Finite Elements in Fluids—Volume 2

Finite Elements in Fluids—Volume 2

Mathematical Foundations, Aerodynamics and Lubrication

Edited by

R. H. Gallagher
Civil Engineering, Cornell University, Ithaca, N.Y.

J. T. Oden
Engineering Mechanics, University of Texas

C. Taylor
Civil Engineering, University College, Swansea

and

O. C. Zienkiewicz
Civil Engineering, University College, Swansea

A Wiley–Interscience Publication

JOHN WILEY & SONS

London · New York · Sydney · Toronto

Library of Congress Cataloging in Publication Data

International Conference on the Finite Element Method in Flow Analysis, University College of Wales, 1974.
Finite elements in fluids.

'A Wiley–Interscience publication.'
CONTENTS: v. 1. Viscous flow and hydrodynamics.—
v. 2. Mathematical foundations, aerodynamics, and lubrication.
1. Fluid dynamics—Congresses. 2. Finite element method—Congresses. I. Gallagher, Richard H., ed.
II. Title.

QA911.153 1974 532′.051 74–13573

ISBN 0 471 29046 7

Printed in Great Britain by J. W. Arrowsmith Ltd., Winterstoke Road, Bristol BS3 2NT

Contributing Authors

J. ALLOUARD — *Bureau d'Études, Franlab,
Rueil-Malmaison,
France.*

A. J. BAKER — *Bell Aerospace Co.,
Buffalo, New York, 14240,
U.S.A.*

F. BARWELL — *Department of Mechanical Engineering,
University College of Wales,
Swansea, Wales, U.K.*

T. J. M. BOYD — *School of Mathematics and Computer Science,
University of North Wales,
Bangor, LL57 2UW, Wales, U.K.*

G. F. CAREY — *Department of Aeronautics and Astronautics,
University of Washington,
Seattle, Washington, 98195,
U.S.A.*

J. F. COUDERT — *Institut Français du Pétrole,
Rueil-Malmaison,
France.*

G. DE VRIES — *Department of Mechanical Engineering,
Calgary University,
Calgary, Alberta,
Canada.*

B. A. FINLAYSON — *Department of Chemical Engineering,
University of Washington,
Seattle, Washington, 98195,
U.S.A.*

J. H. HORLOCK — *S.R.C. Turbomachinery Laboratory,
Engineering Department,
Cambridge University,
Whetstone, London N20.*

K. H. HUEBNER — *General Motors Research Laboratories,
Warren, Michigan,
U.S.A.*

R. D. Milne — *Department of Engineering Mathematics,*
University of Bristol,
Bristol, U.K.

H. Norrie — *Department of Mechanical Engineering,*
Calgary University,
Calgary, Alberta,
Canada.

J. T. Oden, — *Department of Aerospace Engineering and Engineering Mechanics,*
The University of Texas,
Austin, Texas, 78712,
U.S.A.

H. J. Perkins — *GEC Power Engineering Laboratories,*
Whetstone, London N20, U.K.

S. F. Shen, — *Department of Aerospace and Mechanical Engineering,*
Cornell University,
Ithaca, New York, 14850,
U.S.A.

L. C. Wellford — *Department of Civil Engineering,*
University of Southern California,
University Park,
Los Angeles, California, 90007,
U.S.A.

J. R. Whiteman — *School of Mathematical Studies,*
Brunel University,
Uxbridge, UB8 3PH, U.K.

Preface

The finite element method, which was first devised as a procedure for structural analysis, has come to be recognized as an effective analysis tool for a wide range of physical problems. Among these are problems in the field of flow analysis, a subject which is herein interpreted to encompass not only the flow of fluids but also heat flow and lubrication phenomena.

The application of the finite element method to flow analysis problems is a relatively recent development but, nevertheless, a significant literature on the topic has already emerged. Also, a formidable amount of research and application work, involving many aspects of the subject, was in progress by early 1972, the point in time when it was decided to initiate arrangements for a Symposium on finite element methods in flow problems. In planning the Symposium the decision was therefore made to provide a forum for three different types of paper. One type was to be the invited state-of-the-art review of a major aspect of finite element flow analysis. Another type was to feature the in-depth treatment of a particular problem. Finally, there was the paper that outlined work completed, or in progress, on a particular problem.

The International Symposium on the Finite Element Method in Flow Problems was organized along the lines described above and was held at University College of Wales, Swansea, Jan 7–11, 1974. Over 70 papers were presented in the three categories and, despite difficulties in transportation and labour relations that prevailed at the time, a capacity attendance of 250 participants was recorded. A proceedings, available at the time of the Conference was published by the University of Alabama Press at Huntsville under the title, *Finite Element Methods in Flow Problems.* This contained full-length papers in the third category described above and extended abstracts of papers in the first two categories.

Papers in the first two categories are found in this book and in the companion, Finite Elements in Fluids. Volume 1, Viscous Flow and Hydrodynamics.

Although the division of the contents of these two books is accurately characterized by the topic designations (viscous flow, hydrodynamics, mathematical foundations, aerodynamics, lubrication) it would also be possible to assign the designations 'engineering applications' and 'theoretical basis' to them. Exceptions to both modes of designation can be readily identified by the reader.

The present volume is composed of three major segments, comprising papers with an orientation towards 'mathematical foundations' (Chapters 1–7), 'aerodynamics' (Chapters 8–10), and 'lubrication' (Chapters 11–12), plus a separate chapter on the special topic of hydrodynamic stability studies. The first two headings do not represent hard-and-fast categorizations of the papers contained therein. The papers by Baker (Chapter 4) and Allouard and Coudert (Chapter 7) have close identification with the 'viscous flow' and 'wave phenomena' classifications, respectively, of the companion volume and are in fact written by practitioners. The mode of presentation, however, is more consistent with the mathematical rigour of work intended for the present volume. Also, the chapter by Perkins and Horlock (Chapter 8) stands in its own right as a review of practical analysis tools for turbomachinery. Yet, the details of this topic have an alliance with aerodynamic analysis.

In describing the contents of the 'mathematical foundations' segment, it is appropriate to call attention to the introductory chapter of Volume 1, 'Why Finite Elements?', by Zienkiewicz. As we noted in the Preface to Volume 1, the subject matter of that chapter is equally relevant to the contents of the present volume.

It was also noted in the Preface of the companion volume that the Galerkin approach, a component of the more general concept of weighted residuals, forms the basis for many of the formulative efforts in finite element flow analysis. Finlayson, whose efforts have done much to organize and extend the weighted residual concept, opens this book with a chapter that reviews the concept and summarizes work in its application to flow problems. The focus is at first on non-finite element procedures, then upon finite element applications, and finally a promising new finite element scheme is presented in some detail.

The principal motivation for studies on the mathematical side of the finite element method has been in the development of proofs of convergence and errors bounds. Flow problems are frequently described by time-dependent differential equations of the diffusion and convection type. Wellford and Oden direct their attention to these in Chapter 2 and produce an exposition on techniques for determining convergence properties and error estimates for finite element approximation of an important class of time-dependent problems.

Norrie and de Vries, in Chapter 3, explore a new approach to the analysis of non-linear problems by the finite element method, an approach they term the 'pseudo-functional' method. This method can be regarded as a Galerkin method used with a particular approximation scheme. Its effectiveness is illustrated by applications to problems of inviscid compressible flow and viscous incompressible flow.

Baker, in Chapter 4, develops formulations and solution procedures that are closely allied to those presented in Chapters 2–4 of Volume 1, in that the problem of viscous incompressible flow is addressed. Milne (Chapter 5) examines, via the finite element method, a problem that exists in many practical flow analysis circumstances, that of the unbounded region. Another specific problem of wide practical importance is that of singularities, and this is taken up by Whiteman in Chapter 6. To complete this segment of the book Allouard and Coudert take up, in Chapter 7, the analysis of transient non-linear waves by means of spline interpolating functions. This is an approach closely allied to the conventional finite element method and is regarded by many as a component aspect of the general finite element theory.

It has already been observed that Chapter 8, by Perkins and Horlock, covers a topic that may appear peripheral to the subject of 'aerodynamics'. The topic of flow analysis for turbomachinery is extremely broad and these authors, by examining and categorizing the practical and physical aspects of it and by reviewing past efforts in alternative analytical methods, set the stage for future contributions by the finite element method. Both Chapters 9 and 10, by Carey and Shen, respectively, directly address the matter of aerodynamic analysis by the finite element method. Carey contributes a detailed development of the variational formulation for compressible flow and transforms this into a finite element representation in terms of a stream function. Close attention is given to the strategy for solution of the resulting non-linear algebraic equations. Shen, who brings to bear a background of significant contributions to aerodynamic analysis by use of classical solution procedures, presents a far-ranging examination of the present and future role of finite element analysis in this field. Finite element procedures and numerical methods in the specific topic of airfoil analysis are described as are pitfalls which lie ahead for attempts to solve such problems which involve shock wave phenomena by the finite element method. This chapter makes clear that many problems which have presented formidable challenges to alternative solution approaches can be expected to pose similar challenges to the finite element method.

The papers by Barwell (Chapter 11) and Huebner (Chapter 12) complement each other. Barwell examines the component physical aspects of the subject of lubrication, describes how analytical models have emerged for these over the years, and identifies where these are deficient. Recent developments in analytical methods of all types are summarized. Huebner also furnishes a perspective on the physical aspects of lubrication and the basic governing equations but, in his chapter, the focus is upon accomplishments in the solution of these equations by the finite element method. The paper contributes an extensive state-of-the-art review of this topic.

The paper by Boyd (Chapter 13) describes the timely study of the hydromagnetic stability of plasmas that is vital in modern thermonuclear fusion research with particular reference to TOKAMAK systems.

Contents

Chapter 1

Weighted Residual Methods and their Relation to Finite Element Methods in Flow Problems

B. A. Finlayson

1.1 Introduction

Weighted residual methods are contrasted to finite elements methods for applications to flow problems. We first distinguish between principles which will generate the approximate solution and the choice of trial functions. For example, the Galerkin method is sometimes equivalent to the variational method, regardless of the choice of trial functions. The orthogonal collocation method is also seen to be equivalent to the Galerkin, and hence variational, method because the trial functions are orthogonal polynomials. This makes the orthogonal collocation method a discrete form of Galerkin's method in special cases. The importance of collocation-type methods is that quadratures need not be evaluated; several examples are given where this is advantageous.

The main difference between weighted residual methods and finite element methods is in the choice of trial functions or shape functions. Traditionally, weighted residual methods have used trial functions which are defined over the entire domain, whereas finite element methods have used shape functions defined over an element, with elements joined together to cover the entire domain. The various polynomials that have been applied to one-dimensional problems are described and their truncation errors listed. The method of orthogonal collocation on finite elements is then introduced by using as shape functions orthogonal polynomials on finite elements. This method has the advantage that quadratures need not be evaluated in non-linear problems. Conditions are given when it is equivalent to the Galerkin method and applications to time-dependent and two-dimensional problems are outlined. Finally the use of the residual allows for the automatic placement of elements, so that the elements are smallest in regions dictated by the solution. Such a scheme is not feasible in the finite element method when linear shape functions are used, since the residual is not defined.

To illustrate these features of the various methods, we consider applications to several flow problems. Entry-length calculations and boundary layer flows characteristically have solutions with large gradients and the boundary layer flows often have singularities. These properties must be taken into account by efficient weighted residual methods while finite element techniques can be applied in a straightforward (though sometimes inefficient) manner. Flow through packed beds of spheres is shown to have an 8-fold symmetry which must be satisfied exactly for an efficient solution. Weighted residual methods can use trial functions obeying this symmetry, thereby achieving a $2^8 = 256$-fold reduction in the number of unknowns compared to finite element techniques. However, the trial functions utilizing this symmetry in weighted residual methods must be specially constructed for each problem. Applications in the field of petroleum reservoir analysis give useful insight into comparisons of different piecewise trial functions, as well as computing-time comparisons to finite difference methods. The non-linearities associated with the flow of non-Newtonian fluids add difficulties for efficient computing. In the finite element method, linear shape functions are especially convenient (because then the viscosity is constant over an element) but, for more accurate solutions, higher order polynomials have been used for Newtonian fluids. Such shape functions require recalculating the integrals in the finite element method each time the viscosity changes, as it does in each iteration for non-Newtonian fluids. Finally, flows with free boundaries utilize the convenient ability of finite element methods to handle diverse geometries.

1.2 Weighted residual principles

A variational principle is often used to generate the working equations in the finite element method. For a non-linear equation

$$\mathscr{D}(u(\mathbf{x})) = f(\mathbf{x})$$

application of the variational principle leads to the Euler equation of the form

$$\int_\Omega \delta u[\mathscr{D}(u(\mathbf{x})) - f(\mathbf{x})]\,\mathrm{d}\Omega = 0 \tag{1.1}$$

Application of the variational method, for a trial function

$$u(\mathbf{x}) = \sum_{k=1}^{N} c_k u_k(\mathbf{x})$$

where the c_k are constants, gives

$$\int_\Omega u_j[\mathscr{D}(\textstyle\sum c_k u_k) - f(\mathbf{x})]\,\mathrm{d}\Omega = 0 \tag{1.2}$$

The same equations arise from an application of the Galerkin method, when the residual, $N(\sum c_k u_k) - f$, is made orthogonal to the trial function, u_i. Thus there is always a Galerkin method which is equivalent to the variational method.

The variational method is usually applied without performing the integration by parts to obtain Equation 1.1 or 1.2. The trial function then need not be as continuous, since lower order derivatives appear. For example, the slow flow of a generalized Newtonian fluid through an irregularly shaped duct is governed by the variational principle (see Reference 1, p. 274)

$$\Pi(u) = \int_A \left[\int_0^{\mathrm{II}} \eta(\mathrm{II}')\,\mathrm{d}\mathrm{II}' + \mathbf{w}\,.\,\mathbf{e}_{(z)}\frac{\partial p}{\partial z} \right] \mathrm{d}A$$

$$\mathrm{II} = d_{ij}d_{ij}$$

$$d_{ij} = \tfrac{1}{2}(w_{i,j} + w_{j,i})$$

where w_i is the ith component of velocity w, $w_{i,j} \equiv \partial w_i/\partial x_j$, and $\mathbf{e}_{(z)}$ is the unit vector in the $z(2)$ (or 3)-direction. The variation with respect to $\mathbf{w}$ yields

$$\delta\Pi = \int_A \left[2\delta w_{i,j} d_{ij}\eta(\mathrm{II}) + \delta\mathbf{w}\,.\,\mathbf{e}_{(z)}\frac{\partial p}{\partial z} \right] \mathrm{d}A \tag{1.3}$$

while the Euler equation is from

$$\delta\Pi = \int_A \delta w_i \left[-2(d_{ij}\eta(\mathrm{II}))_{,j} + \delta_{3i}\frac{\partial p}{\partial z} \right] \mathrm{d}A + \int_{\Gamma_1} 2\eta(\mathrm{II})\,\delta w_i\,\mathrm{d}_{ij} n_j\,\mathrm{d}\Gamma \tag{1.4}$$

It is clear that the second derivative of w_i must exist if the trial functions are to be substituted into the Galerkin equation, Equation 1.4, since $2d_{ij,j} = w_{i,jj} + w_{j,ij}$, while in the variational formulation, Equation 1.3, only first derivatives of velocity appear. Furthermore, when the variational principle is a positive definite one, the equations resulting from Equation 1.3 will be symmetric and positive definite, thus leading to computational advantages. A further advantage of the variational principle is that natural boundary conditions are easily handled. The natural boundary conditions are contained in Equation 1.3 as $d_{ij}n_j = 0$ on Γ_1 while in the Galerkin formulation, the natural boundary conditions must be adroitly combined with the residual to obtain Equation 1.4 (see Reference 1, pp. 30, 150).

It is thus clear that a variational principle is useful, if one exists. The existence question is answered definitively using Fréchet derivatives as outlined in Reference 1, Chapter 9. Using that formalism, for example, we can prove that the steady state Navier–Stokes equations do not have a variational principle (Reference 1, Section 8.6) unless the inertial terms are zero or $\mathbf{w} \times (\nabla \times \mathbf{w}) = 0$. Thus, to include inertial terms we, of necessity, must use the Galerkin method. We can, however, use the formulation as in Equation 1.3.

$$\delta\Pi = \int_A \left[2\delta w_{i,j} d_{ij} \eta(\mathrm{II}) + \delta w_i \rho w_j w_{i,j} + \delta \mathbf{w} \cdot \mathbf{e}_{(z)} \frac{\partial p}{\partial z} \right] \mathrm{d}A = 0 \qquad (1.5)$$

with the inertial terms added so as to obtain the correct Euler equation, as in Equation 1.4. Then, while the matrix arising in the finite element method will no longer be positive definite or symmetric, Equation 1.5 still only contains first derivatives, so that trial functions can be used which have no second derivative. Thus, even if the Galerkin method is applied, it is useful to understand variational principles. We note in passing that quasi-variational principles and the local potential method are also equivalent to the Galerkin method, and indeed these special constructs often obscure the mathematical basis for the approximate solution (see Reference 1, Chapter 10; Reference 2).

The orthogonal collocation method is an important method because of its relation to the Galerkin method and because of the extensions described below which combine it with finite element ideas. In a collocation method we set the residual to zero at a set of collocation points, e.g.

$$\mathscr{D}\left(\sum_{k=1}^{N} c_k N_k(\mathbf{x}) \right) - f(x) \bigg|_{\mathbf{x}=\mathbf{x}_i} = 0 \qquad i = 1, \ldots, N \qquad (1.6)$$

and solve the resulting equations for c_k. In orthogonal collocation methods (Reference 3; Reference 1, Chapter 5) the trial functions are orthogonal polynomials. For illustrative purposes consider a one-dimensional problem, with $\mathscr{D}$ a second-order ordinary differential equation

$$\mathscr{D}\left(\frac{d^2u}{dx^2}, \frac{du}{dx}, u(x) \right) - f(x) = 0 \qquad 0 < x < 1 \qquad (1.7)$$

and appropriate boundary conditions at $x = 0$ and $x = 1$. The expansion function is taken as

$$u(x) = b + cx + x(1 - x) \sum_{i=1}^{N} a_i P_{i-1}(x) \qquad (1.8)$$

where $P_{i-1}(x)$ is an $(i - 1)$th order polynomial in x made orthogonal to all lower order polynomials using

$$\int_0^1 P_j(x) P_i(x)\, \mathrm{d}x = 0 \qquad i = 0, 1, \cdots, j - 1$$

Now it is clear that if the coefficients b, c, $\{a_i\}$ are known, the function $u(x)$ can be found for any x, since $P_{i-1}(x)$ are known functions. Conversely, if $u(x)$ is known at a set of $N + 2$ collocation points, then the coefficients

$b, c, \{a_i\}$ may be determined. The collocation points are taken as the N roots to $P_N(x) = 0$, which will be in $0 < x < 1$, and the end-points $x = 0, 1$. Derivatives of u, du/dx and d^2u/dx^2, can be written in terms of $b, c, \{a_i\}$, which are known in terms of $u(x_i)$, the function u at the collocation points. Thus it is possible to define the derivatives at the collocation points (Reference 1, p. 100)

$$u_i = u(x_i), \quad \left.\frac{\mathrm{d}u}{\mathrm{d}x}\right|_{x_j} = \sum_{i=1}^{N+2} A_{ji}u_i, \quad \left.\frac{\mathrm{d}^2u}{\mathrm{d}x^2}\right|_{x_j} = \sum_{i=1}^{N+2} B_{ji}u_i \tag{1.9}$$

The orthogonal collocation method applied to Equation 1.7 then yields, similar to Equation 1.6,

$$\mathscr{D}\left(\sum_{i=1}^{N+2} B_{ji}u_i, \sum_{i=1}^{N+2} A_{ji}u_i, u_j, x_j\right) - f(x_j) = 0 \qquad j = 2, \ldots, N+1 \tag{1.10}$$

with two more equations for the boundary conditions at $x_1 = 0$ and $x_{N+2} = 1$. Now comes the important part. Because we have used orthogonal polynomials we can define a quadrature scheme

$$\int_0^1 f(x)\,\mathrm{d}x = \sum_{i=1}^{N+2} W_i^{(N+2)} f(x_i) \tag{1.11}$$

which is exact when $f(x)$ is a polynomial of degree $2N + 1$ or less (Reference 1, p. 105). We can use this result to show that the collocation equations, Equation 1.10, are equivalent to the Galerkin equations for certain linear operators.

First assume the differential equation is of the form

$$(a_1 + a_2x)\frac{\mathrm{d}^2u}{\mathrm{d}x^2} + a_3\frac{\mathrm{d}u}{\mathrm{d}x} - f(x) = 0 \tag{1.12}$$

with the boundary conditions $u(0) = a_4$, $u(1) = a_5$ and $f(x)$ an Nth order polynomial in x. The properties of the quadrature formula, Equation 1.11, can be used to show that the orthogonal collocation and Galerkin methods lead to identical results when the same orthogonal polynomials are used as trial functions in both methods (Reference 1, p. 135). If the problem is more general than Equation 1.12, i.e. the residual is not a polynomial or is a polynomial of degree greater than N, or is non-linear, then the exact correspondence fails. However, in those cases we can still say that the Galerkin method, using an approximate quadrature scheme, is equivalent to the orthogonal collocation method. It is the use of orthogonal polynomials, and the corresponding collocation points, that allows this result, and this is made particularly evident in the convergence results cited below.

Both variational and Galerkin methods require integration of the trial functions over the domain or finite elements. Such procedures are possible *a priori* for certain types of non-linearities, like the $\rho\mathbf{u}\,.\,\nabla\mathbf{u}$ term in Equation 1.5. However, other terms, like $\delta w_{i,j}d_{ij}\eta(\text{II})$ in Equation 1.5, may be complicated enough to prohibit the integration at the start of the calculation, when no solution is known. Then the integrals must be recomputed throughout the iterative calculation and this increases the computation time drastically. The collocation method avoids this problem, because no integrals are evaluated, and the orthogonal collocation method is sometimes equivalent to the Galerkin method anyway. This advantage of the orthogonal collocation method is explored below.

In the subdomain method we use in place of Equation 1.2

$$\int_{\text{Vol}_j} [\mathscr{D}(\textstyle\sum c_k u_k) - f(\mathbf{x})]\,\mathrm{dVol} = 0,$$

thereby setting to zero the average residual in each subdomain Vol_j. Other methods are also possible by choosing other weighting functions in Equation 1.2 in place of u_j.[1,4]

1.3 Trial functions

In order to apply one of the methods of weighted residuals it is necessary to expand the unknown solution in a trial function. The analyst has a choice of two broad classes of trial functions. One class uses functions which are continuous and defined over the entire domain, while the other class divides the domain into subdomains or finite elements and uses trial functions defined only on the elements. The first class will be referred to here as global expansions, for ease in exposition.

The chief advantage of the global orthogonal collocation method is the very rapid convergence as the number of terms in the expansion is increased. For example, one study for an ordinary differential equation[5] found that the error was proportional to $(1/N)^{1.72N}$ where N was the number of collocation points. As N changes from 5 to 6, then, the error decreases by a factor of 100. By contrast, in a finite difference calculation of $O(\Delta x^2)$, where $N = 1/\Delta x$ is the number of grid points, a change of N from 5 to 6 leads to only 1·4 factor improvement in the error, or if the method is of $O(\Delta x^4)$ we get a 2·1 factor improvement. A theoretical study of global polynomial trial functions used with a variational principle showed that the error should decrease as $O(1/N)^{t-1}$ when the number of trial functions is increased (thus increasing the highest degree of polynomial considered).[6] In this expression $N + 1$ is the degree of polynomial and t is thc continuity of the exact solution, and the proof requires $N + 1 \geqslant t$. In the example cited, faster convergence than this was found presumably because the trial functions were functions of x^2 and had certain symmetry properties already included. The rapid convergence of the global methods makes possible a reduction in the number of collocation points, N, for a given accuracy and usually a marked reduction in computation time. However, the global expansions encounter difficulties when the solution has steep gradients in parts of the domain. In those situations a finite element type of approach is preferred, because elements can be bunched near the regions of steep gradients.

It is thus desirable to examine convergence results for piecewise polynomials to see what guidelines they give. The excellent paper by Ciarlet and coworkers[6] treats one-dimensional boundary value problems of the type

$$\sum_{j=0}^{n} (-1)^{j+1} D^j[p_j(x) D^j u(x)] = f(x, u(x)) \qquad 0 < x < 1$$

which have a corresponding variational principle. In addition to the convergence results for global polynomials, referred to above, they present results for various piecewise polynomials. The domain, $0 \leqslant x \leqslant 1$, is divided into a number of elements and a polynomial trial function is used on each element. The polynomials are listed in Table 1.1 together with the convergence results. We have a wide variety of choices available and the power, p, on the rate of convergence, expressed as $(1/M)^p$, depends on the degree of polynomial in the element. The more accurate methods have a higher p, necessitating a higher degree polynomial. The results of Ciarlet and coworkers[6] refer to variational methods whereas DeBoor and Swartz[7] use the collocation method on the subdivision of elements. They also show that the rate of convergence is increased in the collocation method on finite elements when the collocation points are Gaussian quadrature points rather than uniformly distributed. The Gaussian quadrature points correspond to the collocation points in orthogonal collocation on finite elements, as described below. The various choices of polynomials are compared in Table 1.2 depending on the degree of polynomial and the continuity of the trial function. Clearly there is overlap among the methods, even though the formulation and methods of solution are different.

Douglas and DuPont[8] treat time-dependent one-dimensional problems of the type

$$c(x, t, u)\frac{\partial u}{\partial t} = a(x, t, u)\frac{\partial^2 u}{\partial x^2} + b\left(x, t, u, \frac{\partial u}{\partial x}\right)$$

with a collocation method using cubic polynomials on each element. They, too, find that the rate of convergence is increased (to Δx^4) if the collocation points are the Gaussian quadrature points, whereas the convergence goes as Δx^2 for collocation points distributed uniformly on each element. Douglas[9] extends this work for linear problems when the polynomial is of degree $N + 1$. Gaussian quadrature points are used (i.e. orthogonal collocation on finite elements) to find that the convergence goes as Δx^{N+2} globally and Δx^{2N} at the collocation points.

Table 1.1 Comparison of properties of various piecewise polynomials*

Name	Degree of polynomial	Continuity of trial function†	Number of unknowns per element (after continuity)	Rate of convergence‡ as $M \to \infty$
Smooth Hermite§ $m = 1, 2, \ldots$	$2m - 1$	$m - 1$	m	$\left(\frac{1}{M}\right)^{2m-1}$
Non-smooth Hermite§ $m \geqslant 2k; k = 1, 2, \ldots$	$m - 1$	$k - 1$	$m - k$	$\left(\frac{1}{M}\right)^{2m-1}$
Spline functions§ $m = 2, 3, \ldots$	$2m - 1$	$2m - 2$	1	$\left(\frac{1}{M}\right)^{2m-2}$
Orthogonal collocation‖ on finite elements (second-order equations) $N = 1, 2, \ldots$	$N + 1$	1	N	at end of elements: $\left(\frac{1}{M}\right)^{2N}$ interior to elements: $\left(\frac{1}{M}\right)^{N+2}$

* Deduced from Reference 6§ and Reference 7‖; both for second-order ordinary differential equations.
† For example, if entry is t, then approximate solution $u \in \zeta^t$.
‡ t is the continuity of the exact solution; M is the number of subdomains.
§ Variational method.
‖ Orthogonal collocation method.

Table 1.2 Piecewise polynomials in one dimension

First entry is the degree of polynomial.
Second entry is the highest derivative of the solution which is continuous over the whole domain.

				Smooth Hermite ↓	Non-smooth Hermite ↓	↓	↓	
Spline ←	7	5	3	1	2	3	4	→
	6	4	2	0	0	0	0	
Orthogonal collocation on finite elements			2	3	4	5	6	→
			1	1	1	1	1	
				5	6	7	8	→
				2	2	2	2	
				7	8	9	10	→
				3	3	3	3	

1.4 Orthogonal collocation on finite elements in one dimension

Consider the ordinary differential equation, Equation 1.10, to be solved on $0 < x < 1$ together with boundary conditions at $x = 0, 1$. We divide the domain $0 < x < 1$ into NE elements by placing the dividing points at x_l, $l = 1, \ldots, NE + 1$, with $x_1 = 0$ and $x_{NE+1} = 1$. Within each element we define a new variable $s^l = (x - x_l)/\Delta x_l$, $\Delta x_l = x_{l+1} - x_l$, and place interior collocation points at the roots to $P_N(s) = 0$, where P_N is a shifted Legendre polynomial defined on $0 \leqslant s \leqslant 1$. Within each element the variable s^l goes from zero to one. Applying the usual procedure of orthogonal collocation[1] we write the differential equation at the interior collocation points in terms of the value of the solution at the collocation points in the same element. Thus we get

$$\mathscr{D}\left(\frac{1}{\Delta x_l^2}\sum_{i=1}^{N+2} B_{ji}u_i, \frac{1}{\Delta x_l}\sum_{i=1}^{N+2} A_{ji}u_i^l, u_j^l, x_l + s_j^l\Delta x_l\right) - f(x + s_j^l\Delta x_l) = 0 \tag{1.15}$$

$$j = 2, \ldots, N + 1;\ l = 1, \ldots, NE$$

where the matrices B and A are the same ones arising in Equation 1.10. At the division between elements we require continuity of the first derivative.

$$\Delta x_{l+1}\sum_{i=1}^{N+2} A_{N+2,i}u_i^l - \Delta x_l \sum_{i=1}^{N+2} A_{1i}u_i^{l+1} = 0$$

Two additional equations are written for the boundary conditions. For use on a computer we line up the variables u_i^l as one long vector u_k by defining $u_k = u_{(N+1)(l-1)+j}$, where l goes from 1 to NE and j goes from 1 to $N+1$, except for $l = NE$ when j goes to $N + 2$ to get $u_{(N+1)NE+1} = u(x = 1)$. If the problem is linear we can then write the set of equations as

$$\mathbf{AAu} = \mathbf{F} \tag{1.16}$$

and the matrix **AA** has the block diagonal structure shown in Figure 1.1.

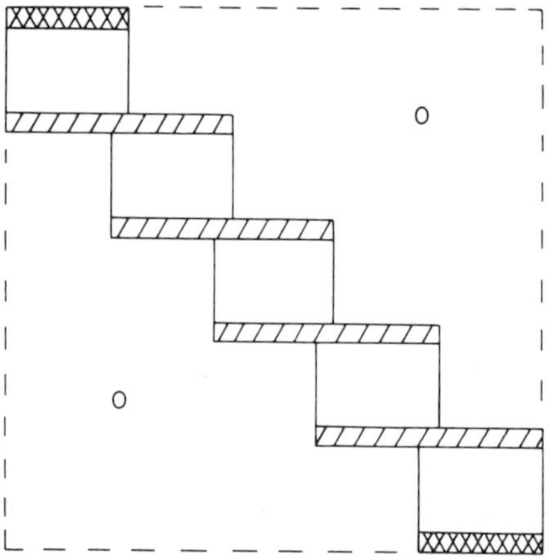

Figure 1.1 Matrix arising in method of orthogonal collocation on finite elements. The only non-zero entries are inside the solid blocks. The cross-hatched areas come from boundary conditions, the slashed areas come from continuity of the first derivative at the element boundaries and the other areas come from the differential equation

There is only one element of overlap, so that we need store only the blocks along the diagonal, as a three-dimensional array $AA(l, i, j)$ with $l = 1, \ldots, NE$ and $AA(l, N + 2, N + 2) = AA(l + 1, 1, 1)$. Such matrices can be inverted using an efficient LU decomposition routine written by Graham Carey of the University of Washington, and the result can be stored on top of AA, in the same block diagonal structure. If the problem is non-linear a simple approach is to collect all the non-linear terms in **F** in Equation 1.16 and solve the equations iteratively, $\mathbf{AAu}^{k+1} = \mathbf{F}^k$. If **AA** is fixed (independent of k) the LU decomposition is carried out only once, and the iterations are performed with fore-and-aft sweeps, thus saving computation time. If this iteration scheme is not convergent then Newton-Raphson can be applied to obtain linear equations of the form Equation 1.16, and the matrix **AA** is inverted at each iteration.

As mentioned above, DeBoor and Swartz[7] have shown that for second-order equations the rate of convergence of the solution and first derivative at the end-points of each element, $s^l = 0$ or 1, goes as Δx^{2N}, where Δx is the uniform element size, while the error decreases as Δx^{N+2} as $\Delta x \to 0$ for the solution at the interior collocation points. It is clear that this scheme can lead to very high order rates of convergence and, indeed, computations bear this out. Of course, as in any finite element method, the element sizes need not be uniform.

Once the calculation is completed, the residual can also be computed. As in any weighted residual method, the residual is the differential equation, evaluated for the approximate solution. It is zero at the interior collocation points of each element, since the solution was found by solving Equation 1.15. In between the collocation points, however, it is non-zero. Calculations have shown that as the approximate solution is refined the residual approaches zero at more and more points and the mean-squared residual also approaches zero (see, for example, Reference 5). Sometimes an upper bound on the error in the solution can be found as a function of the mean-squared residual, even when no exact solution is available.[5] Thus we want the residual to be small, and for the exact solution it is zero. After an orthogonal collocation solution is found on finite elements we can examine the residual. This is most easily done at the end-point of each element, although the interpolation polynomial can be used to find the solution, and hence the residual, anywhere in the domain. Additional elements can then be added in places where the residual is large and the calculation is repeated. In this way we end up with the elements in exactly the place they are needed. Application of this idea to a chemical engineering problem led to very rapid convergence as well as different element sizes whose ratio was as high as 2000.[10]

When using orthogonal collocation on finite elements it is also possible to show equivalence to the Galerkin method. For example, expand the solution as a polynomial within each element

$$\phi_l(s) = b_l + c_l s + s(1 - s) \sum_{i=1}^{N} a_{li} P_{i-1}(s) \tag{1.17}$$

We require continuity of the solution and first derivative at the ends of the element.

$$\phi_l(1) = \phi_{l+1}(0) \quad \text{or} \quad b_l + c_l = b_{l+1} \tag{1.18}$$

$$\Delta x_{l+1} \left.\frac{d\phi_l}{ds}\right|_{s=1} = \Delta x_l \left.\frac{d\phi_{l+1}}{ds}\right|_{s=0} \tag{1.19}$$

or

$$c_l - \sum_{i=1}^{N} a_{li} P_{i-1}(1) = c_{l+1} + \sum_{i=1}^{N} a_{l+1,i} P_{i-1}(0) \tag{1.20}$$

These equations effectively give b_l and c_l. The a_{li} are found by making the residual, here just called $\mathscr{D}(s)$, orthogonal to the weighting function $\partial\phi_l/\partial a_{li} = s(1 - s)P_{i-1}(s)$ for each i.

$$\int_0^1 s(1 - s)P_{i-1}(s)\mathscr{D}(s)\,\mathrm{d}s = 0 \qquad i = 1, \ldots, N$$

Apply the quadrature formula, which is exact if $\mathscr{D}$ is a polynomial in s of degree N or less, to obtain

$$\sum_{j=1}^{N+2} W_j^{(N+2)} s_j(1 - s_j)P_{i-1}(s_j)\mathscr{D}(s_j) = 0$$

This gives effectively $\mathscr{D}(s_j) = 0$ at $j = 2, \ldots, N + 1$, and the conditions Equations 1.18 and 1.20 make the function and first derivative continuous at the element boundary. Thus the Galerkin method requires the collocation equations be satisfied. The reverse argument also holds. Thus, whenever the residual is of degree N in s and the trial function of degree $N + 1$ in s is used, Equation 1.17, the Galerkin and orthogonal collocation methods on finite elements are identical.

There is one further feature of polynomial expansion on finite elements which should be examined: symmetry conditions. To illustrate the point, let us use the differential equation

$$\frac{\mathrm{d}^2u}{\mathrm{d}x^2} + \frac{a-1}{x}\frac{\mathrm{d}u}{\mathrm{d}x} = f(u, x)$$

$$\left.\frac{\mathrm{d}u}{\mathrm{d}x}\right|_{\substack{x=0 \\ u(1)=u}} = 0$$

where $a = 1, 2, 3$ for planar, cylindrical or spherical geometry. If we expand the solution in a power series, $u = \sum_{i=0}^{\infty} a_i x^i$, the boundary condition at $x = 0$ requires $a_1 = 0$. *If* we can prove that when $a_1 = 0$ then $a_3, a_5, \ldots,$ etc. are zero, too, then the solution can be expanded in even powers of x. This is done in Reference 1 and Reference 3, Section 5.1. The net effect of this simplification is that the number of terms needed in the expansion is reduced by a factor of 2. To achieve a trial function of order $2N$ in x we need only N polynomials in x^2 and N interior collocation points. If this simplification is not made, then we must use $2N - 1$ interior collocation points. Now in the finite element expansions the differential equation will be transferred to

$$\frac{1}{\Delta x_l^2}\frac{\mathrm{d}^2u}{\mathrm{d}s^2} + \frac{a-1}{s\Delta x_l + x_l}\frac{\mathrm{d}u}{\mathrm{d}s} = f(u, s\Delta x_l + x_l)$$

and the term $\mathrm{d}u/\mathrm{d}s|_{s=0} = 0$ affects the solution in only the first element. The solution is not a symmetric polynomial in s, even though it is in x, since $x^2 = (s\Delta x_l + x_l)^2 = s^2\Delta x^2 + 2x_l\Delta x_l s + x_l^2$. Thus, by going to any finite element expansion for problems whose solution is symmetric we automatically double the number of terms we must include to achieve the same level of approximation. The same results hold in problems whose solution includes only odd powers of x. As the number of dimensions increases the effect of symmetry is more pronounced. If symmetry exists in a three-dimensional problem, changing to a finite element approach (whether orthogonal collocation on finite elements or some other) will require 8 times as many points to achieve the same degree of polynomials. This is not to say we need 8 times as many points to achieve the same accuracy, since the possibility of placing small elements where needed may make possible improved accuracy with lower order polynomials. However, the symmetry of a problem is clearly an important consideration in choosing a method, and we illustrate this below for flow in a packed bed.

1.5 Orthogonal collocation (global and on finite elements) in two-dimensional problems

For two-dimensional problems the orthogonal collocation method is a straightforward extension of the method for one-dimensional problems. For a rectangular domain the trial function is the two-dimensional analogue of Equation 1.17.

$$u(x, y) = \left[b_1 + b_2 x + x(1 - x) \sum_{i=1}^{NX} a_i P_{i-1}(x) \right] \times \left[c_1 + c_2 y + y(1 - y) \sum_{i=1}^{NY} d_i P_{i-1}(y) \right]$$

The derivatives appearing in the differential equation are evaluated at the collocation points using Equation 1.9 in each direction.

For example, the equations for the fully-developed flow of a generalized non-Newtonian fluid in a rectangular duct are (considering only one quadrant of the duct)

$$0 = -\frac{\partial p}{\partial z} + \frac{\partial}{\partial x}\left(\eta(\mathrm{II})\frac{\partial w}{\partial x}\right) + \frac{\partial}{\partial y}\left(\eta(\mathrm{II})\frac{\partial w}{\partial y}\right)$$

or

$$0 = -\frac{\partial p}{\partial z} + \eta(\mathrm{II})\left(\frac{\partial^2 w}{\partial x^2} + \frac{\partial^2 w}{\partial y^2}\right) + \frac{\partial w}{\partial x}\frac{\partial \eta}{\partial x} + \frac{\partial w}{\partial y}\frac{\partial \eta}{\partial y}$$

$$w = 0 \quad \text{on } x = a, \text{ all } y$$

$$w = 0 \quad \text{on } y = b, \text{ all } x$$

$$\frac{\partial w}{\partial x} = 0 \quad \text{at } x = 0, \text{ all } y$$

$$\frac{\partial w}{\partial y} = 0 \quad \text{at } y = 0, \text{ all } x$$

To examine the symmetry we see if the transformations $(w, x, y) \leftrightarrow (w, -x, -y)$ or $(w, x, y) \leftrightarrow (w, -x, y)$ change the differential equation. Since $2\text{II} = (\partial w/\partial x)^2 + (\partial w/\partial y)^2$ the equations are unchanged by these transformations, and the solution is thus a function of x^2 and y^2, not x and y. Thus we use symmetric polynomials

$$w(x, y) = \left[b + (1 - x^2) \sum_{i=1}^{NX} a_i P_i(x^2)\right]\left[c + (1 - y^2) \sum_{i=1}^{NY} d_i P_i(y^2)\right]$$

where the polynomials $P_i(x^2)$ are of order i in x^2 and are defined by

$$\int_0^1 (1 - x^2) P_i(x^2) P_j(x^2)\,\mathrm{d}x = 0, \qquad i = 0, 1, \ldots, j - 1.$$

The trial function automatically satisfies the derivative condition along $x = 0$ and $y = 0$. The dimensions are also changed so that x and y both go from 0 to 1. The orthogonal collocation formulation of the problem is then

$$\begin{aligned} 0 = \frac{\partial p}{\partial z} &+ \eta(\text{II}_{ij})\left[\frac{1}{a^2} \sum_{k=1}^{NX+1} BX_{ik} w_{kj} + \frac{1}{b^2} \sum_{k=1}^{NY+1} BY_{jk} w_{ik}\right] \\ &+ \frac{1}{a^2}\left[\sum_{k=1}^{NX+1} AX_{ik} \eta(\text{II}_{kj})\right]\left[\sum_{k=1}^{NX+1} AX_{ik} w_{kj}\right] \\ &+ \frac{1}{b^2}\left[\sum_{k=1}^{NY+1} AY_{jk} \eta(\text{II}_{ik})\right]\left[\sum_{k=1}^{NY+1} AY_{jk} w_{ik}\right] \end{aligned}$$

with $w(x_i, y_j) = w_{ij}$, $i = 1, \ldots, NX$, $y = 1, \ldots, NY$ and

$$w_{i,NY+1} = w_{NX+1,j} = 0 \quad \text{all } i, j.$$

For non-linear, two-dimensional chemical reactor calculations we have used a Newton–Raphson method to solve non-linear algebraic equations similar to these,[11] but alternating direction methods are also possible.[12]

If the aspect (b/a) ratio is very large, the velocity profile will be very flat in the y direction. Then the global trial function would require a large number of terms, or NY, to approximate the solution. This is a situation when the method of orthogonal collocation on finite elements is useful. In the y direction, then, we use finite elements. To solve the algebraic equations, we use an alternating direction method: solve a succession of problems in the x direction, while keeping y fixed, and then solve a succession of problems involving the inversion of matrices as shown in Figure 1.1. To do this we need to linearize the equations for each iteration, which can be done by evaluating η at the previous iteration, or by using a Newton–Raphson method.

If the non-Newtonian fluid is such that flat velocity profiles are expected in both x and y direction, then orthogonal collocation on finite elements would be used in each direction. Again an alternating direction would be used in order to permit the storage-saving features of the matrix shown in Figure 1.1. If the profile were not expected to be flat (say a Newtonian fluid with $b/a \sim 1$), then the global orthogonal collocation method would be used because of the $2^2 = 4$-fold reduction in number of collocation points when we include the symmetry of the problem.

1.6 Orthogonal collocation (global and finite elements) in time-dependent problems

Orthogonal collocation methods can be used for time-dependent problems, too, and reduce the problem to sets of ordinary differential equations in time. These equations can be solved using any one of several techniques for integration of ordinary differential equations as discussed in Reference 1, Chapter 5 and References 5, 13–15. Also, analogues of all the modifications of the Crank–Nicolson method for finite difference methods are applicable, while for two-dimensional problems alternating-direction methods are also possible.

Orthogonal collocation on finite elements also leads to ordinary differential equations in time. Explicit methods of integration lead to stability limitations related to Δx_l^2, as in finite difference methods. Implicit methods lead to matrices of the form of Figure 1.1 to invert at each time step (or perhaps only once if it is constant from one time to the next). Two-dimensional, time-dependent problems can be solved with alternating direction implicit methods.

1.7 Applications

A series of applications are considered to illustrate the comparison of global and finite element trial functions.

1.7.1 Entry-length problems

As a fluid moves into a duct, a boundary layer is formed near the walls and the thickness grows as the fluid moves down the duct. The solution to this problem exhibits large gradients, making useful finite element methods which can bunch the elements near the walls. We review first the global techniques and then consider finite element calculations.

In the momentum integral method, first presented by Schiller,[16] the boundary layer equations for a flat plate are integrated over the thickness of the boundary layer. A velocity profile is assumed and the boundary layer thickness can be calculated as a function of length down the duct by solving an ordinary differential equation. Campbell and Slattery[17] showed that more accurate results are achieved if the pressure drop is calculated from the kinetic energy balance including viscous dissipation. However, for some non-Newtonian fluids, even this approach does not give accurate results (see Reference 1, p. 87). These methods are limited in usefulness because the accuracy is sometimes poor and cannot be estimated and a convergent process has not been defined to refine the calculation. In addition they are restricted to a few regular geometries, such as circular tubes or a duct between flat plates.

A convergent mathematical scheme for the flow problem can be applied as presented by Fleming and Sparrow.[18] They write the solution as the sum of the fully developed solution and an entry-region solution. The fully developed solution is found by using trial functions which satisfy the differential equation, which is easy to do since the differential equation for Newtonian fluids is Poisson's equation. Then the boundary conditions are applied by applying collocation on the boundary of the domain, which can be irregular. They use more collocation points than there are constants to fit in the trial function and so use a least squares collocation condition: minimize the sum of the collocation residuals on the boundary. The entry-region solution is then found from the solution to,

$$\varepsilon(z)U\frac{\partial w^*}{\partial z} = \Lambda(z) + \nu\left[\frac{\partial^2 w^*}{\partial x^2} + \frac{\partial^2 w^*}{\partial y^2}\right]$$

where U is the mean velocity and the $\varepsilon(z)$ and $\Lambda(z)$ functions are found in the solution. Thus the solution includes inertial effects (with an approximation) and viscous terms in the axial direction, $\partial^2 w/\partial z^2$, are also neglected. The solution is expressed as an infinite series in h_i,

$$w^* = \sum c_i h_i e^{-\mu_i \xi}$$

$$\nabla^2 h_i + \mu_i h_i = 0, \qquad \frac{1}{\mu_i A}\oint \frac{\partial h_i}{\partial h}\,dS = 1$$

Again the trial function satisfies the differential equation (with an as yet undetermined eigenvalue). The eigenvalue is found iteratively to satisfy the normalization condition and the boundary conditions are satisfied by a least squares collocation condition on the boundary. Since the entry-region solution is expressed as a series, the three-dimensional problem is reduced to a series of two-dimensional eigenvalue problems. Arbitrarily shaped ducts can be handled, although calculations are presented only for rectangular and triangular ducts. To achieve good accuracy near the inlet from 12 to 100 terms are required in the eigenfunction expansion, depending on the problem.

Atkinson and coworkers[19] solved the problem of slow flow of a Newtonian fluid using a finite element method. The problem was solved in terms of the stream function, so that second derivatives appeared in the functional, and quadratic shape functions were used on triangular elements. The number of elements needed (as well as the number of eigenfunctions used by Fleming and Sparrow) depends on how close to the entrance one wants to resolve the velocity profile. For flow in a pipe the finite element calculation used 451 mesh points, in both radial and axial directions, but the closest profile to the inlet presented by Atkinson and coworkers is at length $z = 0{\cdot}1R$, where R is the pipe radius. By contrast, the expansion method of Fleming and Sparrow could be evaluated as close as $z = 0{\cdot}0006Re \,.\, D_h$ for rectangular ducts, using 100 eigenfunctions, where Re is the Reynolds number and D_h is the hydraulic diameter. For equilateral triangular ducts only 12 terms are needed. For an entry-length heat transfer problem for flow between parallel plates, Tay and Davis[20] used the finite element method with linear shape functions on triangles. They calculated to within $0{\cdot}0005Pe \,.\, h$, where Pe is the Peclet number and h is the thickness between plates; but the calculations still showed some error even for $21 \times 25 = 525$ mesh points.

Both global and finite element methods are applicable to arbitrary geometries perpendicular to the duct axis, but the eigenfunction techniques are probably more accurate for a given number of trial functions.

1.7.2 Boundary layer flows

In boundary layer flows the computations are complicated by the singularities of the solution and the semi-infinite domain. All the applications of the Method of Weighted Residuals known to the author have been with trial functions defined over the entire boundary layer rather than over finite elements, although the entry-length calculations cited above indicate finite element calculations might be possible.

One advantage of global expansion methods is that if the type of singularity is known it can sometimes be transformed away. As mentioned by Thompson,[21] and illustrated by him, if the solution varies wildly, an analytical transformation may be used to make the bad behaviour algebraically explicit; the new unknowns are then smooth. Some of the analyses of laminary boundary layers are attempts to do this.

In the method of integral relations (see, for example, Reference 1, p. 78) the boundary layer equations are integrated from the solid boundary to the edge of a subdomain. The subdomains usually are parallel to the shape of the body. The equations are then satisfied in subdomains, or strips running parallel to the body creating the boundary layer. In one sense the method is a finite element method in the transverse direction. The ordinary differential equations thus generated are integrated along the length of the body. As pointed out by Melnik and Ives[22] this method leads to equations which, if linearized, have both positive and negative eigenvalues. Furthermore, the magnitude of the eigenvalues increases rapidly as the number of strips increases. This makes the calculations difficult and the number of subdomains is usually limited to five or less. Melnik and Ives[22] indicate that an increase in accuracy is possible if the location of the strips is not uniform, but the edges of the strip are placed at the roots to Chebyshev polynomials. They recommend for engineering calculations that two strips, so placed, are sufficient. For supersonic flows similar methods are applicable, but Holt and Ndefo[23] indicate they are too expensive. Chushkin,[24] on the other hand, maintains the scheme is efficient for supersonic flows about conical bodies, including three-dimensional cases.

Galerkin methods are also possible and have been developed by Dorodnitsyn[25,26] and Bethel.[27,28] The boundary conditions on velocity must be approached asymptotically so that the choice of trial functions may be difficult. To avoid this problem, a transformation is made

$$\theta(x, u) = (\partial u/\partial \eta)^{-1}$$

where the variables are θ—new dependent variable, x—lengthwise coordinate, u—lengthwise velocity, η—transverse coordinate. The full equations are summarized in Reference 1, p. 79. This method is efficient and often four term solutions are comparable in accuracy to 100 term finite difference solutions (Reference 1, p. 80). Unfortunately, it is not possible to ensure the completeness of the trial functions, owing to the transformation, and reversed flow formulations require different trial functions. Indeed for accelerating flows the trial function is $\theta(x, \mu) = \sum c_j(x)P_j(u)/(1 - u)$, where $P_j(u)$ is a polynomial in u. If the singularity in u is not handled correctly, the result may not converge to the solution. Thus, without considerable *a priori* knowledge about the solution, the accuracy cannot be guaranteed. This is both an advantage, if one is doing many calculations for a class of problems, and a disadvantage if one is doing many different kinds of problems.

The question of the proper singularity in Bethel's method is an important one. For free shear layers Stoy[29] applies the method and finds the nonlinear equations difficult to solve. The method was criticized by Wortman and Franks[30] as not being appropriate for an accurate solution and Stoy, in his reply, agreed. These conclusions apply to the type of trial function used and its singularity, rather than to the Method of Weighted Residuals as a whole.

To avoid this problem, MacDonald[31] applies a Crocco transformation in which the new independent variables become x and u while the new dependent variable is $\theta = (\partial u/\partial \eta)^2$. In this formulation there are no singularities and simple polynomial expansions in u can be used. The trial function is substituted in the differential equation, it is multiplied by a weighting function and the integrated result is set to zero. This process leads to nonlinear ordinary differential equations; if the problem admits a similarity solution, algebraic equations result.

Bossel[32] uses exponential trial functions, without making any transformation, and in compressible flow calculations[33] found that 5 terms in the series compared in accuracy to 200 grid-point finite difference calculations. The calculations were faster per time step than finite difference calculations but required more steps due to the stiffness of the equations. Jaffe and Thomas[34] introduced yet another variation by applying quasilinearization to the basic equations and solving the resulting linear equations using a series of Chebyshev polynomials. We note that the same equations result if the solution is expanded in the polynomials and then the Newton-Raphson method is applied. For the Falkner–Skan equation their method required about 10 times fewer terms than were needed in finite difference methods, but the computation times were comparable. In many ways this approach is comparable to applying orthogonal collocation and the corresponding generalization of applying orthogonal collocation on finite elements has not yet been applied.

Another interesting problem for which the Method of Weighted Residuals has been used to good advantage is the treatment of laminar flow in strongly curved tubes.[35] At high Reynolds number, boundary layers are formed, and these are treated using the integral methods described in the previous section. For lower Reynolds number the velocities are expanded in a Fourier series in the angular variable. The trial function is substituted into the equations and the coefficients of $\sin n\phi$ and $\cos n\phi$ are set to zero. This is a Fourier series method, but can also be viewed as a collocation method with Fourier series expansion functions. The ordinary differential equations in r are then solved. In this way the rather formidable two-dimensional, three-velocity-component, flow problem is reduced to manageable proportions.

It is clear from this summary that applications to boundary layer flows using global trial functions have been successful in varying degrees. Finite element methods are feasible, as illustrated by other papers in this book. Certainly the orthogonal collocation method on finite elements is one candidate for calculation, with larger elements far from the body to handle the infinite domain and perhaps generating differential equations in the lengthwise direction. Whether finite element methods are competitive in terms of efficiency of computation (accuracy, computation time and storage) remains an open question.

1.7.3 Flow in packed beds

One of the strengths of the finite element method is its adaptability to different and complicated geometries. This feature is especially useful when one wants to solve similar problems, differing only in geometry, such as in design studies. If the problem has a great deal of symmetry, however, global methods are able to reduce the number of trial functions needed by taking that symmetry into account. Finite element methods, however, can take the symmetry into account (by setting derivatives to zero, for example) only in the elements nearest the line of symmetry. An example of the efficiency of global methods is presented for the slow flow of a Newtonian fluid through a bed of solid spheres packed in a simple cubic array.

Sørensen[36] solved the three-dimensional flow problem given by

$$\mu \nabla^2 \mathbf{v} = -\nabla p$$

with $\mathbf{v} = 0$ on the boundary of the spheres. The equations and geometry, and hence the solution, exhibit the symmetry shown in Table 1.3. The average flow is in the z direction, which is parallel to one orientation of the spheres.

The three-dimensional velocity is expressed as

$$\mathbf{v} = \nabla F \times \nabla G$$

which automatically satisfies the continuity equation.

Table 1.3 Symmetry conditions for flow in simple cubic packed bed*

Transformation	Functions which are unaffected	Functions which change sign
(1) $x \to -x$	v_y, v_z, p	v_x
(2) $y \to -y$	v_x, v_z, p	v_y
(3) $z \to -z$	$v_z, p \to 2p_0 - p(x, y, -z)$	v_x, v_y
Other transformations		
(4) $(x, y) \to (y, x)$	$v_x \to v_y$ v_z, p unchanged	
Periodicity conditions		
(5–7) $\mathbf{v}(x, y, z) = \mathbf{v}(x + 2i, y + 2j, z + 2k)$ $p(x, y, z) = p(x + 2i, y + 2j, z + 2k) - 2k\,\Delta p$ $i, j, k =$ any integer values		

* The spatial dimensions have been non-dimensionalized in terms of the sphere radius.

The symmetry is included in the trial functions by a change of coordinates. The coordinate ξ_2 is given by

$$\xi_2 = 2\frac{\sum_{\pm j=1,2,\ldots} j(1 - 1/P_j)}{\sum_{\pm j=0,1,\ldots} (1 - 1/P_j)}$$

$$P_j = \prod_i \prod_{\pm k=0,1,\ldots} [1 - \exp(-w1 d_{ijk})][1 - \exp(-w2 d_{ijk})]$$

$$\pm i = 1, 3, 5, \ldots, j \text{ odd}$$

$$\pm i = 0, 2, 4, \ldots, j \text{ even}$$

$$(d_{ijk} + 1)^2 = (x - 2i)^2 + (y - 2j)^2 + (z - 2k)^2$$

Three other coordinate lines are obtained from the same expressions using various interchanges of indices. The pair (ξ_1, ξ_2) give coordinate lines which are nearly perpendicular and cover the region (in the x–y plane) between the four spheres. The other pair (ξ_3, ξ_4) curve in the opposite sense and both grids are needed to obtain the full symmetry.

The trial functions are written as Fourier series

$$F_1 = \xi_1/2 \qquad G_2 = \xi_2/2$$

$$G_1 = \xi_2/2 + 2\sum C_m \cos(j\pi\xi_1)\sin(i\pi\xi_2)\cos(k\pi z)$$

$$F_2 = \xi_1/2 + 2\sum C_m \sin(i\pi\xi_1)\cos(j\pi\xi_2)\cos(k\pi z)$$

$$\mathbf{v}_{12} = \nabla F_1 \times \nabla G_1 + \nabla F_2 \times \nabla G_2$$

and this construction for F and G keeps the problem linear in the unknowns, $\{C_m\}$. A similar expression is written to obtain $\mathbf{v}_{34}$ and to obtain the full symmetry we use

$$\mathbf{v} = \mathbf{v}_{12} + \mathbf{v}_{34}$$

The collocation points are spaced somewhat uniformly between the spheres in each x–y plane. The residuals are set to zero and the equations solved. The pressure drop was calculated by evaluating the viscous dissipation, with very good accuracy. For a total number of unknowns of 30, 42 and 54 the pressure drop differed by only $\pm 0{\cdot}04\,\%$. Using that answer as the exact result, the calculations for $n = 2$ had an error of $0{\cdot}7\,\%$ while $n = 8$ gave $0{\cdot}4\,\%$ error. The fact that $1\,\%$ accuracy can be achieved with only 2 terms is a result of including the eight-fold symmetry in the trial functions. If a finite element or finite difference calculation were used we would need about 2^8 times as many mesh points or grid points. This is a dramatic illustration of the power of global trial functions in special cases.

1.7.4 Newtonian fluids flowing past spheres and cylinders

Flow past spheres and cylinders at finite Reynolds numbers is a problem which has been solved many times using different methods. The Galerkin method has been applied to flow past a sphere using a trial function which contains the Stokes solution as a special case[37]

$$\psi(r, \theta) = P_1(r)\sin^2\theta + P_2(r)\sin^2\theta\cos\theta$$

With only two terms in the θ direction, good accuracy could not be expected. At $Re = 100$ the solution was compared to a finite difference solution (using 60 annular grid points and 78 radial points). The Galerkin method using the crude velocity profile gave reasonable accuracy of average quantities, such as drag coefficient, but pointwise values of separation angle, pressure profile on the sphere, etc., were not well determined.

Other authors have expanded the solution in Fourier series (see Reference 38, for cylinders) and Legendre polynomials in $\cos\theta$ (see References 39 and 40 for spheres and 41 for cylinders). The trial function is substituted into the differential equation and the coefficient of each angular function in the trial function is set to zero. This can be interpreted as a collocation method in the θ direction. All these authors then used various finite difference methods to solve the set of ordinary differential equations in the r direction. The calculations could not be taken to high Reynolds number because the number of modes needed increased dramatically with Reynolds number. For the sphere,[39] 16 modes were necessary for very accurate solutions at $Re = 40$. For the cylinder[41] 80 modes were used with $Re = 100$. Collins and Dennis[41] also solved the problem of an impulsively started cylinder and, for small times, transformed the problem using boundary layer coordinates to obtain accurate solutions.

The author and one of his students (Ronald Andermann) have applied orthogonal collocation on finite elements to flow past a sphere. The problem was formulated in terms of the stream function and vorticity.

$$\sin\theta\left[\frac{\partial\psi}{\partial r}\frac{\partial}{\partial\theta}\left(\frac{\zeta}{r^2\sin^2\theta}\right) - \frac{\partial\psi}{\partial\theta}\frac{\partial}{\partial r}\left(\frac{\zeta}{r^2\sin^2\theta}\right)\right] = \frac{2}{Re}E^2\zeta$$

$$r\sin\theta\zeta = E^2\psi$$

$$E^2 = \frac{\partial^2}{\partial r^2} + \frac{\sin\theta}{r^2}\frac{\partial}{\partial\theta}\left(\frac{1}{\sin\theta}\frac{\partial}{\partial\theta}\right)$$

In the θ direction each variable was expanded in sine functions, while the r direction was divided up into elements and orthogonal collocation was applied in each element. The equations were solved iteratively using an alternating direction method. For a given stream function, the vorticity was found from the first equation by successive sweeps in the θ and then the r direction. When sweeping in the θ direction (constant r) the inertial terms involving $\partial\zeta/\partial\theta$ were put on the side with $\partial^2\zeta/\partial\theta^2$ and when sweeping in the r direction the $\partial\zeta/\partial r$ and $\partial^2\zeta/\partial r^2$ terms were used as unknowns in the iteration. When this problem had converged, the next equation was solved for stream function, for the given vorticity. This was also done using an alternating direction method. In this way the storage requirements for the matrices to invert were no larger than needed to solve ordinary differential equations using orthogonal collocation on finite elements. The calculations are not yet complete, but the results to date indicate the method gives very accurate answers; the practical upper limit in Reynolds number, however, has not yet been determined.

The finite element method has been applied to the flow past a cylinder for Reynolds numbers up to 100 by Taylor and Hood.[42] Their recommended procedure leaves the equations in terms of the primary variables, velocity and pressure, and keeps the time derivatives in order to integrate to steady state. The Galerkin method was used on elements with trapezoidal bases and apparently quadratic shape functions.* Taylor and Hood found that the non-linear inertial terms had to be included in the linear iteration scheme, rather than simply be evaluated at the old iteration, in order to have stable calculations. Unfortunately, for the cylinder problem the detailed features like drag coefficient and separation angle were not reported, so that comparisons cannot be made concerning accuracy. The computation time was large, being greater than 100 minutes for $Re = 100$ on an ICL 1905E computer. The work does, however, show the efficacy of the finite element method for Reynolds numbers this high. This paves the way for other applications, like flow past banks of cylinders, where the ease of handling irregular geometries will be especially useful.

1.7.5 Reservoir engineering problems

In the section on trial functions, considerable discussion was given to convergence results for various types of polynomials for one-dimensional problems. The same polynomials can be used in two dimensions, making the methods finite elements on square bases. Applications like this have been made in the field of reservoir engineering problems and the results have been compared to finite difference results, which is the usual method of calculation. The comparisons give some interesting results.

Price and Varga[43] solved a linear time-dependent problem in one dimension, for parameters giving rise to a moving wave front in the solution. They used the Galerkin method with piecewise linear, cubic and quintic trial functions, as well as one calculation with piecewise linear functions everywhere except near the wave front where quintic polynomials were used. The results showed that the computation time was much less than that for finite difference calculations, sometimes 10 times less, and that the improvement was best when solutions were desired of high accuracy.

The next comparison is for a non-linear problem, but still for one space dimension and time. Culham and Varga[44] used smooth linear and cubic Hermite polynomials, non-smooth cubic polynomials and cubic spline polynomials. These trial functions were used along with the Galerkin method. Cubic spline functions were also used with a collocation method. The results were compared based on the pointwise error, but a bias was introduced into the results. The finite difference results were compared based on the error at the grid points, whereas the Galerkin results and collocation results were based on the error throughout the entire domain. The convergence results cited above indicate that the solution in the orthogonal collocation method is more accurate at the collocation points than

* The shape functions are defined for another problem in the same paper, but not for flow past a cylinder.

globally (see, particularly, Reference 9), so that the comparison of the spline collocation method may be comparing the least accurate part of the collocation solution with the most accurate part of the finite difference solution. The Galerkin results may not be discriminated against, since the rate of convergence applies globally. The authors also compared four different methods of handling the time-dependent calculations. They found that the Galerkin results were always inferior to the finite difference results when comparing computation time to achieve a specified accuracy. The cubic spline collocation results were more competitive, but were faster than the finite difference results only for the most accurate solutions determined. The reason given for the dramatic difference in conclusion reached in this non-linear problem, compared to the linear problem, was that a large fraction of the computation time was to evaluate quadratures to be used in the Galerkin method. Since the problem was non-linear, these quadratures had to be recomputed at each time step and the entire method became inefficient. We note also that cubic splines give convergence as Δx^2 whereas cubic polynomials in orthogonal collocation gives convergence as Δx^4.

Two-dimensional calculations are reported by McMichael and Thomas[45] for the full scale, three phase production of an oil field (non-linear equations, parabolic in time, elliptic in two space dimensions). They used both linear and cubic polynomials in each direction on finite elements with square bases. They concluded that the Galerkin method was feasible to apply and in many cases gave results which were superior to those from finite difference schemes and gave more realistic profiles. For a given time step the Galerkin method requires significantly more machine time, but larger time steps are allowed; the net effect was that the finite difference results took from $\frac{1}{2}$ to $\frac{6}{5}$ times as much computer time as the Galerkin method. The Galerkin method also gave less numerical dispersion than the finite difference methods.

These results suggest that for certain non-linear problems the finite element method may not be competitive with finite difference methods in terms of computation time, because of excessive time spent calculating integrals in the Galerkin method. This same problem does not arise in the method of orthogonal collocation on finite elements because the collocation method is used, but the rate of convergence is the same as the Galerkin method when Gaussian quadrature points are used. Consequently this is a type of problem for which the marriage of orthogonal collocation and finite elements may provide a solution to an important difficulty.

1.7.6 Non-Newtonian fluids

When the fluid is no longer Newtonian, the viscosity is a function of the velocity gradient and the integrals in a variational principle (if one exists) or the Galerkin method become difficult to evaluate (e.g. Equation 1.4). Some recent finite element work has tackled these problems using a viscosity which was constant within an element, but which varied from element to element.[46–48] In these cases linear shape functions were used for velocity and pressure (or quadratic functions for stream function) so that the velocity gradients were constant within triangular elements. Then the viscosity function is actually constant within an element. The calculations of Taylor and Hood,[42] however, used quadratic trial functions for Newtonian fluids. If these shape functions were used for non-Newtonian problems, the viscosity would vary over the element, the stiffness matrices would have to be recalculated at every iteration and a finite element solution might run into the same problem found in the reservoir engineering calculations: the finite element may not be competitive with finite difference methods in terms of computation time for a given accuracy.

Another feature of interest is more general constitutive equations. The power law fluid is a common one and has

$$\boldsymbol{\tau} = -2K(2d_{ij}d_{ji})^{n-1/2}\mathbf{d}$$

Polymer flows are usually viscoelastic, however, leading to constitutive equations of the form, e.g.

$$(1 + \lambda_1 F)\boldsymbol{\tau} = 2\mu_0(1 + \lambda_2 F)\mathbf{d}$$

$$F\boldsymbol{\tau} = \frac{\partial \boldsymbol{\tau}}{\partial t} + \mathbf{u}\,.\,\nabla\boldsymbol{\tau} + \mathbf{w}\,.\,\boldsymbol{\tau} - \boldsymbol{\tau}\,.\,\mathbf{w} - (1 + \varepsilon)(\mathbf{d}\,.\,\boldsymbol{\tau} + \boldsymbol{\tau}\,.\,\mathbf{d}) + \tfrac{2}{3}(1 + \varepsilon)(\mathbf{d}\,.\,\boldsymbol{\tau})\mathbf{1}$$

The applicability of finite element methods (or any other method) has yet to be shown, but the possibility of using both stress and velocity shape functions in the finite element method may be useful.

Global trial functions have been applied to non-Newtonian fluids, particularly for flow around spheres and flow through ducts (see Reference 1 for references), but rarely are the solutions done numerically and continued until numerical convergence was obtained. The method of orthogonal collocation on finite elements might be a good compromise for these problems: by avoiding the calculation of integrals but retaining some of the convenient features of finite element methods.

1.7.7 Flow with free boundaries

When a free boundary exists, on which the normal and viscous stresses must be balanced between the two fluids, the calculation is complicated by the unknown position of the boundary. The finite element method is ideally suited to such problems: by assuming the shape of the boundary, solving the problem and then correcting the shape and repeating the calculation, the location of the boundary can be determined. The advantage of the finite element method in handling diverse geometries is especially useful here. Thompson and coworkers[46] treated a power law fluid squeezed between two flat plates, with the fluid exposed to air in between the plates. Chan and Larock[49] applied the finite element method to an inviscid, irrotational fluid out of an orifice. The viscous problem is much more complicated, of course, but the iterative method of finding the shape of the jet was verified. The author is currently applying the method of orthogonal collocation on finite elements to problems involving both non-Newtonian fluids and free boundaries.

1.8 Conclusion

The classical Method of Weighted Residuals, in which the trial functions are defined over the entire domain, and the finite element method, which uses variational or Galerkin principles but utilizes trial functions defined over finite elements, are similar in many respects. The finite element method is preferred when there are rapid changes of the solution, or complicated geometries. In other situations the Method of Weighted Residuals, with global trial functions, is preferred for its very rapid convergence and savings in both storage and computation time. In addition the analytic form of the solution is convenient in many situations. The method of orthogonal collocation on finite elements is a combination of both methods. By using a collocation principle, the quadrature evaluation is eliminated, giving rise to time savings in non-linear problems. By using orthogonal collocation, the method converges as fast as a Galerkin method (in certain cases). So far the method has only been applied to problems with some symmetry, so that diverse geometries have not yet been handled. Orthogonal collocation on finite elements provides a bridge between methods using global and finite element trial functions and the limits of the method are yet to be defined. A broad outlook of weighted residual methods and finite element methods leads to interesting and useful comparisons and interrelations which reveal the advantages of each approach.

Acknowledgement

This work was supported by the National Science Foundation under grant GK-12517.

References

1. B. A. Finlayson, *The Method of Weighted Residuals and Variational Principles*, Academic Press, New York, 1972.
2. B. A. Finlayson and L. E. Scriven, 'On the search for variational principles', *Int. J. Heat Mass Transfer*, **10**, 799–821 (1967).
3. J. V. Villadsen and W. E. Stewart, 'Solution of boundary-value problems by orthogonal collocation', *Chem. Eng. Sci.*, **22**, 1483–1501 (1967).

4. B. A. Finlayson and L. E. Scriven, 'The method of weighted residuals—a review', *Appl. Mech. Rev.*, **19**, 735–748 (1966).
5. N. B. Ferguson and B. A. Finlayson, 'Error bounds for approximate solutions to nonlinear ordinary differential equations', *AIChE J.*, **18**, 1053–1059 (1972).
6. P. G. Ciarlet, M. H. Schultz and R. S. Varga, 'Numerical methods of high-order accuracy for nonlinear boundary value problems. I. One dimensional problem', *Num. Math.*, **9**, 394–430 (1967).
7. C. DeBoor and B. Swartz, 'Collocation of Gaussian points', *SIAM J. Numer. Anal.*, **10**, 582–606 (1973).
8. J. Douglas, Jr. and T. Dupont, 'A finite element collocation method for quasilinear parabolic equations', *Math. Comp.*, **27**, 17–28 (1973).
9. J. Douglas, Jr., 'A superconvergence result for the approximate solution of the heat equation by a collocation method', in *The Mathematical Foundations of the Finite Element Method with Applications to Partial Differential Equations*, A. K. Aziz (ed.), Academic Press, New York, 1972, pp. 475–490.
10. G. F. Carey and B. A. Finlayson, 'Orthogonal collocation on finite elements' (unpublished).
11. L. C. Young and B. A. Finlayson, 'Axial dispersion in nonisothermal packed bed chemical reactors', *IEC Fund.*, **12**, 412–422 (1973).
12. W. F. Ames, *Nonlinear Partial Differential Equations in Engineering*, Academic Press, New York, 1965.
13. B. A. Finlayson, 'Packed bed reactor analysis by orthogonal collocation', *Chem. Eng. Sci.*, **26**, 1081–1091 (1971).
14. L. Lapidus and J. H. Seinfeld, *Numerical Solution of Ordinary Differential Equations*, Academic Press, New York, 1971.
15. F. B. Hildebrand, *Introduction to Numerical Analysis*, McGraw-Hill, New York, 1956.
16. L. Schiller, 'Die Entwichlung der laminaren Geschwendigkeitsverteilung und ihre Bedeutung für Zähigkeitmessungen', *Z. Angew. Math. Mech.*, **2**, 96–106 (1922).
17. W. D. Campbell and J. C. Slattery, 'Flow in the entrance of a tube', Trans. ASME, Series D, *J. Basic Eng.*, **85**, 41–46 (1963).
18. D. P. Fleming and E. M. Sparrow, 'Flow in the hydrodynamic entrance region of ducts of arbitrary cross section', Trans ASME, Series C, *J. Heat Transfer*, **91**, 345–354 (1969).
19. B. Atkinson, M. P. Brocklebank, C. C. H. Card and J. M. Smith, 'Low Reynolds number developing flows', *AIChE J.*, **15**, 548–553 (1969).
20. A. O. Tay and G. De Vahl Davis, 'Application of the finite element method to convection heat transfer between parallel planes', *Int. J. Heat. Mass Transfer*, **14**, 1057–1069 (1971).
21. B. W. Thompson, 'Some semi-analytical methods in numerical fluid dynamics', *Proceedings of the Second International Conference on Numerical Methods in Fluid Dynamics, Lecture Notes in Physics*, Vol. 8, Springer-Verlag, New York, 1971, pp. 73–77.
22. R. E. Melnik and D. C. Ives, 'Subcritical flows over two-dimensional airfoils by a multistrip method of integral relations', *Proceedings of the Second International Conference on Numerical Methods in Fluid Dynamics*, Sept. 15–19, 1970, Lecture Notes in Physics, Vol. 8, Holt, M. (editor), Springer-Verlag, New York, 1971, pp. 243–251.
23. M. Holt and D. E. Ndefo, 'A numerical method for calculating steady unsymmetrical supersonic flow past cones,' *J. of Computational Physics*, **5**, 463–486 (1970).

24. P. I. Chushkin, 'Supersonic flows about conical bodies', *J. of Computational Physics*, **5**, 572–586 (1970).
25. A. A. Dorodnitsyn, 'General method of integral relations and its application to boundary layer theory', *Adv. Aeronaut. Sci.*, **3**, 207–219 (1962).
26. A. A. Dorodnitsyn, 'Exact numerical methods in the boundary-layer theory', in *Fluid Dynamic Transactions*, Vol. 1, W. Fiszdon (ed.), Pergamon Press, Oxford.
27. H. E. Bethel, 'On a convergent multi-moment method for the laminar boundary layer equations', *Aeronautical Quarterly*, **18**, 332–353 (1967); errata, **19**, 402 (1968).
28. H. E. Bethel, 'An improved reversed-flow formulation of the Galerkin–Kantorovich–Dorodnitsyn multi-moment integral method', *Aeronaut. Q.* **20**, 191–202 (1969).
29. R. L. Stoy, 'Method of weighted residuals applied to free shear layers', *AIAA J.*, **8**, 1527–1528 (1970).
30. A. Wortman and W. J. Franks, Comments on 'Method of weighted residuals applied to free shear layers', *AIAA J.*, **9**, 2303–2304 (1971).
31. D. A. MacDonald, 'Solution of the incompressible boundary layer equations via the Galerkin–Kantorovich technique', *J. Inst. Maths. Applics.*, **6**, 115–130 (1970).
32. H. H. Bossel, 'Use of exponentials in the integral solution of the parabolic equations of boundary layer, wake, jet, and vortex flows', *J. Computational Physics*, **5**, 359–382 (1970).
33. N. K. Mitra and H. H. Bossel, 'Compressible boundary-layer computation by the method of weighted residuals using exponentials', *AIAA J.*, **9**, 2370–2377 (1971).
34. N. A. Jaffe and J. Thomas, 'Application of quasi-linearization and Chebyshev series to the numerical analysis of the laminar boundary-layer equations', *AIAA J.*, **8**, 483 (1970).
35. R. J. Nunge and T.-S. Lin, 'Laminar flow in strongly curved tubes', *AIChE J.*, **19**, 1280–1281 (1973).
36. J. P. Sørensen, 'Solution of transport problems by use of trial function techniques', Ph.D. thesis, University of Wisconsin, 1972.
37. A. E. Hamielec, T. W. Hoffman and L. L. Ross, 'Numerical solution of the Navier-Stokes equations for flow past spheres with finite radial mass efflux at the surface', *AIChE J.*, **13**, 212 (1967).
38. R. L. Underwood, 'Calculation of incompressible flow past a circular cylinder at moderate Reynolds numbers', *J. Fluid .Mech.*, **37**, 95–114 (1969).
39. S. C. R. Dennis and J. D. A. Walker, 'Calculation of the steady flow past a sphere at low and moderate Reynolds numbers', *J. Fluid Mech.*, **48**, 771–789 (1971).
40. S. C. R. Dennis and J. D. A. Walker, 'Numerical solutions for time-dependent flow past an impulsively started sphere', *Phys. Fluids*, **15**, 517–525 (1972).
41. W. M. Collins and S. C. R. Dennis, 'Flow past an impulsively started circular cylinder', *J. Fluid Mech.*, **60**, 105–127 (1973).
42. C. Taylor and P. Hood, 'A numerical solution of the Navier–Stokes equations using the finite element technique', *Computers and Fluids*, **1**, 73–100 (1973).
43. H. S. Price and R. S. Varga, 'Error bounds for semidiscrete Galerkin approximations of parabolic problems with applications to petroleum reservoir mechanics', in *Numerical Solution of Field Problems in Continuum Physics*, Vol. II, G. Birkoff and R. S. Varga (eds.), American Mathematical Society, Providence, R.I., 1970.
44. W. E. Culham and R. S. Varga, 'Numerical methods for time dependent nonlinear boundary value problems', *Soc. Pet. Eng. J.*, **11**, 374–388 (1971).
45. C. L. McMichael and G. W. Thomas, 'Reservoir simulation by Galerkin's method', *Soc. Pet. Eng. J.*, **13**, 125–138 (1973).

46. E. G. Thompson, L. R. Mack and F. S. Lin, 'Finite-element method for incompressible slow viscous flow with a free surface', *Devel. in Mechanics*, **5**, 93–111 (1969).
47. K. Palit and R. T. Fenner, 'Finite element analysis of slow non-Newtonian channel flow', *AIChE J.*, **18**, 628–633 (1972).
48. K. Palit and R. T. Fenner, 'Finite element analysis of two-dimensional slow non-Newtonian flows., *AIChE J.*, **18**, 1163–1170 (1972).
49. S. T. K. Chan and B. E. Larock, 'Fluid flows from axisymmetric orifices and valves', *J. Hydraulics Div., Proc. ASCE*, **99**, 81–97 (1973).

Chapter 2

Accuracy and Convergence of Finite Element/Galerkin Approximations of Time-Dependent Problems with Emphasis on Diffusion

L. C. Wellford Jr. and J. T. Oden

2.1 Introduction

In very recent times, several distinct methods have emerged for studying the central mathematical questions surrounding finite element schemes for time-dependent problems: accuracy, convergence and stability. The ultimate utility of the method in applications to evolution problems hinges on the answers to these questions. The present paper is largely an exposition aimed at analysing a number of techniques for arriving at error estimates of finite element/finite difference approximations of time-dependent diffusion problems. However, much of what we consider can be applied to finite element approximations of all types of linear boundary/initial-value problems.

In the exposition to follow, we describe a number of basic techniques for determining rates-of-convergence of approximations of a class of linear time-dependent problems. We choose to categorize these techniques as follows:

(1) Semigroup theoretic estimates
(2) Energy methods
(3) L_2 methods
(4) Other methods

(1) It is a widely known result in the theory of partial differential equations that for a broad class of evolution equations the fundamental solution operator is a member of a semigroup of operators (see, for example, References 6 and 7). By using certain basic properties of semigroups, error estimates can often be easily obtained, particularly in those cases in which it is possible to transform the problem into a one-parameter family of elliptic problems. This makes the problem of estimating the spatial rate-of-convergence relatively straightforward. The rate-of-convergence of the temporal approximations can be established by exploiting certain properties of fundamental

solutions. For example, if the fundamental solution is a member of a semigroup, then it has a matrix exponential form which can be approximated using the Padé matrix-approximation theory. The use of semigroups in difference approximations of time-dependent problems has been discussed by several authors (e.g. Peetre and Thomee[8]; Widlund[9]). Some features of the methods we discuss were used by Babuska and Aziz in collaboration with Fix.[1]

(2) In most mathematical models of physical problems a norm can be developed which is equivalent to the total energy in the system. Normally, the study of convergence of various Galerkin approximations in an appropriate energy norm is a very natural undertaking. This is due to the fact that the weak forms of most boundary-value problems of mathematical physics can be interpreted physically in terms of changes in energy. The use of energy methods can be found in the finite difference literature (see, for example, Reference 18), and variants have been used by Fujii[11] and Oden and Fost[12] for finite element analyses.

(3) A fairly extensive literature has accumulated in recent years on Galerkin approximations of the diffusion equation in which L_2 estimates in spatial variations and L_2 or L_∞ estimates are obtained in the temporal variations. This 'L_2 theory' has been largely developed by Douglas and Dupont,[13,14,15] Varga,[16] Wheeler[17] and others. Interestingly enough, the methods do not involve energy-error estimates and enable one to go directly to stronger L_2 estimates.

(4) There are, of course, several other techniques in use for studying finite element approximations of time-dependent problems. The projection methods of Thomee,[18] for example, make use of the fact that the finite element technique produces a system of difference equations in R_n. By using standard difference concepts and projections from R^n back into the space V in which the original problem is posed, error estimates for certain finite element approximations can be obtained. Alternatively, certain parabolic problems can be shown to have coercive properties under an appropriate choice of norm (see, for example, Lions and Magenes[19]). Thus, 'elliptic type' error estimates can be obtained, as shown by Cella and Cecchi.[20] It is only a matter of interpretation as to whether or not these 'projection' and 'coercive operator' techniques do not actually belong to the semigroup and the energy methods described previously.

As in elliptic theory, the study of convergence of finite element approximations rests firmly on certain results from interpolation theory. For this reason we discuss, in the section following this introduction, certain features of interpolation theory which are essential for our investigation. Next, in Section 2.3, we describe finite element/Galerkin models of a general class of diffusion problems, and in Sections 2.4, 2.5 and 2.6 we obtain error estimates for these models using the semigroup theory, energy methods and L_2 methods, respectively.

2.2 Finite element approximation and interpolation

Consider a Hilbert space $\mathscr{H}$ whose elements are functions $u(\mathbf{x})$ of points $\mathbf{x} = (x_1, x_2, \ldots, x_n)$ in some bounded domain Ω of R^n. In subsequent sections, the context shall make clear the specific properties of $\mathscr{H}$, but for the moment we need only assume that it is endowed with an inner product, (u_1, u_2). A finite element model of Ω (and $\mathscr{H}$) is another region $\hat{\Omega}$ which is partitioned into a finite number E of disjoint open sets Ω_e called finite elements:

$$\hat{\Omega} = \bigcup_{e=1}^{E} \bar{\Omega}_e; \qquad \Omega_e \cap \Omega_f = 0 \quad \text{if } e \neq f \tag{2.1}$$

Here $\bar{\Omega}_e$ is the closure of Ω_e. Within each element we identify a set of local basis functions $\psi_N^{\boldsymbol{\alpha}(e)}(\mathbf{x})$ which have the property that

$$\begin{aligned} D^{\boldsymbol{\beta}}\psi_N^{\boldsymbol{\alpha}(e)}(\mathbf{x}_f^M) = \delta^{\boldsymbol{\beta}\boldsymbol{\alpha}}\delta_N^M\delta_f^e; \qquad & \psi_N^{\boldsymbol{\alpha}(e)}(\mathbf{x}) = 0 \qquad \mathbf{x} \neq \Omega_e \\ \boldsymbol{\alpha}, \boldsymbol{\beta} \in Z_+^n; \qquad & M, N = 1, 2, \ldots, N_e; \\ e, f = 1, 2, \ldots, E; \quad & |\boldsymbol{\alpha}| \leqslant k \end{aligned} \tag{2.2}$$

Here $\mathbf{x}_f^M$ is a nodal point labelled M in element Ω_f, $\delta^{\boldsymbol{\beta}\boldsymbol{\alpha}}$, δ_N^M, δ_f^e, are Kronecker deltas, N_e is the number of nodes in element Ω_e, and we have used multi-index notation, i.e. $\boldsymbol{\alpha}$ and $\boldsymbol{\beta}$ are ordered n-tuples of non-negative integers, $\boldsymbol{\alpha} = (\alpha_1, \alpha_2, \ldots, \alpha_n)$, and the following conventions are used:

$$\begin{aligned} D^{\boldsymbol{\alpha}}u(\mathbf{x}) \equiv \frac{\partial^{|\boldsymbol{\alpha}|}u(\mathbf{x})}{\partial x_1^{\alpha_1}\partial x_2^{\alpha_2}\ldots\partial x_n^{\alpha_n}}; \qquad & |\boldsymbol{\alpha}| \equiv \alpha_1 + \alpha_2 + \cdots + \alpha_n \\ \mathbf{x}^{\boldsymbol{\alpha}} \equiv x_1^{\alpha_1}x_2^{\alpha_2}\ldots x_n^{\alpha_n}; \qquad & \delta^{\boldsymbol{\alpha}\boldsymbol{\beta}} = \delta^{\alpha_1\beta_1}\delta^{\alpha_2\beta_2}\ldots\delta^{\alpha_n\beta_n} \end{aligned} \tag{2.3}$$

The *local* representation of a function in terms of the basis functions $\psi_N^{\boldsymbol{\alpha}(e)}(\mathbf{x})$ is of the form

$$V_e(\mathbf{x}) = \sum_{|\boldsymbol{\alpha}| \leqslant k} \sum_{N=1}^{N_e} a_{\boldsymbol{\alpha}(e)}^N \psi_N^{\boldsymbol{\alpha}(e)}(\mathbf{x}); \qquad a_{\boldsymbol{\alpha}(e)}^N = D^{\boldsymbol{\alpha}}u_e(\mathbf{x}_e^N) \tag{2.4}$$

and the *global* representation is of the form

$$V(\mathbf{x}) = \bigcup_{e=1}^{E} V_e(\mathbf{x}_e) = \sum_{|\alpha| \leqslant k} \sum_{\Delta=1}^{G} A_{\boldsymbol{\alpha}}^{\Delta}\chi_{\Delta}^{\boldsymbol{\alpha}}(\mathbf{x}) \tag{2.5}$$

Here $\chi_{\Delta}^{\boldsymbol{\alpha}}(\mathbf{x})$ are global basis functions given by

$$\chi_{\Delta}^{\boldsymbol{\alpha}}(\mathbf{x}) = \bigcup_{e=1}^{E} \sum_{N=1}^{N_e} \overset{(e)}{\Omega}{}_{\Delta}^{N} \psi_N^{\boldsymbol{\alpha}(e)}(\mathbf{x}_e) \tag{2.6}$$

where $\overset{(e)}{\Omega}{}_{\Delta}^{N}$ defines a Boolean transformation of the disconnected system of elements into the connected model Ω (i.e. $\overset{(e)}{\Omega}{}_{\Delta}^{N} = 1$ if node N of Ω_e coincides with node $\mathbf{x}^{\Delta}$ of Ω and $\overset{(e)}{\Omega}{}_{\Delta}^{N} = 0$ if otherwise).

Suppose $\Omega = \hat{\Omega}$. Then the set of functions $\{\chi_\Delta^\alpha(\mathbf{x})\}_{\Delta=1}^G$; $|\alpha| \leqslant k$ defines a finite-dimensional subspace of $\mathscr{H}$ which we shall denote $\mathscr{S}_h(\Omega)$. Here h is the mesh parameter of the finite element mesh (i.e. if $h_e = \text{dia}\,(\Omega)_e$, $h = \max(h_1, h_2, \ldots, h_E)$). We define $\mathscr{S}_h(\Omega)$ more precisely subsequently. For economy in notation, we shall re-label the global basis functions $\chi_\Delta^\alpha(\mathbf{x})$ as $\phi_N(\mathbf{x})$ $N = 1, 2, \ldots, N_0$ (i.e. we introduce an automorphism $a_N^{\Delta\alpha}: \mathscr{S}_h(\Omega) \to \mathscr{S}_h(\Omega)$). Then each global representation is of the form

$$V(\mathbf{x}) = \sum_{N=1}^{N_0} A^N \phi_N(\mathbf{x}) \tag{2.7}$$

Returning now to the space $\mathscr{H}$, consider a typical element u $(= u(\mathbf{x}))$. The pair $\{(\cdot, \cdot), \{\phi_N\}_{N=1}^{N_0}\}$ define an orthogonal projection $Q_h: \mathscr{H} \to \mathscr{S}_h(\Omega)$ such that

$$Q_h u \equiv W = \sum_{N=1}^{N_0} (u, \phi^N)\phi_N(\mathbf{x}) \tag{2.8}$$

where $\phi^N(\mathbf{x}) \equiv \sum_M (\phi_N, \phi_M)^{-1}\phi_M(\mathbf{x})$. The function

$$E(\mathbf{x}) \equiv u(\mathbf{x}) - Q_h u(\mathbf{x}) \tag{2.9}$$

is referred to as the (pointwise) *interpolation error* of the finite element approximation $W(\mathbf{x}) = Q_h u(\mathbf{x})$. Its properties depend explicitly on the properties of the subspace $\mathscr{S}_h(\Omega)$.

To appreciate the importance of the interpolation error in finite element/Galerkin theory, consider the abstract boundary-value problem, find $u \in \mathscr{H}$ such that

$$(\mathscr{P}(u), v) = (f, v) \qquad \forall v \in \mathscr{M} \tag{2.10}$$

where $\mathscr{P}: \mathscr{H} \to \mathscr{M}$ is a linear operator. The finite element/Galerkin approximation of the solution u of (2.10) is the function $U \in \mathscr{S}_h$ such that

$$(\mathscr{P}(U), V) = (f, V) \qquad \forall V \in \mathscr{S}_h(\Omega) \tag{2.11}$$

The function

$$e(\mathbf{x}) \equiv u(\mathbf{x}) - U(\mathbf{x}) \tag{2.12}$$

is the (pointwise) *approximation error* of $U(\mathbf{x})$. Since $U(\mathbf{x}) \in \mathscr{S}_h(\Omega)$, there is a mapping $\Pi_h: \mathscr{H} \to \mathscr{S}_h(\Omega)$ such that $\Pi_h u = U$; but Π_h is not an orthogonal projection into $\mathscr{S}_h(\Omega)$ relative to $\{(\cdot, \cdot), \{\phi_N\}\}$. Indeed, by setting $v = V$ in (2.10) and subtracting (2.11) from the result, we see that $(\mathscr{P}e, V) = 0$; thus, $\mathscr{P}e$ is orthogonal to $\mathscr{S}_h$.

The function

$$\mathscr{E}(\mathbf{x}) = \Pi_h u(\mathbf{x}) - Q_h u(\mathbf{x}) = U(\mathbf{x}) - Q_h u(\mathbf{x}) \tag{2.13}$$

is referred to as the *projection error*. In most of the developments to follow, we show that it is possible to bound certain norms of the projection error by the corresponding norm of the interpolation error; i.e. we derive relationships of the type

$$\|\mathscr{E}\|_{\mathscr{H}} = C(h)\|\mathscr{E}\|_{\mathscr{H}} \tag{2.14}$$

Since $e = E - \mathscr{E}$, use of the triangle inequality gives

$$\|e\|_{\mathscr{H}} \leqslant (1 + C(h))\|E\|_{\mathscr{H}} \tag{2.15}$$

Thus, (*i*) if the coefficient $(1 + C(h))$ remains bounded as $h \to 0$, and (*ii*) if $\|E\|_{\mathscr{H}} \to 0$ as $h \to 0$, we have proved convergence of the method in the $\|\cdot\|_{\mathscr{H}}$ norm. Criterion (*i*) is a question of *stability* of the approximation while (*ii*) is a question of *consistency*. The latter question depends explicitly on $\mathscr{H}$ and the properties of $\mathscr{S}_h(\Omega)$, so that the convergence of the method is connected to the interpolation error in a fundamental way.

In many instances, the space $\mathscr{H}$ is a Sobolev space $H^m(\Omega)$, the elements of which are functions whose partial derivatives of order $\leqslant m$ are square integrable on Ω. The inner product in $H^m(\Omega)$ is then

$$(u, v)_m = \int_\Omega \sum_{|\alpha| \leqslant m} D^\alpha u D^\alpha v \, \mathrm{d}\Omega \tag{2.16}$$

and the norm is

$$\|u\|_m^2 = \int_\Omega \left(\sum_{|\alpha| \leqslant m} D^\alpha u \right)^2 \mathrm{d}\Omega \tag{2.17}$$

where $\mathrm{d}\Omega = \mathrm{d}x_1 \, \mathrm{d}x_2 \ldots \mathrm{d}x_n$. In such cases we construct the finite element subspaces $\mathscr{S}_h(\Omega)$ so as to have the following properties:

(1) For every $u \in H^m(\Omega)$, there is a constant C such that $\|Q_h u\|_m \leqslant C\|u\|_m$ for all $h \geqslant 0$.

(2) If $p(\mathbf{x})$ is a polynomial of degree $\leqslant k$,

$$Q_h p(\mathbf{x}) = p(\mathbf{x})$$

(3) Let $h \to 0$ uniformly (i.e., for each refinement of the mesh, let the radius ρ_e of the largest sphere that can be inscribed in Ω_e be proportional to h_e). Then there is a constant K independent of h such that

$$\|E\|_n \leqslant K h^{k+1-n} |u|_{k+1} \tag{2.18}$$

for $n \leqslant m$, where $|u|_{k+1}$ is the semi-norm

$$|u|_m^2 \equiv \int_\Omega \left(\sum_{|\alpha| \leqslant m} D^\alpha u \right)^2 \mathrm{d}\Omega \tag{2.19}$$

Interpolation results such as (2.18) were derived by Strang,[21] Ciarlet and Raviart[22] and others. We shall henceforth assume that the *spatial* interpolation spaces $\mathscr{S}_h(\Omega)$ have properties (2.1)–(2.3).

2.3 Finite element approximation of the diffusion equation

We now consider a class of time-dependent problems characterized by equations of the form

$$\frac{\partial u(\mathbf{x}, t)}{\partial t} + A(\mathbf{x})u(\mathbf{x}, t) = f(\mathbf{x}, t) \qquad \mathbf{x} \in \Omega;\ t \in (0, T]$$

$$D^{\alpha}u(\mathbf{x}, t) = 0, \qquad \mathbf{x} \in \partial\Omega;\ |\boldsymbol{\alpha}| \leqslant m - 1$$

$$u(\mathbf{x}, 0) = u_0(\mathbf{x}), \qquad x \in \Omega \tag{2.20}$$

where A is the $2m$th order differential operator

$$A(\mathbf{x}) \equiv \sum_{|\boldsymbol{\alpha}|,|\boldsymbol{\beta}| \leqslant m} (-1)^{|\boldsymbol{\alpha}|} D^{\boldsymbol{\alpha}} A_{\boldsymbol{\alpha\beta}}(\mathbf{x}) D^{\boldsymbol{\beta}} \tag{2.21}$$

and the coefficients $A_{\boldsymbol{\alpha},\boldsymbol{\beta}}(\mathbf{x})$ are such that A is m elliptic (strongly elliptic), i.e. there exists a sesquilinear form

$$a(u, v) \equiv (Au, v) = \int_{\Omega} \sum_{|\boldsymbol{\alpha}|,|\boldsymbol{\beta}| \leqslant m} A_{\boldsymbol{\alpha\beta}}(\mathbf{x}) D^{\boldsymbol{\alpha}} u D^{\boldsymbol{\beta}} v \, d\Omega \tag{2.22}$$

such that there are positive constants μ_0 and μ_1 for which

$$a(u, u) \geqslant \mu_0 \|u\|_m^2 \tag{2.23}$$

and

$$a(u, v) \leqslant \mu_1 \|u\|_m \|v\|_m \tag{2.24}$$

We then replace (2.20) by the equivalent (weaker) variational problem,

$$\left(\frac{\partial u(t)}{\partial t}, v\right)_0 + a(u(t), v) = (f(t), v)_0$$

$$(u(\,.\,, 0), v)_0 = (u_0, v)_0, \qquad \forall v \in H_0^m(\Omega)$$

$$\forall t \in (0, T] \tag{2.25}$$

where $H_0^m(\Omega)$ is the Sobolev space of $H^m(\Omega)$ functions with compact support in Ω.

Now consider a Galerkin approximation of (2.24) which involves seeking a function $U(\mathbf{x}, t) \in \mathscr{S}_h(\Omega) \times C^1(0, T)$ such that

$$\left(\frac{\partial U(t)}{\partial t}, V\right)_0 + a(U(t), V) = (f(t), V)_0 \qquad t \in (0, T]$$

$$(U(\,.\,, 0), V)_0 = (u_0, V)_0 \qquad \forall V \in \mathscr{S}_h(\Omega) \tag{2.26}$$

Since a continuous dependence on t is still assumed, $U(\mathbf{x}, t)$ is referred to as a *semidiscrete* Galerkin approximation. Now, the coefficients A^N in (2.7) are functions of time t. Thus, introducing (2.7) into (2.26) we obtain a system of first-order differential equations for the specific coefficients $A^N(t)$ corresponding to the finite element approximation:

$$\sum_{M=1}^{N_0} C_{NM}\dot{A}^M(t) + \sum_{M=1}^{N_0} K_{NM}A^M(t) = f_N(t)$$

$$\sum_{M=1}^{N_0} C_{NM}A^M(0) = l_N \tag{2.27}$$

Here

$$C_{NM} = (\phi_N, \phi_M)_0; \qquad K_{NM} = a(\phi_N, \phi_M); \qquad f_N = (f, \phi_N)_0;$$
$$l_N = (u_0, \phi_N)$$

and $\dot{A}^M(t) = \mathrm{d}A^M(t)/\mathrm{d}t$.

In practical calculations, we introduce the partition P of $[0, T]$ composed of the set $\{t_0, t_1, \ldots, t_R\}$ where $0 = t_0 < t_1 < \cdots < t_R = T$ with $t_{n+1} - t_n = \Delta t$; and we introduce the sequence $\{U^n\}_{n=0}^R$ to denote the value of the function $U(t) \in \mathscr{S}_h(\Omega) \times C^1(0, T)$ at the time points of partition P. Thus $\{U^n\}_{n=0}^R = \{U(t_n)\}_{n=0}^R$. Then we construct a family of *finite difference/Galerkin* approximations associated with parameter θ $(0 \leqslant \theta \leqslant 1)$ which represent solutions to the following equation:

$$(\delta_t U^n, V)_0 + (1 - \theta)a(U^{n+1}, V) + \theta a(U^n, V) = (f(t), V)_0 \qquad t \in (0, T]$$

$$(U^0, V)_0 = (u_0, V)_0 \qquad \forall V \in \mathscr{S}_h(\Omega) \tag{2.28}$$

where δ_t denotes the forward difference operator, i.e. $\delta_t U^n = (U^{n+1} - U^n)/\Delta t$. Evaluation of (2.28) with the finite element approximation (2.7) leads to a system of algebraic equations for the coefficients $A_N^n = A^N(t_n)$:

$$\sum_{M=1}^{N_0} [C_{NM} + \Delta t(1 - \theta)\delta_{NQ}K_{QM}]A_M^{n+1} = \sum_{M=1}^{N_0} [C_{NM} - \Delta t\theta\delta_{NQ}K_{QM}]A_M^n + \Delta t f_N$$

$$\sum_{M=1}^{N_0} C_{NM}A_M^0 = l_N \tag{2.29}$$

2.4 Error estimates for the diffusion equation using semigroup theoretic results

The calculation of error estimates for (2.25) can be embedded in the theory of the approximation of elliptic partial differential equations by using semigroup tools. In particular, the traditional characterization of the semigroup through the resolvent operator can be used to transform the parabolic diffusion problem into a one-parameter family of elliptic problems. The error in the spatial discretization for the parabolic problem can then be related to errors encountered in modelling elliptic problems. On the other hand, errors due to the discretization in time can be determined through the exponential function representation for the semigroup and its relationship to the Padé approximations.

Initially, we must define the components of the error of the approximation scheme. $u(\mathbf{x}, t)$ is the exact solution to (2.25); $U(\mathbf{x}, t)$ is the solution to the semidiscrete Galerkin approximation (2.26); and $U^n(\mathbf{x})$ is the solution to the finite difference/Galerkin approximation (Equation 2.28) at time $t = n\Delta t$. We introduce the definitions,

$$e(\mathbf{x}, n\,\Delta t) = u(\mathbf{x}, n\,\Delta t) - U^n(\mathbf{x}) = \text{approximation error} \tag{2.30}$$

$$\sigma(\mathbf{x}, n\,\Delta t) = u(\mathbf{x}, n\,\Delta t) - U(\mathbf{x}, n\,\Delta t) = \text{semidiscrete approximation error} \tag{2.31}$$

$$\tau(\mathbf{x}, n\,\Delta t) = U(\mathbf{x}, n\,\Delta t) - U^n(\mathbf{x}) = \text{temporal approximation error} \tag{2.32}$$

Then, for any choice of norm on $\mathbf{x}$,

$$\begin{aligned} \|e(t)\| &= \|u(\,.\,, n\Delta t) - U^n(\,.\,) + U(\,.\,, n\Delta t) - U(\,.\,, n\Delta t)\| \\ &= \|\sigma + \tau\| \leqslant \|\sigma\| + \|\tau\| \end{aligned} \tag{2.33}$$

The evolution problem (Equation 2.20) is reminiscent of the linear dynamical system

$$\dot{\mathbf{q}} - \mathbf{A}\mathbf{q} = \mathbf{f}, \qquad \mathbf{q}(0) = \mathbf{q}_0$$

the solution of which is

$$\dot{\mathbf{q}} = e^{\mathbf{A}t}\mathbf{q}_0 + \int_0^t e^{\mathbf{A}(t-s)}\mathbf{f}(s)\,\mathrm{d}s$$

For simplicity, let us assume that $\mathbf{f}(s) = 0$. Then the matrix $\mathbf{E}(t)$ given by

$$\mathbf{E}(t) = e^{\mathbf{A}t}$$

is called the *fundamental solution operator* and

$$\mathbf{q}(t) = \mathbf{E}(t)\mathbf{q}_0$$

The operator $\mathbf{E}(t)$ is a member of a multiplicative semigroup, $\mathscr{G}$; i.e., if $\mathbf{E}_1(t)$ and $\mathbf{E}_2(t)$ are in $\mathscr{G}$, then the semigroup properties,

$$\begin{aligned} &\text{(i) } \mathbf{E}_1 \,.\, \mathbf{E}_2 \in \mathscr{G} \qquad \text{(closure)} \\ &\text{(ii) } \mathbf{E}_1 \,.\, (\mathbf{E}_2 \,.\, \mathbf{E}_3) = (\mathbf{E}_1 \,.\, \mathbf{E}_2) \,.\, \mathbf{E}_3 \qquad \text{(associativity)} \end{aligned} \tag{2.34}$$

are satisfied. In fact, we also have the important properties,

$$\begin{aligned} &\mathbf{E}(t)\mathbf{E}(s) = \mathbf{E}(t + s) \\ &\lim_{t \to 0} \mathbf{E}(t) = \mathbf{I} \end{aligned} \tag{2.35}$$

How does one construct the fundamental solution operator $\mathbf{E}(t)$ from a given matrix $\mathbf{A}$? Let $\mathscr{L}(\mathbf{q}(t)) = \bar{\mathbf{q}}$ denote the Laplace transform of $\mathbf{q}(t)$. Then

$$s\bar{\mathbf{q}} - \mathbf{q}(0) = \mathbf{A}\bar{\mathbf{q}}$$

Thus, if s is not an eigenvalue of $\mathbf{A}$,

$$\bar{\mathbf{q}} = (s\mathbf{I} - \mathbf{A})^{-1}\mathbf{q}_0$$

$$\mathbf{E}(t) = \mathscr{L}^{-1}(s\mathbf{I} - \mathbf{A})^{-1} \tag{2.36}$$

The operator $(s\mathbf{I} - \mathbf{A})^{-1}$ is called the resolvent operator $\mathbf{R}(\mathbf{x}, s)$. To ensure that the inverse Laplace transform exists, we assume that the operator $\mathbf{A}$ is the infinitesimal generator of a strongly continuous semigroup. The Hille–Yosida–Phillips theorem (Friedman)[23] guarantees that if the operator $\mathbf{A}$ is the infinitesimal generator of a strongly continuous semigroup, there exist real numbers M and ω such that for every $R_e s > \omega$, s is in the resolvent set of $\mathbf{A}$ (i.e. $(s\mathbf{I} - \mathbf{A})^{-1}$ exists) and

$$\|\mathbf{R}(\mathbf{x}, s)^n\| \leqslant \frac{M}{(s - \omega)^n} \qquad n = 1, 2, 3, \ldots \tag{2.37}$$

Thus if we select the contour Γ in the complex plane so that if $s \in \Gamma$, s is in the resolvent set, $R_e s > \beta$, and $R_e s = \beta$ as $s \to \infty$ where β is a small negative constant, then the integrand in (2.36) is bounded in accordance with (2.37) as s increases. Thus the Laplace transform (2.36) exists. We choose β so that $|\beta| < \mu_0$.

Now a similar situation is encountered in the use of the weak formulation of partial differential equations of the type (2.25). In this case we define a fundamental solution operator $\mathbf{E}(\mathbf{x}, t)$ such that

$$u(\mathbf{x}, t) = \mathbf{E}(\mathbf{x}, t)u_0(\mathbf{x})$$

Now we take the Laplace transform of the weak parabolic partial differential equation (2.25) and set $f(t) = 0$, for convenience. Then

$$\mathscr{L}\left[\left(\frac{\partial u(t)}{\partial t}, v\right)_0\right] + \mathscr{L}[a(u(t), v)] = 0 \qquad \forall v \in H_0^m(\Omega) \tag{2.38}$$

Now

$$\begin{aligned}
\mathscr{L}\left[\left(\frac{\partial u(t)}{\partial t}, v\right)_0\right] &= \int_\Omega \int_0^t e^{-st} \frac{\partial u(\mathbf{x}, t)}{\mathrm{d}t} v(\mathbf{x})\,\mathrm{d}t\,\mathrm{d}\Omega \\
&= s\int_\Omega \bar{u}(\mathbf{x}, s)v(\mathbf{x})\,\mathrm{d}\Omega - \int_\Omega u_0(\mathbf{x})v(\mathbf{x})\,\mathrm{d}\Omega \\
&= s(\bar{u}(s), v)_0 - (u_0, v)_0
\end{aligned} \tag{2.39}$$

and

$$\mathscr{L}[a(u(t), v)] = a(\bar{u}(s), v) \tag{2.40}$$

Hence introducing (2.39) and (2.40) into (2.38), we obtain the boundary-value problem

$$s(\bar{u}(s), v)_0 + a(\bar{u}(s), v) = (u_0, v)_0 \qquad \forall v \in H_0^m(\Omega) \tag{2.41}$$

Similarly, if U is the Galerkin approximation to the exact solution, then we define the approximate fundamental solution operator $\mathbf{E}_h(\mathbf{x}, t)$ so that

$$U(\mathbf{x}, t) = \mathbf{E}_h(\mathbf{x}, t)u_0(\mathbf{x}) \tag{2.42}$$

Now if we take the Laplace transform of the semidiscrete Galerkin equation (2.26) and set $f(t) = 0$ for convenience, we obtain an approximate boundary-value problem for (2.41).

$$s(\bar{U}(s), V)_0 + a(\bar{U}(s), V) = (u_0, V)_0 \qquad \forall V \in \mathscr{S}_h(\Omega) \tag{2.43}$$

Subtracting (2.43) from (2.41) and defining the transformed approximation error $\bar{\sigma}(s)$ by $\bar{u}(s) - \bar{U}(s) = \mathscr{L}[\sigma(t)] = \mathscr{L}[u(t) - U(t)]$, we get

$$s(\bar{\sigma}(s), V)_0 + a(\bar{\sigma}(s), V) = 0 \qquad \forall V \in \mathscr{S}_h(\Omega) \tag{2.44}$$

Thus it is clear that if we can estimate the approximation error $\bar{\sigma}(s)$ for the boundary-value problem (2.41), then we can deduce the approximation error for problem (2.25) by using the inverse Laplace transform, (2.36):

$$\sigma(t) = \mathscr{L}^{-1}[\bar{\sigma}(s)] = \mathscr{L}^{-1}[\bar{u}(s) - \bar{U}(s)] \tag{2.45}$$

The operator A satisfies the m elliptic condition (2.23). Thus the operator $\mathbf{A} + R_e s\mathbf{I}$ occurring in (2.41) also satisfies an m elliptic condition since

$$\begin{aligned} ((\mathbf{A} + R_e s\mathbf{I})v, v)_0 &= (\mathbf{A}v, v)_0 + R_e s(v, v)_0 \\ &= a(v, v) + R_e s(v, v)_0 \\ &= a(v, v) + R_e s\|v\|_0^2 \\ &\geqslant a(v, v) + \beta\|V\|_m^2 \\ &\geqslant \mu_2\|V\|_m \end{aligned} \tag{2.46}$$

Thus the boundary value problem (2.41) is strongly elliptic for each $s \in \Gamma$. Equation 2.46 leads to the following error estimate.

Theorem 2.1 If $\bar{U}(s)$ is the Galerkin approximation in $\mathscr{S}_h$ to $\bar{u}(s)$ the solution of problem (2.41), then

$$\|\bar{u}(s) - \bar{U}(s)\|_m \leqslant C_1 h^{k+1-m}|\bar{u}(s)|_{k+1} \tag{2.47}$$

Proof: From the m ellipitic property of operator $A + R_e s\mathbf{I}$ (Equation 2.46) we have that if $\bar{U}^*(s)$ is an arbitrary element of $\mathscr{S}_h$

$$\begin{aligned}\mu_2\|\bar{u}(s) - \bar{U}(s)\|_m^2 &\leqslant R_e s(\bar{u}(s) - \bar{U}(s), \bar{u}(s) - \bar{U}(s))_0 \\ &\quad + a(\bar{u}(s) - \bar{U}(s), \bar{u}(s) - \bar{U}(s)) \\ &= R_e s(\bar{u}(s) - \bar{U}(s), \bar{u}(s) - \bar{U}^*(s) \\ &\quad + \bar{U}^*(s) - \bar{U}(s))_0 + a(\bar{u}(s) - \bar{U}(s), \\ &\quad \bar{u}(s) - \bar{U}^*(s) + \bar{U}^*(s) - \bar{U}(s)) \\ &= R_e s(\bar{u}(s) - \bar{U}(s), \bar{u}(s) - \bar{U}^*(s))_0 \\ &\quad + a(\bar{u}(s) - \bar{U}(s), \bar{u}(s) - \bar{U}^*(s))\end{aligned} \tag{2.48}$$

To obtain the last result, we have used the identity

$$R_e s(\bar{u}(s) - \bar{U}(s), \bar{U}^*(s) - \bar{U}(s))_0 + a(\bar{u}(s) - \bar{U}(s), \bar{U}^*(s) - \bar{U}(s)) = 0$$

obtained by taking the real part of (2.44) and setting $V = \bar{U}^*(s) - \bar{U}(s)$. Now using the Schwarz inequality (2.24) in (2.48)

$$\begin{aligned}\mu_2\|\bar{u}(s) - \bar{U}(s)\|_m^2 &\leqslant R_e s M_1\|\bar{u}(s) - \bar{U}(s)\|_0\|\bar{u}(s) - \bar{U}^*(s)\|_0 \\ &\quad + \mu_1\|\bar{u}(s) - \bar{U}(s)\|_m\|\bar{u}(s) - \bar{U}^*(s)\|_m \\ &\leqslant (R_e s M_1 M_2 + \mu_1)\|\bar{u}(s) - \bar{U}(s)\|_m\|\bar{u}(s) - \bar{U}^*(s)\|_m\end{aligned} \tag{2.49}$$

Thus, simplifying (2.49), we obtain

$$\|\bar{u}(s) - \bar{U}(s)\|_m \leqslant \frac{(R_e s M_1 M_2 + \mu_1)}{\mu_2}\|\bar{u}(s) - \bar{U}^*(s)\|_m \tag{2.50}$$

Now let $\bar{U}^*(s)$ be the arbitrary element of $\mathscr{S}_h$ which interpolates $\bar{u}(s)$. Then we see from the interpolation result (2.18) that

$$\begin{aligned}\|\bar{u}(s) - \bar{U}(s)\|_m &\leqslant \frac{(R_e s M_1 M_2 + \mu_1)}{\mu_2} k h^{k+1-m}|\bar{u}(s)|_{k+1} \\ &= C_1 h^{k+1-m}|\bar{u}(s)|_{k+1}\end{aligned}$$

This completes the proof.

Thus the error estimate for the semidiscrete approximation (2.26) to (2.25) can be determined.

Theorem 2.2 If $u(\mathbf{x}, n\,\Delta t)$ is the solution to (2.25) at time $t = n\,\Delta t$ and $U(\mathbf{x}, n\,\Delta t)$ is the solution to (2.26) at time $t = n\,\Delta t$, then

$$\begin{aligned}\|\sigma(\mathbf{x}, n\,\Delta t)\|_m &= \|u(\mathbf{x}, n\,\Delta t) - U(\mathbf{x}, n\,\Delta t)\|_m \\ &\leqslant C_2 h^{k+1-m}|u(x, n\,\Delta t)|_{k+1}\end{aligned} \tag{2.51}$$

Proof: Using the transform (2.45) between the error involved in the approximation of Equations 2.25 and 2.41 the result (2.51) follows

$$\begin{aligned}\|\sigma(\mathbf{x}, n\,\Delta t)\|_m &= \|u(\mathbf{x}, n\,\Delta t) - U(\mathbf{x}, n\,\Delta t)\|_m \\ &= \left\| \frac{1}{2\pi_i} \int_\Gamma (\bar{u}(s) - \bar{U}(s))\, e^{st}\, \mathrm{d}s \right\|_m \\ &\leqslant \frac{C_1}{2\pi_i} h^{k+1-m} \int_\Gamma |\bar{u}(s)|_{k+1}\, e^{st}\, \mathrm{d}s \\ &= C_2 h^{k+1-m} |u(\mathbf{x}, t)|_{k+1}\end{aligned}$$

Now let $E_h(\mathbf{x}; t, \tau)$ be the *fundamental solution operator* associated with the semidiscrete Galerkin approximation (2.26) and $E_h^{\Delta t,\theta}(\mathbf{x}; (v+1)\,\Delta t, v\,\Delta t)$ be the fundamental solution operator (often called the *amplification matrix*) associated with the finite difference–Galerkin approximation (2.38). Then

$$U(\mathbf{x}, (v+1)\,\Delta t) = E_h(\mathbf{x}; (v+1)\,\Delta t, v\,\Delta t) U(\mathbf{x}, v\,\Delta t) \tag{2.52}$$

and

$$U^{n+1}(\mathbf{x}) = E_h^{\Delta t,\theta}(\mathbf{x}; (v+1)\,\Delta t, v\,\Delta t) U^n(\mathbf{x}) \tag{2.53}$$

The operators E_h and $E_h^{\Delta t,\theta}$ have the semigroup property (2.35). Thus

$$E_h(\mathbf{x}; t, \tau) = E_h(\mathbf{x}; t, s) E_h(\mathbf{x}; s, \tau) \tag{2.54}$$

$$\begin{gathered}E_h^{\Delta t,\theta}(\mathbf{x}; v\,\Delta t, n\,\Delta t) = E_h^{\Delta t,\theta}(\mathbf{x}; v\,\Delta t, \gamma\,\Delta t) E_h^{\Delta t,\theta}(\mathbf{x}; \gamma\,\Delta t, \eta\,\Delta t) \\ \eta \leqslant \gamma \leqslant v\end{gathered} \tag{2.55}$$

Now the semidiscrete problem (2.26) is assumed to be well-posed and the finite difference/Galerkin problem (2.28) is assumed to be stable (depends continuously on the initial data). Thus the fundamental solution operator and the amplification matrix are bounded

$$\|E_h(\mathbf{x}; (v-1)\,\Delta t, 0) U(\mathbf{x})\|_m \leqslant C_3 \|U(\mathbf{x})\|_m \qquad \begin{gathered} v = 0, 1, \ldots, N \\ \forall U \in \mathscr{S}_h(\Omega)\end{gathered} \tag{2.56}$$

$$\|E_h^{\Delta t,\theta}(\mathbf{x}; n\,\Delta t, v\,\Delta t) V(\mathbf{x})\|_m \leqslant C_4 \|V(\mathbf{x})\|_m \qquad \begin{gathered} 0 \leqslant v \leqslant n \leqslant N \\ \forall V \in \mathscr{S}_h(\Omega)\end{gathered} \tag{2.57}$$

Equation 2.57 implies that

$$\begin{aligned}\|E_h^{\Delta t,\theta}(\mathbf{x}; n\,\Delta t, v\,\Delta t)\| &= \sup_{V \in \mathscr{S}_h(\Omega)} \frac{\|E_h^{\Delta t,\theta}(\mathbf{x}; n\,\Delta t, v\,\Delta t) V(\mathbf{x})\|_m}{\|V(\mathbf{x})\|_m} \\ &\leqslant C_4 \qquad 0 \leqslant v \leqslant n \leqslant N\end{aligned} \tag{2.58}$$

Consider the semidiscrete finite element/Galerkin equation (2.27). Solving (2.27) for the value of the solution vector $A^M(\nu\,\Delta t)$ based on the 'initial value' of $A^M((\nu - 1)\,\Delta t)$ and setting $f_n = 0$

$$A^M(\nu\,\Delta t) = E_{M\beta}A^\beta((\nu - 1)\,\Delta t) \tag{2.59}$$

where $E_{M\beta} = e^{-\Delta t B_{M\beta}}$ and $B_{M\beta} = C^{-1}_{M\eta}K_{\eta\beta}$. The matrix $E_{M\beta}$ is a discrete approximation for the fundamental solution operator E_h. Similarly, solving the finite difference/Galerkin equation (2.29) for the value of A^ν_M in terms of $A^{\nu-1}_M$ and setting $f_n = 0$, we obtain

$$A^\nu_M = F^\theta_{M\beta}A^{\nu-1}_\beta \tag{2.60}$$

where

$$F^\theta_{M\beta} = [\delta_{M\gamma} + \Delta t(1 - \theta)\delta_{MN}B_{N\gamma}]^{-1}[\delta_{\gamma\beta} - \Delta t\theta\delta_{\gamma\alpha}B_{\alpha\beta}]$$

The operator $F^\theta_{M\beta}$ in (2.60) is a discrete analogue for $E^{\Delta t}_h, \theta$.

A useful approximation for the term $E_{M\beta}$ in (2.59) can be obtained through the Padé approximations. The Padé approximations $R_{p,q}(\Delta t B_{M\beta})$ are a rational matrix approximation for $e^{-\Delta t B_{M\beta}}$ defined by

$$\begin{aligned} R_{p,q}(\Delta t B_{M\beta}) &= n_{p,q}(\Delta t B_{M\beta})[d_{p,q}(\Delta t B_{M\beta})]^{-1} \\ &= e^{-\Delta t B_{M\beta}} + \mathrm{O}(\Delta t B_{M\beta})^r \end{aligned} \tag{2.61}$$

where

$$n_{p,q}(\Delta t B_{M\beta}) = \sum_{k=0}^{q} (p + q - k)![(p + q)!k!(q - k)!]^{-1}(-\Delta t B_{M\beta})^k$$

$$d_{p,q}(\Delta t B_{M\beta}) = \sum_{k=0}^{p} (p + q - k)![(p + q)!k!(p - k)!]^{-1}(\Delta t B_{M\beta})^k$$

$$r = p + q + 1$$

From (2.59), (2.60) and (2.61) we see that the following relationships hold

$$\begin{aligned} R_{0,1} &= F^1_{M\beta} = E_{M\beta} + \mathrm{O}(\Delta t)^2 \\ R_{1,0} &= F^0_{M\beta} = E_{M\beta} + \mathrm{O}(\Delta t)^2 \\ R_{1,1} &= F^{\frac{1}{2}}_{M\beta} = E_{M\beta} + \mathrm{O}(\Delta t)^3 \end{aligned} \tag{2.62}$$

Here the choice of θ corresponds to the forward difference, backward difference and Crank–Nicolson schemes, respectively. The notation indicates that each term of the matrix $F^\theta_{M\beta}$ can be expressed as the sum of the corresponding term in the matrix $E_{M\beta}$ plus a term of order of magnitude Δt^r. Equation (2.62) implies that

$$\begin{aligned} E_{M\beta} - F^1_{M\beta} &= \mathrm{O}(\Delta t)^2 \\ E_{M\beta} - F^0_{M\beta} &= \mathrm{O}(\Delta t)^2 \\ E_{M\beta} - F^{\frac{1}{2}}_{M\beta} &= \mathrm{O}(\Delta t)^3 \end{aligned} \tag{2.63}$$

Now

$$(E_h - E_h^{\Delta t,\theta})(\cdot) = \sum_{M=1}^{N} \phi_M(E_{M\beta} - F_{M\beta}^{\theta})g^{\beta\alpha} \int_{\Omega} (\cdot)\phi_\alpha \, d\Omega \tag{2.64}$$

Using (2.63) and (2.64), we find that

$$\begin{aligned} \|E_h(\mathbf{x}; v\,\Delta t, (v-1)\,\Delta t) - E_h^{\Delta t,1}(\mathbf{x}; v\,\Delta t, (v-1)\,\Delta t)\| &\leqslant C_5\,\Delta t^2 \\ \|E_h(\mathbf{x}; v\,\Delta t, (v-1)\,\Delta t) - E_h^{\Delta t,0}(\mathbf{x}; v\,\Delta t, (v-1)\,\Delta t)\| &\leqslant C_6\,\Delta t^2 \\ \|E_h(\mathbf{x}; v\,\Delta t, (v-1)\,\Delta t) - E_h^{\Delta t,\frac{1}{2}}(\mathbf{x}; v\,\Delta t, (v-1)\,\Delta t)\| &\leqslant C_7\,\Delta t^3 \end{aligned} \tag{2.65}$$

Based on the previous results, the temporal error estimate theorem can be introduced. We will prove the temporal error estimate theorem only for the forward difference approximation ($\theta = 1$). Error estimates for other temporal operators can be derived in similar fashion.

Theorem 2.3 Let $U(\mathbf{x}, n\,\Delta t)$ be the solution to the semidiscrete Galerkin equation (2.26) at time $t = n\,\Delta t$ and $U^n(\mathbf{x})$ be the solution to the forward differenced Galerkin approximation, (2.28) with $\theta = 1$, at time $t = n\,\Delta t$. In addition let (2.56), (2.58) and (2.65) hold. Then the temporal approximation error is

$$\|\tau(\mathbf{x}, n\,\Delta t)\|_m = \|U(\mathbf{x}, n\,\Delta t) - U^n(\mathbf{x})\|_m \leqslant C_8\,\Delta t \|U^0(\mathbf{x})\|_m \tag{2.66}$$

Proof: Using the semigroup property, (2.54) and (2.55)

$$\begin{aligned} \|\tau(\mathbf{x}, n\,\Delta t)\|_m &= \|U(\mathbf{x}, n\,\Delta t) - U^n(\mathbf{x})\|_m \\ &= \|(E_h(\mathbf{x}; n\,\Delta t, 0) - E_h^{\Delta t,1}(\mathbf{x}; n\,\Delta t, 0))U^0(\mathbf{x})\|_m \\ &= \Bigg\| \sum_{v=1}^{n} E_h^{\Delta t,1}(\mathbf{x}; n\,\Delta t, v\,\Delta t)[E_h^{\Delta t,1}(\mathbf{x}; v\,\Delta t, (v-1)\,\Delta t) \\ &\qquad - E_h(\mathbf{x}; v\,\Delta t, (v-1)\,\Delta t)]E_h(\mathbf{x}; (v-1)\,\Delta t, 0)U^0(\mathbf{x}) \Bigg\|_m \\ &\leqslant \sum_{v=1}^{n} \|E_h^{\Delta t,1}(\mathbf{x}; n\,\Delta t, v\,\Delta t)\| \, \|E_h^{\Delta t,1}(\mathbf{x}; v\,\Delta t, (v-1)\,\Delta t) \\ &\qquad - E_h(\mathbf{x}; v\,\Delta t, (v-1)\,\Delta t)\| \, \|E_h(\mathbf{x}; (v-1)\,\Delta t, 0)U^0(\mathbf{x})\|_m \end{aligned}$$

And using (2.56), (2.58) and (2.65)

$$\begin{aligned} \|\tau(\mathbf{x}, n\,\Delta t)\|_m &\leqslant nC_4C_5\,\Delta t^2 C_3 \|U^0(\mathbf{x})\|_m \\ &\leqslant C_8\,\Delta t \|U^0(\mathbf{x})\|_m \end{aligned}$$

The total error estimate of the approximation is given in the following theorem.

Theorem 2.4 If the components of the error of the approximation are given by (2.30), (2.31) and (2.32), then

$$\|e(\mathbf{x}, n\,\Delta t)\|_m \leqslant C_2 h^{k+1-m} |u(\mathbf{x}, n\,\Delta t)|_{k+1} + C_8\,\Delta t \|U^0(\mathbf{x})\|_m \tag{2.67}$$

Proof: From (2.33)

$$\|e(\mathbf{x}, n\,\Delta t)\|_m \leqslant \|\sigma(\mathbf{x}, n\,\Delta t)\|_m + \|\tau(\mathbf{x}, n\,\Delta t)\|_m$$

But introducing (2.51) and (2.66)

$$\|e(\mathbf{x}, n\,\Delta t)\|_m \leqslant C_2 h^{k+1-m}|u(\mathbf{x}, n\,\Delta t)|_{k+1} + C_8\,\Delta t\|U^0(\mathbf{x})\|_m$$

2.5 Error estimates for the diffusion equation using the energy method

We often find it mathematically convenient and physically appealing to seek an error estimate that indicates the error in energy of an approximation. This estimate of the error in energy of the approximation requires the definition of an appropriate energy norm. For the diffusion problem the most natural energy norm can be constructed using the bilinear form $a(u, v)$ connected with the operator **A** (2.22). The bilinear form $a(u, v)$ satisfies the m ellipticity condition, (2.23). Thus effectively, an inner product space $\mathscr{H}_A$ is introduced in association with the operator **A** and the inner-product $a(u, v)$. The norm associated with this inner-product is called the *energy norm*,

$$\|u\|_A = \sqrt{a(u, u)} \qquad u \in \mathscr{H}_A \tag{2.68}$$

the properties of which depend intrinsically on the operator A. The energy space $\mathscr{H}_A$ is *topologically equivalent* to the Sobolev space $H^m(\Omega)$, i.e. positive constants γ_0 and γ_1 exist such that

$$\gamma_0\|u\|_m \leqslant \|u\|_A \leqslant \gamma_1\|u\|_m \tag{2.69}$$

In the subsequent developments we find it convenient to replace the energy norm $\|\cdot\|_A$ by the Sobolev norm $\|\cdot\|_m$ using (2.69).

Now suppose $u|_n$ is the exact solution to (2.25) evaluated at time point $t = n\,\Delta t$ and U^n is the solution to the finite difference/Galerkin equation (2.28) at time point $t = n\,\Delta t$. We seek an estimate in energy for the difference between $u|_n$ and U^n. In this section we consider only the case in which the temporal operator is the forward difference operator in time. Thus we set $\theta = 1$ in (2.28). Now let W^n be an arbitrary function in the subspace $\mathscr{S}_h(\Omega)$. Then we define the error components

$$\begin{aligned} e^n &= u|_n - U^n \\ \mathscr{E}^n &= W^n - U^n \\ E^n &= u|_n - W^n \\ \varepsilon^n &= \left.\frac{\partial u}{\partial t}\right|_n - \delta_t u|_n \end{aligned} \tag{2.70}$$

We can now cite a theorem which specifies the behaviour of the error.

Theorem 2.5 Let e^n, $\mathscr{E}^n$, E^n and ε^n be the error components defined in (2.70). Then

$$a(\mathscr{E}^n, \mathscr{E}^n) = -(\delta_t e^n, \mathscr{E}^n)_0 - a(E^n, \mathscr{E}^n) - (\varepsilon^n, \mathscr{E}^n)_0 \tag{2.71}$$

Proof: Subtracting (2.28) with $\theta = 1$ from (2.25) evaluated at $t = n\,\Delta t$

$$\left(\frac{\partial u}{\partial t}\bigg|_n - \delta_t U^n, V\right)_0 + a(u|_n - U^n, V) = 0 \qquad V \in \mathscr{S}_h(\Omega) \tag{2.72}$$

where $\partial u/\partial t|_n$ denotes $\partial u/\partial t$ evaluated at time point $t = n\,\Delta t$. Rewriting (2.72) in the form

$$\left(\frac{\partial u}{\partial t}\bigg|_n - \delta_t u|_n + \delta_t u|_n - \delta_t U^n, V\right)_0 + a(u|_n - W^n + W^n - U^n, V) = 0$$

$$\forall V \in \mathscr{S}_h(\Omega)$$

and introducing (2.70), we get

$$(\varepsilon^n + \delta_t e^n, V)_0 + a(E^n + \mathscr{E}^n, V) = 0 \qquad \forall V \in \mathscr{S}_h(\Omega)$$

Now setting $V = \mathscr{E}^n$ and using the bilinearity of $(.\,,.)$ and $a(.\,,.)$, we obtain (2.71).

The error component $\mathscr{E}^n$ can be estimated using the following theorem.

Theorem 2.6 If the error components e^n, $\mathscr{E}^n$, E^n and ε^n are related by (2.71), then there exist positive constants μ_0, γ_1, μ_1 and M_2 such that

$$\mu_0 \|\mathscr{E}^n\|_m \leqslant \gamma_1^2 \|\mathscr{E}^n\|_m + 2\mu_1 \|E^n\|_m + 2M_2 \|\varepsilon^n\|_0 \tag{2.73}$$

Proof: The proof of this theorem is similar to the one used in developing Lemma 3 in Reference 12.

Now by expanding u in a Taylor's series about $t = n\,\Delta t$, introducing the result into (2.70(4)) and taking the L_2 norm, we see that

$$\|\varepsilon^n\|_0 \leqslant c\,\Delta t |||u|||_2 \tag{2.74}$$

where $|||u|||_2^2 = \int_\Omega (\partial^2 u/\partial t^2)^2 \,\mathrm{d}\Omega$.

Then the estimate of the approximation error e^n is contained in the following theorem.

Theorem 2.7 If the error components e^n, $\mathscr{E}^n$, E^n and ε^n are defined by (2.70) and $\mathscr{E}^n$ satisfies (2.73), then there exist positive constants M_4, M_5, M_6 and M_7 such that

$$\|e^n\|_m \leqslant M_5 \|E^n\|_m + M_4 \|\varepsilon^n\|_0 \tag{2.75}$$

and

$$\|e^n\|_m \leqslant M_6 h^{k+1-m} |u|_{k+1} + M_7\,\Delta t |||u|||_2 \tag{2.76}$$

Proof: The proof of this theorem uses the triangle inequality and is similar to Theorem 2 in Reference 13.

2.6 Error estimates for the diffusion equation using L_2 methods

The derivation of L_2 error estimates for finite element/Galerkin models for the diffusion equation follows very naturally from the interpolation theory developed in Section 2.2. The key point in the derivation of the L_2 error estimate is establishing a bound on the approximation error for all time points based on the pointwise error estimate in time. Traditionally this has been carried out through a discrete version of Gronwall's inequality.[25] The method used here to derive error estimates for the finite element/Galerkin solution circumvents the complex arguments involved in using the discrete Gronwall's inequality.

We consider the solution of the weak parabolic problem (2.25) with $f(t) = 0$. The exact solution to this problem has been characterized by Goldstein.[26]

$$u \in C^1[0, T; H^m(\Omega)] \tag{2.77}$$

However if the coefficients $A_{\alpha\beta}$ in (2.21) are such that $A_{\alpha\beta} \in C^\infty(\Omega)$ and $D^\gamma(A_{\alpha\beta}) \in L_\infty(\Omega)$ for all $\boldsymbol{\alpha}$, $\boldsymbol{\beta}$ and γ, then

$$u \in C^\infty[0, T; H^m(\Omega)] \tag{2.78}$$

We assume in this section that the coefficients $A_{\alpha\beta}$ are constants so that (2.78) holds. However, this choice is not basic to the method of derivation of error estimates to be introduced here. In fact, the methods used here are valid as long as the regularity of the exact solution u is such that

$$\frac{\partial^2 u}{\partial t^2} \in L_2[0, T; H^m(\Omega)]$$

Let us define the space $\overline{PC}^1[0, T; \mathscr{S}_h(\Omega)]$. If $PC^1[0, T; \mathscr{S}_h(\Omega)]$ is the space of functions which are continuous with piecewise continuous derivatives between the points of the partition P of $[0, T]$ introduced in Section 2.3, then $\overline{PC}^1[0, T; \mathscr{S}_h(\Omega)]$ is the subspace of $PC^1[0, T; \mathscr{S}_h(\Omega)]$ of functions which are piecewise linear; $\mathscr{S}_h(\Omega)$ is the finite dimensional subspace of $H^m(\Omega)$ described in Section 2.2. We seek an approximate Galerkin solution U^n to (2.25) using the finite difference/Galerkin approximation (2.28) with $\theta = 1$ (this corresponds to the forward difference in time case). The solution of this problem is fully equivalent to the solution of the following problem. Find a $U \in \overline{PC}^1[0, T; \mathscr{S}_h(\Omega)]$ such that

$$\left(\frac{\partial U(t)}{\partial t}, V\right)_0^{t=n\,\Delta t} + a(U(t), V)^{t=n\,\Delta t} = 0 \qquad \forall V \in \overline{PC}^1[0, T; \mathscr{S}_h(\Omega)]$$

$$n = 0, \ldots, R$$

$$(U(\cdot, 0), V)_0 = (u_0, V_0) \tag{2.79}$$

where the notation indicates that a collocation is performed at time points $t = n\,\Delta t$, $n = 0, \ldots, R$. Our subsequent developments pertain to (2.79).

Now let $W(t)$ be that element of $\overline{PC}^1[0, T; \mathscr{S}_h(\Omega)]$ which interpolates u in the sense of (2.8), and let $U^n = U(n\,\Delta t)$. We can define the error components of the approximation by

$$\begin{aligned} e &= u(t) - U(t) \\ \mathscr{E}_1 &= W(t) - U(t) \\ \mathscr{E}_2 &= W(t) - U^n \\ E &= u(t) - W(t) \end{aligned} \tag{2.80}$$

Then the following theorems describe the behaviour of the error.

Theorem 2.8 *If* $u(t)$ is the exact solution to (2.25), $U(t)$ is the approximate solution to (2.25) defined by (2.79), $W(t)$ is the interpolant of $u(t)$ defined by (2.8), and the error components are defined by (2.80), then

$$\left(\frac{\partial \mathscr{E}_1}{\partial t}, \mathscr{E}_1\right)_0 + a(\mathscr{E}_1, \mathscr{E}_1) = -\left(\frac{\partial E}{\partial t}, \mathscr{E}_1\right)_0 - a(E, \mathscr{E}_1) - a(\mathscr{E}_2 - \mathscr{E}_1, \mathscr{E}_1)$$
$$n\,\Delta t \leqslant t \leqslant (n+1)\,\Delta t \tag{2.81}$$

Proof: Subtracting (2.79) from (2.25)

$$\left(\frac{\partial u(t)}{\partial t}, V(t)\right)_0 - \left(\frac{\partial U(t)}{\partial t}, V(t)\right)_0^{t=n\,\Delta t} + a(u(t), V(t)) - a(U(t), V(t))^{t=n\,\Delta t} = 0$$
$$\forall V \in \overline{PC}^1[0, T; \mathscr{S}_h(\Omega)] \tag{2.82}$$

But $(\partial U(t)/\partial t, V(t))_0^{t=n\,\Delta t} = (\partial U(t)/\partial t, V(t))$ for $n\,\Delta t \leqslant t \leqslant n + 1(\Delta t)$, since $U(t)$ is piecewise linear in t. Thus

$$\left(\frac{\partial u(t)}{\partial t} - \frac{\partial U(t)}{\partial t}, V(t)\right)_0 + a(u(t) - U^n, V(t)) = 0 \qquad \forall V \in \overline{PC}^1[0, T; \mathscr{S}_h(\Omega)] \tag{2.83}$$

Rewriting (2.83)

$$\left(\frac{\partial u(t)}{\partial t} - \frac{\partial W(t)}{\partial t} + \frac{\partial W(t)}{\partial t} - \frac{\partial U(t)}{\partial t}, V(t)\right)_0 + a(u(t) - W(t) + W(t) - U^n, V(t))$$
$$= 0 \qquad \forall V \in \overline{PC}^1[0, T; \mathscr{S}_h(\Omega)] \tag{2.84}$$

Then introducing (2.80) into (2.84) and setting $V = \mathscr{E}_1$

$$\left(\frac{\partial \mathscr{E}_1}{\partial t}; \mathscr{E}_1\right) + a(\mathscr{E}_2, \mathscr{E}_1) = -\left(\frac{\partial E}{\partial t}, \mathscr{E}_1\right)_0 - a(E, \mathscr{E}_1) \tag{2.85}$$

The result (2.81) follows directly from (2.85) by splitting up the second term on the left-hand side.

The following theorem describes the variation of $\mathscr{E}_1$ with time:

Theorem 2.9 Let the hypotheses of Theorem 2.8 hold, then

$$\frac{1}{2}\frac{\mathrm{d}}{\mathrm{d}t}\|\mathscr{E}_1\|_0^2 \leqslant \frac{1}{4\eta}\left\|\frac{\partial E}{\partial t}\right\|_0^2 + \eta\|\mathscr{E}_1\|_0^2 + \frac{\mu_1^2}{4\mu_0}\left\|\frac{\partial U(t)}{\partial t}(t - n\,\Delta t)\right\|_m^2 \qquad (2.86)$$
$$n\,\Delta t \leqslant t < (n+1)\,\Delta t$$

Proof: $\mathscr{E}_2 - \mathscr{E}_1 = U(t) - U^n = \partial U(t)/\partial t(t - n\,\Delta t)$ for $n\,\Delta t \leqslant t < (n+1)\,\Delta t$. Thus using the Cauchy–Schwarz inequality (2.24)

$$a(\mathscr{E}_2 - \mathscr{E}_1, \mathscr{E}_1) = a\left(\frac{\partial U(t)}{\partial t}(t - n\,\Delta t), \mathscr{E}_1\right)$$
$$\leqslant \mu_1\left\|\frac{\partial U(t)}{\partial t}(t - n\,\Delta t)\right\|_m \|\mathscr{E}_1\|_m \qquad n\,\Delta t \leqslant t < (n+1)\,\Delta t$$

Now using the elementary inequality $ab \leqslant \frac{1}{2}[\varepsilon a^2 + (1/\varepsilon)b^2]$ for $\varepsilon > 0$ (with $a = \mu_1\|\partial U(t)/\partial t(t - n\,\Delta t)\|_m$, $b = \|\mathscr{E}_1\|_m$ and $\varepsilon = 1/2\mu_0$)

$$a(\mathscr{E}_2 - \mathscr{E}_1, \mathscr{E}_1) \leqslant \frac{\mu^2}{4\mu_0}\left\|\frac{\partial U(t)}{\partial t}(t - n\,\Delta t)\right\|_m^2 + \mu_0\|\mathscr{E}_1\|_m^2 \qquad n\,\Delta t < t \leqslant (n+1)\,\Delta t \qquad (2.87)$$

Now by the definition of the interpolation operation (2.8) used to define $W(t) \in \overline{PC}^1[0, T; \mathscr{S}_h(\Omega)]$ and the best approximation property of Hilbert spaces,[27] we have that

$$(E, V)_0 = 0 \qquad \forall V \in \overline{PC}^1[0, T; \mathscr{S}_h(\Omega)] \qquad (2.88)$$

Now for the moment we suppress the dependence on time. If $V \in \overline{PC}^1[0, T; \mathscr{S}_h(\Omega)]$, then for any particular time point, $V \in \mathscr{S}_h(\Omega)$. Now $\mathscr{S}_h(\Omega)$ contains all polynomials of degree $\leqslant k$, and A is the operator of order $2m$ introduced in (2.21). In this development we restrict the subspace $\mathscr{S}_h$ to the class of subspaces which allow interpolation in the sense of (2.8) of derivatives through the $m - 1$th order of the solution u, and we assume that A maps $\mathscr{S}_h(\Omega)$ into $\mathscr{S}_h(\Omega)$. This implies that if $V \in \overline{PC}^1[0, T; \mathscr{S}_h(\Omega)]$, then $AV \in \overline{PC}^1[0, T; \mathscr{S}_h(\Omega)]$. Thus

$$a(E, \mathscr{E}_1) = (E, A\mathscr{E}_1)_0$$
$$= 0 \qquad (2.89)$$

Introducing (2.87) and (2.89) into (2.81) gives

$$\left(\frac{\partial \mathscr{E}_1}{\partial t}, \mathscr{E}_1\right)_0 + a(\mathscr{E}_1, \mathscr{E}_1) \leqslant -\left(\frac{\partial E}{\partial t}, \mathscr{E}_1\right)_0 + \frac{\mu_1^2}{4\mu_0}\left\|\frac{\partial U(t)}{\partial t}(t - n\,\Delta t)\right\|_m^2 + \mu_0\|\mathscr{E}_1\|_m^2$$
$$n\,\Delta t \leqslant t < (n+1)\,\Delta t \qquad (2.90)$$

Now using the definition of the L_2 norm and the m elliptic property of (2.23), we get

$$\frac{1}{2}\frac{\mathrm{d}}{\mathrm{d}t}\|\mathscr{E}_1\|_0^2 + \mu_0\|\mathscr{E}_1\|_m^2 \leqslant -\left(\frac{\partial E}{\partial t}, \mathscr{E}_1\right)_0 + \frac{\mu_1^2}{4\mu_0}\left\|\frac{\partial U(t)}{\partial t}(t - n\,\Delta t)\right\|_m^2 + \mu_0\|\mathscr{E}_1\|_m^2$$

$$n\,\Delta t \leqslant t < (n+1)\,\Delta t \tag{2.91}$$

Thus, cancelling $\mu_0\|\mathscr{E}_1\|_m^2$ on each side of (2.91) and using the inequality $(a, b) \leqslant 1/4\eta\|a\|^2 + \eta\|b\|^2$ to estimate the first term on the right-hand side in (2.91), we obtain (2.86).

The following theorem gives an estimate for $\mathscr{E}_1$ at the discretization points in time.

Theorem 2.10 Let the hypothesis of Theorem 2.8 hold. Then

$$\|\mathscr{E}_1\|_{L^\infty(L_2)} = \sup_{0 \leqslant M \leqslant R} \|\mathscr{E}_1(M\,\Delta t)\|_0$$

$$\leqslant C_4\left[\|\mathscr{E}_1(0)\|_0 + \|E\|_{H^1(L_2)} + \Delta t\left\|\frac{\partial u(t)}{\partial t}\right\|_{L_2(H^m)}\right] \tag{2.92}$$

where, for example, the notation $L_2(H^m)$ indicates the function is in L_2 in the time variable and H^m in the spatial variable.

Proof: Integrating (2.86) from $t = n\,\Delta t$ to $t = (n+1)\,\Delta t$ gives

$$\|\mathscr{E}_1(n+1)\,\Delta t\|_0^2 - \|\mathscr{E}_1(n\,\Delta t)\|_0^2 \leqslant \frac{1}{2\eta}\int_{n\,\Delta t}^{(n+1)\,\Delta t}\left\|\frac{\partial E}{\partial t}\right\|^2 \mathrm{d}t + 2\eta\int_{n\,\Delta t}^{(n+1)\,\Delta t}\|\mathscr{E}_1\|_0^2\,\mathrm{d}t$$

$$+ \frac{\mu_1^2}{2\mu_0}\int_{n\,\Delta t}^{(n+1)\,\Delta t}\left\|\frac{\partial U(t)}{\partial t}(t - n\,\Delta t)\right\|_m^2 \mathrm{d}t \tag{2.93}$$

Now simplifying the last term on the right-hand side

$$\int_{n\,\Delta t}^{(n+1)\,\Delta t}\left\|\frac{\partial U(t)}{\partial t}(t - n\,\Delta t)\right\|_m^2 \mathrm{d}t = \int_{n\,\Delta t}^{(n+1)\,\Delta t}\left\|\frac{\partial U(t)}{\partial t}\right\|_m^2 (t - n\,\Delta t)^2\,\mathrm{d}t$$

$$= \frac{\Delta t^2}{3}\int_{n\,\Delta t}^{(n+1)\,\Delta t}\left\|\frac{\partial U(t)}{\partial t}\right\|_m^2 \mathrm{d}t \tag{2.94}$$

Introducing (2.94) into (2.93) and summing from $n = 0$ to $n = M$, we obtain

$$\|\mathscr{E}_1(M\,\Delta t)\|_0^2 - \|\mathscr{E}_1(0)\|_0^2 \leqslant \frac{1}{2\eta}\int_0^{M\,\Delta t}\left\|\frac{\partial E}{\partial t}\right\|_0^2 \mathrm{d}t + 2\eta\int_0^{M\,\Delta t}\|\mathscr{E}_1\|_0^2\,\mathrm{d}t$$

$$+ \frac{\mu_1^2}{6\mu_0}\Delta t^2\int_0^{M\,\Delta t}\left\|\frac{\partial U(t)}{\partial t}\right\|_m^2 \mathrm{d}t \tag{2.95}$$

But $\|(\partial U/\partial t)(t)\|_m \leqslant C_2 \|\partial u(t)/\partial t\|_m$ for all $t \in [0, M\,\Delta t]$. C_2 is of the form $1 + f(h, \Delta t)$. We do not introduce the explicit expression for f into our equations because it can only result in higher order terms in the error estimate. This result then implies that

$$\|\mathscr{E}_1(M\,\Delta t)\|_0^2 \leqslant \left[\|\mathscr{E}_1(0)\|_0^2 + \frac{1}{2\eta}\left\|\frac{\partial E}{\partial t}\right\|^2_{L_2(L_2)} + \frac{\mu_1^2 C_2^2}{6\mu_0}\Delta t^2 \left\|\frac{\partial u(t)}{\partial t}\right\|^2_{L_2(H^m)}\right] + 2\eta \int_0^T \|\mathscr{E}_1\|_0^2\,\mathrm{d}t \tag{2.96}$$

But the classical Gronwall's inequality[28] implies that if

$$|x(t)| \leqslant a + \int_0^t |x(s)| C\,\mathrm{d}s$$

where a and C are positive constants, then $|x(t)| \leqslant a\,e^{Ct}$. Thus

$$\|\mathscr{E}_1(M\,\Delta t)\|_0^2 \leqslant \left[\|\mathscr{E}_1(0)\|_0^2 + \frac{1}{2\eta}\left\|\frac{\partial E}{\partial t}\right\|^2_{L_2(L_2)} + \frac{\mu_1^2 C_2^2}{6\mu_0}\Delta t^2 \left\|\frac{\partial u(t)}{\partial t}\right\|^2_{L_2(H^m)}\right] e^{2\eta T}$$

or using the embedding result $\|\partial E/\partial t\|_{L_2(L_2)} \leqslant C_3 \|E\|_{H^1(L_2)}$

$$\|\mathscr{E}_1(M\,\Delta t)\|_0 \leqslant C_4\left[\|\mathscr{E}_1(0)\|_0 + \|E\|_{H^1(L_2)} + \Delta t\left\|\frac{\partial u(t)}{\partial t}\right\|_{L_2(H^m)}\right] \tag{2.97}$$

The result (2.92) follows by taking the supremum of (2.97) for all integers M with $0 \leqslant M \leqslant R$. Note that $\overline{PC}^1[0, T; \mathscr{S}_h(\Omega)]$ is a space of piecewise linear functions. Thus since $\mathscr{E}_1 \in \overline{PC}^1[0, T; \mathscr{S}_h(\Omega)]$, $\mathscr{E}_1$ attains its maximum at one of the discretization points in time. Thus the supremum of the $\mathscr{E}_1$ at the discretization points in time is exactly the L_∞ norm.

The approximation error can then be established through the following theorem.

Theorem 2.11 Let the hypotheses of Theorem 2.8 hold. Then

$$\|e\|_{L_2(L_2)} \leqslant C_5\left[\|e(0)\|_0 + \|\mathscr{E}\|_{H^1(L_2)} + \Delta t\left\|\frac{\partial u(t)}{\partial t}\right\|_{L_2(H^m)}\right] \tag{2.98}$$

Proof: Using (2.80),

$$\begin{aligned}\|e\|_{L_2(L_2)} &= \|E\|_{L_2(L_2)} + \|\mathscr{E}_1\|_{L_2(L_2)} \\ &\leqslant \|E\|_{L_2(L_2)} + T^{\frac{1}{2}}\|\mathscr{E}_1\|_{L_\infty(L_2)}\end{aligned} \tag{2.99}$$

19. J. L. Lions and E. Magenes, *Non-Homogeneous Boundary Value Problems and Applications*, Vol. I, Springer-Verlag, New York, 1972.
20. A. Cella and M. M. Cecchi, 'An extended theory for the finite element method', *Variational Methods in Engineering, Vol. I*, C. A. Brebbia and H. Tottenham (Eds.), Southampton University Press, 1973, pp. 74–84.
21. G. Strang, 'Approximation in the finite element method,' *Numerische Mathematik*, **19**, 81–98 (1972).
22. P. G. Ciarlet and P. A. Raviart, 'General Lagrange and Hermite interpolation in R^n with applications to finite element methods', *Archives for Rational Mechanics and Analysis*, **46**, 177–199 (1972).
23. A. Friedman, *Partial Differential Equations*, Holt, Rinehart and Winston, 1969.
24. M. Schultz, *Spline Analysis*, Prentice-Hall, Englewood Cliffs, N.J. 1973.
25. M. Lees, '*A priori* estimates for the solution of difference approximations to parabolic partial differential equations', *Duke Mathematical Journal*, **27**, 297–311 (1960).
26. J. Goldstein, *Semigroups of Operators and Abstract Cauchy Problems*, Tulane University, 1970.
27. J. T. Oden, *Finite Elements of Nonlinear Continua*, McGraw-Hill, New York, 1972.
28. R. Bellman, *Stability Theory of Differential Equations*, McGraw-Hill, New York, 1952.

Chapter 3

Application of the Pseudo-Functional Finite Element Method to Non-Linear Problems

D. H. Norrie and G. de Vries

3.1 Introduction

The variational finite element method is still predominant, although residual and other methods are being increasingly used.[1] In many physical problems, true variational principles have been discovered leading to functionals whose stationary values yield the solution to the problems.[2,3,4] Where a true functional is not available, so-called variational principles have often been devised which allow the solution to be obtained after a lengthy and complex process. The present paper outlines another approach for such problems, making use of a *pseudo-functional*, which has the advantage of simplicity and reduced computation. It is not a true variational method since from another point of view it can be regarded as a Galerkin method used with a particular approximation scheme.

To illustrate the method, several applications to non-linear problems are presented.

3.2 The pseudo-functional method

Consider the domain or field equation for the problem in the form

$$A_1 + A_2 + \cdots + B_1 + B_2 + \cdots = 0 \qquad \text{in } \Omega \tag{3.1}$$

where $A_1, A_2, \ldots, B_1, B_2, \ldots,$ are differential or integral terms involving the dependent and independent variables. Commonly, $A_1 + A_2 + \cdots$ are the linear terms and $B_1 + B_2 + \cdots$ the non-linear terms, but this is not a necessary restriction on the method. The boundary conditions are assumed to comprise a Dirichlet condition on a portion Γ_1 of the boundary and a different condition on the remaining portion Γ_2 of the boundary. It is assumed in the following that the group $A_1, A_2, \ldots,$ usually comprise the terms of major influence in the field equation, with $B_1, B_2, \ldots,$ being of lesser influence.

If there exists a *true or classical variational principle*[3,4] for the problem,* the associated functional is known and the condition for the functional to be stationary leads to the solution of the problem. The true functional will be written in the form

$$\Pi = A + B \tag{3.2}$$

where A and B are functionals related respectively to $A_1, A_2, \ldots,$ and $B_1, B_2, \ldots,$ through the variational calculus.

Consider now the case where the functional

$$\Pi = A \tag{3.3}$$

can be derived from the field equation

$$A_1 + A_2 + \cdots = 0 \qquad \text{in } \Omega \tag{3.4}$$

but when $B_1, B_2, \ldots,$ are added to the left-hand side of Equation 3.4 to give Equation 3.1 no corresponding functional can be found for the resulting field equation. There are two possible cases to be considered:

I Although it has not been found, a true functional for Equation 3.1 does exist.
II A true functional for Equation 3.1 does not exist.

Case I—Functional exists

The pseudo-functional method consists of a process of iteration in which the (unknown) true functional Π is replaced by a pseudo-functional P. The solution which makes the pseudo-functional P stationary, at each level of iteration, successively approaches the true solution associated with the true functional Π. The pseudo-functional P can be written as

$$P = P_1 + P_2 \tag{3.5}$$

where P_1 is identical to A and P_2 is a sum of one or more selected functionals which collectively represent the remaining terms of the true functional together with any surface integrals which may be needed to satisfy the boundary conditions associated with the problem. The selection of the functional P_2 requires consideration of both the domain equation (3.1) and the associated boundary conditions, but does not require knowledge of the remaining terms of the true functional.

Case II—Functional does not exist

Although the mechanics of the pseudo-functional method for this case is the same as for the previous, the rationale is slightly different. If the functional

* A *classical* variational principle exists when there is an explicit functional (variational integral) which attains a stationary value for the solution function satisfying the original domain equation of the problem and the associated boundary equations.

does not exist, the pseudo-functional can simply be regarded as an integral expression whose stationary value at the *final level** of an iterative process is produced by that same numerical solution which satisfies the field equation (3.1) and its associated boundary conditions. The pseudo-functional, as before, has the form

$$P = P_1 + P_2 \tag{3.6}$$

where P_1 is identical to A, as previously, and P_2 is an appropriately chosen set of domain and surface integrals.

In both cases, at each level of iteration, the solution is determined by a Ritz finite element procedure with the pseudo-functional taking the place of the true functional. Convergence is dependent on the choice of the pseudo-functional as well as on the finite element approximation. In the following, only a single dependent variable will be considered but the approach can be extended to multiple variables.

It will now be shown that for both Case I and Case II, the functional P_2 as given by

$$P_2(v) = -\int_\Omega (B_1 + B_2 + \cdots)v \, d\Omega + \int_{\Gamma_2} (\ldots) \, d\Gamma \tag{3.7}$$

when added to P_1 and used in the iterative process yields an appropriate pseudo-functional P. At the mth level of iteration, the pseudo-functional is determined from the following chosen form which is used in a standard Ritz finite element procedure to determine v^m

$$P(v^m) = A(v^m) - \int_\Omega (B_1^{m-1} + B_2^{m-1} + \cdots)v^m \, d\Omega + \int_{\Gamma_2} (\ldots) \, d\Gamma \tag{3.8}$$

where B_1^{m-1}, $B_2^{m-1}, \ldots,$ are found from $B_1, B_2, \ldots,$ using the previously obtained solution v^{m-1} and where the surface integral is determined through consideration of the boundary conditions for the particular problem. If $(A_1, A_2, \ldots)$ are linear then $A(v^m)$ is quadratic or quadratic-linear, and since the second right-hand side term is linear and the third right-hand side term commonly is linear or quadratic-linear, $P(v^m)$ is quadratic-linear. The system matrix equation is therefore linear with a symmetric stiffness matrix. If $(B_1 + B_2 + B_3 + \cdots)$ contains highest-order derivatives of order n, then use of an nth order polynomial shape function gives the necessary derivatives for a given level of iteration from the previous level.

If the process is convergent, $v^{m-1} \to v^m$ and $B_1^{m-1} \to B_1^m$, $B_2^{m-1} \to B_2^m, \ldots,$ and, in the limit, the pseudo-functional P tends to

$$P(v^m) = A(v^m) - \int_\Omega \{B_1^m + B_2^m + \cdots\}v^m \, d\Omega + \int_{\Gamma_2} (\ldots) \, d\Gamma \tag{3.9}$$

* Strictly, in the limiting case, i.e. when the number of iterations tends to infinity.

Since v^m is that function for which $P(v^m)$ has a stationary value, the variational calculus can be used to determine the necessary conditions for the functional $P(v^m)$ to be stationary. These conditions can be shown to be

$$A_1 + A_2 + A_3 + \cdots B_1 + B_2 + \cdots = 0 \qquad \text{in } \Omega \tag{3.10}$$

which is the original domain equation for the problem, and where $v = \lim_{m\to\infty} v^m$, $B_1 = \lim_{m\to\infty} B_1^m$, $B_2 = \lim_{m\to\infty} B_2^m, \ldots$. If the surface integral in Equation 3.7 has been chosen properly, the boundary conditions associated with the variational procedure will also be those for the original problem.

It has been shown above that the pseudo-functional, (3.8), used in a Ritz finite element formulation yields the solution $v = \lim_{m\to\infty} v^m$ to the original problem, (3.1), if the process is convergent. It is expected that the more dominant the effect of P_1, the more rapid the convergence. The method can be extended to problems with several domain equations, in which case several (corresponding) pseudo-functionals must be determined. The previously simple iterative procedure must now be expanded to a coupled iteration scheme. The pseudo-functional approach is of particular value in non-linear problems.

To illustrate the pseudo-functional method and to show how the boundary conditions can be handled, a simple example is given in the Appendix at the end of this chapter.

The iterative method of solution for non-linear problems, outlined here, has been successfully used by the authors in several applications. A slight disadvantage of the procedure is that it is not known whether or not a particular pseudo-functional will yield convergence or not until it has been tried, since a mathematical criterion for convergence is not yet available. If a certain functional does not converge, one has no choice but to modify it. Such a change alters the *stiffness* matrix and may now result in convergence. Some experience with the method assists in choosing an appropriate functional on intuitive grounds. The authors *postulate* that the process will converge if the terms which dominate the physical behaviour of the system are included are those terms in the functional which are not iterated upon but are used only in the minimization procedure.

3.3 Application to compressible flow

The governing (non-linear) field equation for steady, two-dimensional, inviscid, irrotational, compressible flow may be written as

$$\nabla^2\phi - a^{-2}(\phi_x^2\phi_{xx} + 2\phi_x\phi_y\phi_{xy} + \phi_y^2\phi_{yy}) = 0 \qquad \text{in } \Omega \tag{3.11}$$

with boundary conditions

$$\phi = g(x, y) \quad \text{on } \Gamma_1 \quad \text{and} \quad \frac{\partial \phi}{\partial n} = 0 \quad \text{on } \Gamma_2 \qquad (3.12a, 3.12b)$$

where the domain Ω is enclosed by the boundary $\Gamma = \Gamma_1 + \Gamma_2$, and where ϕ is the velocity potential (related to the velocity vector q, by $q = -\nabla\phi$) and where ϕ_x, ϕ_y, ϕ_{xy} are the partial derivatives $\partial\phi/\partial x$, $\partial\phi/\partial y$, $\partial^2\phi/\partial x\, \partial y$, respectively. The symbol a denotes the local speed of sound and satisfies the relation

$$a^2 = A + B(\phi_x^2 + \phi_y^2) \qquad (3.13)$$

where A and B are constants defined by

$$A = a_\infty^2 + \tfrac{1}{2}(\gamma - 1)q_\infty^2, \qquad B = \tfrac{1}{2}(1 - \gamma) \qquad (3.14a, 3.14b)$$

where γ denotes the ratio of specific heats and conditions at infinity being indicated by the subscript ∞.

The solution ϕ can be obtained[5,6] by successive determination of that function v^m from the class of admissible functions which makes the following pseudo-functional stationary

$$P(v^m) = \int_\Omega [\tfrac{1}{2}\{(v_\beta^m)^2 + (v_y^m)^2\} + G^m v^m]\, d\Omega \qquad (3.15)$$

where G^m is computed using

$$G^m = (a^m)^{-2}[(\phi_x^{m-1})^2\phi_{xx}^{m-1} + 2\phi_x^{m-1}\phi_y^{m-1}\phi_{xy}^{m-1} + (\phi_y^{m-1})^2\phi_{yy}^{m-1}] \qquad (3.16)$$

and where a^m is determined from

$$(a^m)^2 = A + B[(\phi_x^{m-1})^2 + (\phi_y^{m-1})^2] \qquad (3.17)$$

It should be noted that the conditions (3.12a, 3.12b) are the principal and natural boundary conditions for this problem.

The flow around a circular cylinder at low Mach numbers was determined using the above iterative scheme together with the Ritz finite element procedure. The element sub-domains were six-node triangles with second-order polynomial interpolation. The result for $M = 0{\cdot}3$ is shown in Figure 3.1(a).

The flow for $M = 0{\cdot}7$ is shown in Figure 3.1(b) although the solution obtained is physically incorrect since the free stream Mach number exceeds the critical Mach number of approximately $M = 0{\cdot}4$. This result, however, indicates that the method is stable under significantly non-linear conditions.

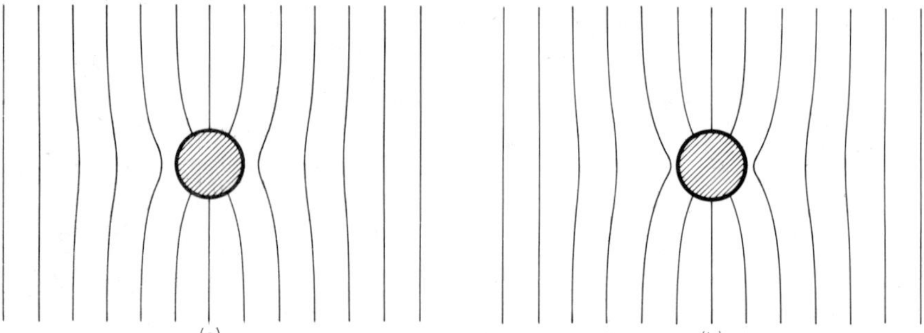

Figure 3.1 Isopotential lines for isentropic compressible flow past a circular cylinder (a) for $M = 0{\cdot}30$, (b) for $M = 0{\cdot}70$

The above finite element approach to compressible flow was first presented in 1970[7] and has since been applied to a number of practical aerodynamic problems by Perriaux[8,9] with good results being obtained even when the flow contained supersonic patches.

3.4 Visco-plastic torsion of a prismatic bar of square cross-section

The section considers the problem of pure torsion of a right prismatic, homogeneous, isotropic, visco-plastic (Bingham) bar of square cross-section. The problem may be formulated in terms of a stress function using a semi-inverse method, similar to Saint Venant's procedure for the elastic torsion problem. This formulation involves the definition of the stress function $\overline{U}(x, y)$, which is continuous and has piecewise continuous first derivatives. The equilibrium equation in terms of the unknown components τ_{xz} and τ_{yz} of the stress tensor becomes

$$\frac{\partial \tau_{xz}}{\partial x} + \frac{\partial \tau_{yz}}{\partial y} = 0 \tag{3.18}$$

which is satisfied by defining the stress function $\overline{U}(x, y)$ through

$$\tau_{xz} = \frac{\partial \overline{U}}{\partial y}, \qquad \tau_{yz} = -\frac{\partial \overline{U}}{\partial x} \tag{3.19a, 3.19b}$$

Writing the strain rates in terms of the stress function $\overline{U}$, it can be shown that the compatibility equation in non-dimensional form reduces to

$$\nabla^2 U - [U_x^2 + U_y^2]^{-\frac{3}{2}}[U_y^2 U_{xx} - 2U_x U_y U_{xy} + U_x^2 U_{yy}] + 2\varepsilon = 0 \tag{3.20}$$

where U is the non-dimensional stress function. The parameter ε gives the ratio of the characteristic viscous shear stress and the plastic shear stress, that is when the viscosity effect is small, $\varepsilon \ll 1$.

It can be shown that the boundary conditions associated with the torsion problem, formulated above, are

$$U = g(x, y) \quad \text{on } \Gamma_1 \quad \text{and} \quad \frac{\partial U}{\partial n} = 0 \quad \text{on } \Gamma_2 \qquad (3.21a, 3.21b)$$

Although a true functional can be shown to exist, the following pseudo-functional can be used

$$P(v^m) = \int_\Omega \left[\tfrac{1}{2}\{(v_x^m)^2 + (v_y^m)^2\} + G^m v^m\right] d\Omega \qquad (3.22)$$

where G^m is defined by

$$G^m = [(v_x^{m-1})^2 + (v_y^{m-1})^2]^{-\frac{3}{2}}[(v_y^{m-1})^2 v_{xx}^{m-1} - 2v_x^{m-1}v_y^{m-1}v_{xy}^{m-1} + (v_x^{m-1})^2 v_{yy}^{m-1}] - 2\varepsilon \qquad (3.23)$$

It is noted that the boundary conditions given in Equations 3.21a and 3.21b are the principal and natural boundary conditions associated with the functional (3.22).

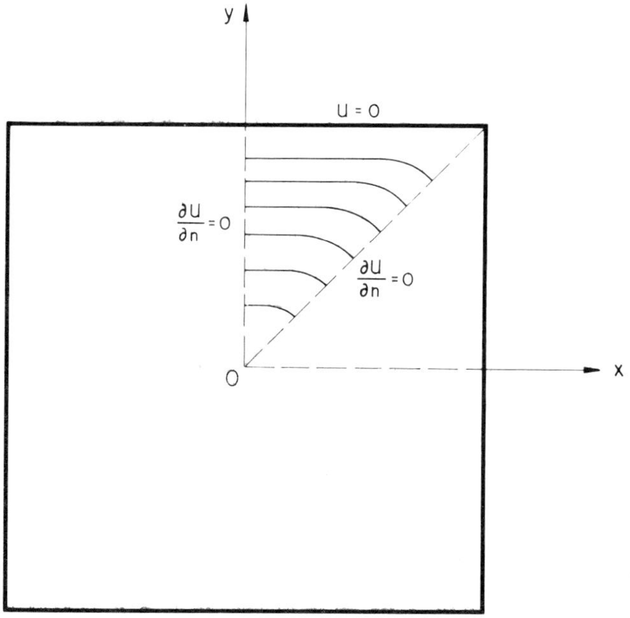

Figure 3.2 Stress trajectories for $\varepsilon = 0{\cdot}50$ (lines of constant U)

The iterative method presented previously was used in conjunction with the Ritz finite element procedure and solutions were obtained for various values of ε. Second-order polynomial approximations on six-node triangles were again used. The stress trajectories (or constant U lines) are shown in Figure 3.2 for the case of $\varepsilon = 0{\cdot}50$. Because of symmetry only one sector of the solution is shown.

Further details of this application are given in Reference 10.

3.5 Application to viscous flow

The governing field equation for steady, viscous two-dimensional flow may be written in non-dimensional form as

$$\nabla^4\psi = \frac{Re}{2}[\psi_x\nabla^2\psi_y - \psi_y\nabla^2\psi_x] \tag{3.24}$$

where ψ is the stream function defined in terms of the x, y components of the velocity vector q, by

$$u = -\frac{\partial\psi}{\partial x}, \qquad v = \frac{\partial\psi}{\partial y} \qquad \text{(3.25a, 3.25b)}$$

and where Re is the Reynolds number defined in terms of the coefficient of viscosity ν and the volume flow rate Q by

$$Re = \frac{2Q}{\nu} \tag{3.26}$$

The boundary conditions are dependent on the particular problem but involve the quantities ψ, ψ_x and ψ_y.

The solution ψ may be obtained[11] by an iterative scheme, similar to those introduced previously, using the following pseudo-functional associated with the governing equation (3.24)

$$P(\psi^m) = \int_\Omega [\{(\psi^m_{xx} - \psi^m_{yy})^2 + 4(\psi^m_{xy})^2\} + G^m\psi^m]\, d\Omega \tag{3.27}$$

where G^m is defined by

$$G^m = -\left[\frac{Re}{2Q}\{\psi_x^{m-1}\nabla^2\psi_y^{m-1} - \psi_y^{m-1}\nabla^2\psi_x^{m-1}\}\right] \tag{3.28}$$

The flow of a viscous fluid through a two-dimensional channel for various Reynolds numbers has been determined using the above iterative scheme and the Ritz finite element procedure. A cubic-order polynomial approximation on a three-node triangle, the Tocher-10 element, was used. Solutions at higher Reynolds numbers were obtained by iterating from those at lower Reynolds numbers. The flow through a straight-sided, two-dimensional diffuser at Reynolds numbers of 20 and 37·5 is shown in Figure 3.3.

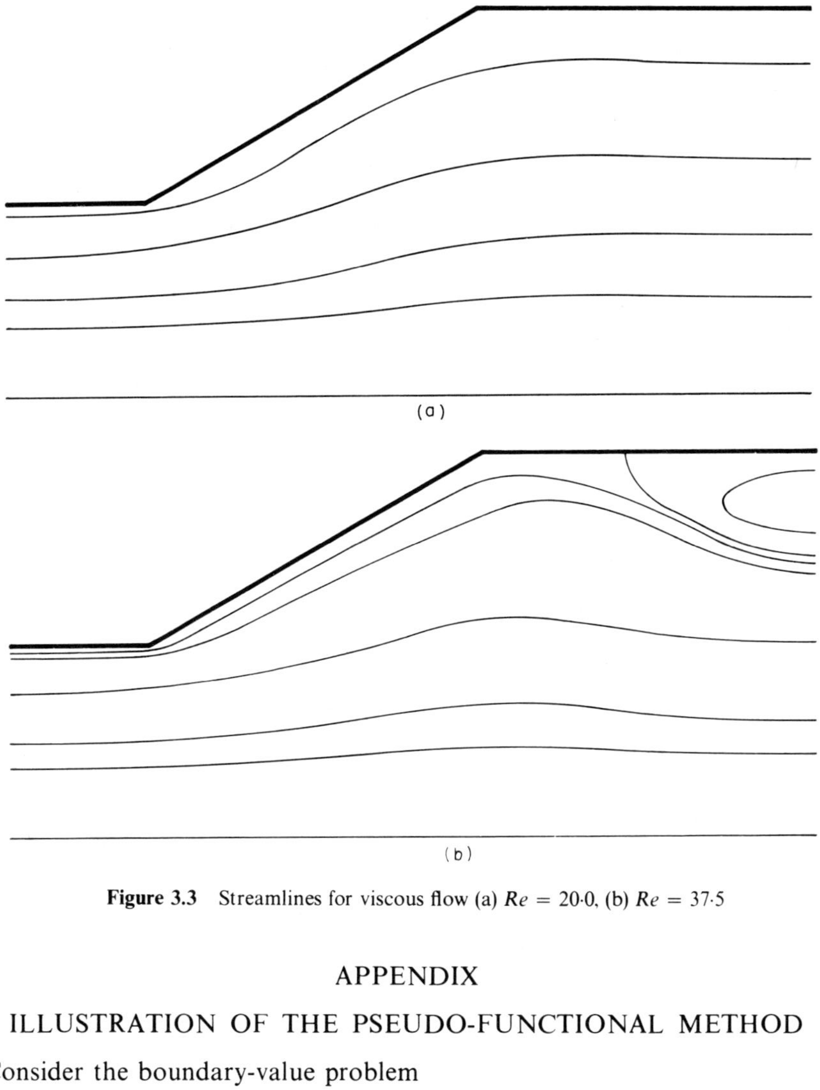

Figure 3.3 Streamlines for viscous flow (a) $Re = 20{\cdot}0$, (b) $Re = 37{\cdot}5$

APPENDIX

ILLUSTRATION OF THE PSEUDO-FUNCTIONAL METHOD

Consider the boundary-value problem

$$\nabla^2\phi = \phi_{xx} + \phi_{yy} = 0 \qquad \text{in } \Omega \tag{3.29}$$

subject to

$$\phi = g(x, y) \qquad \text{on } \Gamma_1, \qquad \frac{\mathrm{d}\phi}{\mathrm{d}n} = 0 \qquad \text{on } \Gamma_2 \tag{3.30}$$

where the domain Ω is enclosed by the boundary $\Gamma = \Gamma_1 + \Gamma_2$, and where ϕ_{xx}, ϕ_{yy} denote the partial derivatives $\partial^2\phi/\partial x^2$, $\partial^2\phi/\partial y^2$, respectively.

Although the true functional for this problem is well known, let it be assumed that only the functional $\int_\Omega \frac{1}{2}v_x^2\, d\Omega$, corresponding to the first term in Equation 3.29 can be determined. Let it further be assumed that in the present case it is known that the term ϕ_{xx} dominates the behaviour of the domain equation (3.29). In the notation of the previous section, it is clear that $A_1 + A_2 + \cdots = \phi_{xx}$, and $B_1 + B_2 + \cdots = \phi_{yy}$. Consider the pseudo-functional of the form

$$P(v) = P_1(v) + P_2(v) = \underbrace{\int_\Omega (\tfrac{1}{2}v_x^2)\, d\Omega}_{P_1} + \underbrace{\int_\Omega (Gv)\, d\Omega + \int_{\Gamma_2} (Hv)\, d\Gamma}_{P_2} \tag{3.31}$$

where v belongs to the class of admissible functions for this problem and where G, H are functions of x, y only and not of v or its derivatives. For $P(v)$ to be stationary when $v = \phi$, the requisite conditions can be shown to be

$$\phi_{xx} - G = 0 \quad \text{in } \Omega \tag{3.32}$$

$$\phi = g(x, y) \quad \text{on } \Gamma_1, \quad \text{and} \quad \phi_x n_x + H = 0 \quad \text{on } \Gamma_2 \tag{3.33}$$

where n_x, n_y are the x, y components of the unit outward normal n to Γ_2.

Comparison of Equations 3.32 and 3.33 with Equations 3.29 and 3.30 shows that if $G(x, y)$, $H(x, y)$ are chosen as

$$G(x, y) = -\phi_{yy} \quad \text{in } \Omega \tag{3.34}$$

$$H(x, y) = \phi_y n_y \quad \text{on } \Gamma_2 \tag{3.35}$$

the solution $v = \phi$ satisfies the governing equations (3.29) and (3.30). Since the solution ϕ is not known, $G(x, y)$, $H(x, y)$ cannot be determined directly. However, Equations 3.34 and 3.35 can be used in an iterative procedure by using the pseudo-functional in the form

$$P(v^m) = \int_\Omega [\tfrac{1}{2}(v_x^m)^2 + G^m v^m]\, d\Omega + \int_{\Gamma_2} H^m v^m\, d\Gamma, \qquad m = 1, 2, \ldots \tag{3.36}$$

where

$$G^m(x, y) = -\phi_{yy}^{m-1} \quad \text{in } \Omega \tag{3.37}$$

$$H^m(x, y) = \phi_y^{m-1} n_y \quad \text{on } \Gamma_2 \tag{3.38}$$

Applying the same variational procedure as before yields

$$\phi_{xx}^m + \phi_{yy}^{m-1} = 0 \quad \text{in } \Omega \tag{3.39}$$

$$\phi^m = g(x, y) \quad \text{on } \Gamma_1, \quad \text{and} \quad \phi_x^m n_x + \phi_y^{m-1} n_y = 0 \quad \text{on } \Gamma_2 \tag{3.40}$$

If the scheme is convergent so that $\phi = \lim_{m \to \infty} \phi^m$, the converged result will be the solution to the original problem.

Acknowledgements

The investigations described in this paper have been financially supported by the National Research Council (Grants Nos. A4192, A7432, C0679, and C0310) and this assistance is recorded with appreciation.

References

1. D. H. Norrie and G. de Vries, *The Finite Element Method—Fundamentals and Applications*, Academic Press, New York, 1973.
2. B. A. Finlayson and L. E. Scriven, 'On the search for variational principles', *International Journal of Heat and Mass Transfer*, **10**, 799–821 (1967).
3. B. A. Finlayson, *The Method of Weighted Residuals and Variational Principles*, Academic Press, New York, 1972.
4. L. G. Napolitano, 'Finite element methods in fluid-dynamics', AGARD-VKI Lecture Series, *Advances in Numerical Fluid Dynamics*, Van Karman Institute for Fluid Dynamics, Brussels, March 5–9, 1973.
5. G. de Vries, G. P. Berard and D. H. Norrie, 'Application of the finite element technique to compressible flow problems', *Mechanical Engineering Department Report No.* 18, University of Calgary, August, 1970.
6. G. de Vries and D. H. Norrie, 'Application of the pseudo-functional finite element method to compressible flow problems', *Mechanical Engineering Department Report No.* 51, University of Calgary, January, 1974.
7. D. H. Norrie and G. de Vries, 'Application of finite element methods in fluid dynamics', in *Numerical Methods in Fluid Dynamics*, edited by J. J. Smolderen, AGARD Lecture Series LS-48, May, 1972.
8. J. Perriaux, 'Two and three dimensional analysis of subsonic compressible potential flows in ducts and around lifting bodies with finite element techniques', in *Finite Element Methods in Flow Problems* edited by J. T. Oden, O. C. Zienkiewicz, R. H. Gallagher and C. Taylor, University of Alabama in Huntsville Press, 1974.
9. J. Perriaux, 'Three dimensional analysis of compressible potential flows with the finite element method', *AMD-BA Internal Report*, Avions Marcel Dassault–Breguet Aviation, 1973.
10. G. de Vries and D. H. Norrie, 'Application of the pseudo-functional finite element method to visco-plastic torsion', *Mechanical Engineering Department Report No.* 46, University of Calgary, October, 1973.
11. R. Balasubramanian, *An Iterative Finite Element Method for Viscous Flow*, M.Sc. Thesis, University of Calgary, January, 1974.

Chapter 4

Finite Element Solution Algorithm for Incompressible Fluid Dynamics

A. J. Baker

4.1 Introduction

Numerical solution of field problems in mechanics has been made possible by the large digital computer. Development of solution procedures for specific disciplines has been highly problem oriented, and little cross-fertilization has occurred that takes advantage of the uniform mathematical description for problems in continuum mechanics. For example, finite difference methods have been almost universally employed for computational fluid mechanics, while the finite element method has found wide acceptance for analysis of complex structural systems. However, these apparently diverse approaches can be viewed in a unified manner as application of specific criteria within the Method of Weighted Residuals (MWR),[1] and satisfaction of the governing differential equations in a weighted average sense. The select choice of weighting and approximation functions renders each approach identifiable, including classical integral approaches like von Karman–Pohlhausen and Integral Method of Strips in boundary layer flow, and the many variations of Galerkin, Kantorovich and collocation methods.

Over the years, the finite element procedure has proven highly adaptable to solution of linear elliptic boundary value problems involving complex boundary conditions applied on irregularly shaped, non-coordinate surface solution domain closures. Extension to the Navier–Stokes equations using variational concepts appeared theoretically difficult without gross assumptions on the concept of a local potential. However, it is now readily verified that a MWR development for a linear equation produces a computational form that is identical to extremization of the equivalent stationary principle. However, since no linearity constraint exists in the MWR derivation, a general theory for finite element solution of non-linear equations is established, especially for specific forms of the Navier–Stokes equations.[2–6]

In this chapter the finite element solution algorithm is established for the two-dimensional Navier–Stokes equations governing the kinematics and thermodynamics of a variable viscosity, constant density fluid. The primitive dependent variables are replaced by a vorticity-stream-function description, which provides a uniformly elliptic differential equation system description for all computational variables. The preferred differential equation systems are established in rectangular, cylindrical and spherical Cartesian coordinate systems. The finite element algorithm is derived for the generalized, non-linear elliptic boundary value problem of mathematical physics and contains no requirements for either computational mesh or solution domain closure regularity. Boundary condition constraints on the normal flux and tangential distribution of all computational dependent variables are routinely piecewise enforceable on domain closure segments arbitrarily oriented with respect to a global reference frame. The intrinsic finite element shapes for one- and two-dimensional domains spanned by linear approximation functions are the line and triangle. The area-coordinate concept of structural mechanics has been utilized to establish a natural coordinate function evaluation of the finite element matrices. The finite element solution algorithm for the characteristic equation system is embodied in the COMOC (Computational Continuum Mechanics) computer program system, specific variants of which have produced solutions in three-dimensional subsonic and supersonic viscous flow fields[3,4] and non-linear transient heat conduction[7] in addition to the Navier–Stokes solutions to be discussed. COMOC has been evaluated for several flow problems to assess solution accuracy, convergence and versatility. Convergence of velocity with discretization is numerically demonstrated for developing flow in a duct. The finite element algorithm is illustrated to predict imbedded recirculation regions for sample problems without resorting to special boundary condition techniques for vorticity. The discussed computed results yield a favourable assessment of the viability of the solution algorithm and its computational embodiment.

4.2 Formulation of governing equations

The description of a fluid dynamical state in hydrodynamics is contained within the solution of the system of coupled, non-linear, second-order partial differential equations describing local conservation of mass, momentum and energy, in conjunction with appropriate specifications of constitutive behaviour and applicable initial and boundary conditions. These equations for a single-component, viscous incompressible fluid are:

$$0 = -[u_i]_{;i} \tag{4.1}$$

$$u_{i,t} = -\left[u_i u_j + \frac{p}{\rho}\delta_{ij} - \tau_{ij}\right]_{;j} \tag{4.2}$$

$$\left(H - \frac{p}{\rho}\right)_{,t} = -\left[u_i H - \tau_{ij}u_j - \frac{k}{\rho}T_{,i}\right]_{;i} \tag{4.3}$$

The dependent variables in Equations 4.1 to 4.3 have their usual interpretation with u_i the vector velocity, p the pressure, ρ the (constant) density, τ_{ij} the stress tensor, H the stagnation enthalpy, k the thermal conductivity, and T the static temperature. Cartesian tensor notation is used;[6] the semicolon denotes the vector derivative of the Cartesian scalar components of a tensor, while the comma denotes the usual gradient operator. Summation convention is employed for repeated subscripts.

The solution of Equations 4.1 to 4.3 requires specification of constitutive relationships between appropriate dependent variables. For laminar flow of a Newtonian fluid, the dynamic relations are contained within Stokes viscosity law, which becomes, for an incompressible fluid:

$$\tau_{ij} \equiv \nu[u_{i;j} + u_{j;i}] \tag{4.4}$$

Before transforming Equations 4.1 to 4.3 to the desired form, it is convenient to non-dimensionalize all variables to extract the useful fluid dynamic parameters. Selecting a characteristic length (L), velocity (U_∞) and uniform density (ρ) as the reference parameters, Equations 4.1 to 4.3 become:

$$0 = -[u_i]_{;i} \tag{4.5}$$

$$u_{i,t} = -\left[u_i u_j + p\delta_{ij} - \frac{1}{Re}\tau_{ij}\right]_{;j} \tag{4.6}$$

$$(H - Ec(p))_{,t} = -\left[u_i H - \frac{Ec}{Re}\tau_{ij}u_j - \frac{1}{Re\,Pr}\nu H_{,i}\right]_{;i} \tag{4.7}$$

The non-dimensional groupings of fluid parameters are defined as:

$$\text{Reynolds number:} \quad Re = \frac{U_\infty L}{\nu_\infty} \tag{4.8}$$

$$\text{Prandtl number:} \quad Pr = \frac{c\nu\rho}{k} \tag{4.9}$$

$$\text{Eckert number:} \quad Ec = \frac{U_\infty^2}{c_p T_\infty} \tag{4.10}$$

The solution of Equations 4.5 to 4.7 is a formidable task. It is desired to restrict the generality of the description to the degree that the concomitant mathematical advantages are applicable to non-trivial problems. The essence of this development is restriction to two independent space dimensions, but in which certain three-dimensional flow fields may exist. The derivations are outlined; the interested reader is referred to Reference 6 for details.

From mathematical physics it is known that a vector field is defined when its divergence and curl are known. The vector field of fluid mechanics is velocity, and Equation 4.5 defines its vanishing divergence. Hence, it is possible to specify this field in terms of the curl of a vector potential function Ψ_i. Little analytical benefit accrues from this specification (as does occur, for example, in Maxwell's equations) in fluid dynamics, due primarily to the non-linearity of Equation 4.6. Significant numerical benefit can occur, however, if either (a) the velocity vector u_i is planar, or (b) a scalar component, u_3 of u_i, is independent of the corresponding variable, x_3. In these cases, the continuity equation can be identically satisfied in terms of the x_3 scalar component of the vector potential Ψ_i, and u_3, as

$$u_i \equiv \frac{1}{J}\varepsilon_{3ij}\Psi_{3;j} + u_3\delta_{i3} \tag{4.11}$$

In Equation 4.11, ε_{3ij} is the Cartesian alternating tensor and J is the determinant of the metric. Restricting attention to spaces spanned by rectangular, cylindrical and/or spherical Cartesian coordinate systems, Equation 4.11 can be advantageously written in terms of a scalar function ψ, and the gradient operator, as:

$$u_i = \frac{1}{r\sin^\alpha\phi}\varepsilon_{3ij}\psi_{,j} + u_3\delta_{i3} \tag{4.12}$$

In Equation 4.12, α is nonzero only for spherical coordinates, when it is unity, r is set equal to unity (and $u_3 \equiv 0$) for rectangular coordinates, and x_3 corresponds to the azimuthal angle (θ) in both cylindrical and spherical coordinates with range $0 \leqslant x_3 \leqslant 2\pi$. It is the reasonable assumption that the space of fluid dynamical problems to be considered can be conveniently spanned by at least one of these Cartesian bases. It is readily shown[6] that Equation 4.12 identically satisfies Equation 4.5 in these coordinate systems.

With the zero divergence of the flow field ensured by identification of this scalar function (stream-function), it remains to obtain the curl of the velocity u_i. Identify the vorticity vector ω_i as:

$$\omega_i \equiv \frac{1}{J}\varepsilon_{ijk}u_{k;j} \tag{4.13}$$

The x_3 scalar component (ω) of ω_i in the three Cartesian bases, is:

$$\omega = \frac{1}{r^\alpha}\varepsilon_{3ij}u_{j;i} \tag{4.14}$$

with α interpreted as before. In the derived solution algorithm, this scalar vorticity component is employed as an auxiliary dependent variable. Substituting Equation 4.12 into Equation 4.14, and using the skew-symmetric properties of the alternating tensor contraction, the following compatibility equation is obtained:

$$\omega = -\frac{1}{r^\alpha}\left[\frac{1}{r \sin^\alpha \phi}\psi_{,k}\right]_{;k} \tag{4.15}$$

In rectangular coordinates, Equation 4.15 is the familiar Poisson equation involving the Laplacian operator. In the other coordinate systems, it lacks certain matric terms of being Laplacian.

The curl of the Navier–Stokes equation ensures the existence of the streamfunction. Since the problem class is restricted from full dimensional generality, determination of the x_3 scalar component of the curl, coupled with the x_3 scalar component, is adequate to determine u_i. From Equation 4.17, obtain

$$\varepsilon_{3ki}\left\{u_{i,t} + \left[u_i u_j - \frac{1}{Re}\tau_{ij}\right]_{;j}\right\}_{;k} = 0 \tag{4.16}$$

Advantage has been taken in Equation 4.16 of the symmetry properties of the alternating tensor to eliminate the appearance of pressure—a prime computational feature of the dependent-variable transformation. Equation 4.16 is rearranged to explicit appearance of the computational variables, using Equations 4.12 and 4.14 to replace the velocity and velocity derivative. Omitting details,[6] Equation 4.16, written in terms of vorticity, streamfunction and x_3 component of velocity u_3, becomes:

$$\begin{aligned} r^\alpha \omega_{,t} = & -e_{3ki}\left[\left(\frac{r^\alpha \omega \psi_{,i}}{r \sin^\alpha \phi}\right) + u_3 \varepsilon_{3il}\left(\frac{\psi_{,l}}{r \sin^\alpha \phi}\right)_{;3}\right]_{;k} \\ & + \frac{1}{Re}\left[\nu(r^\alpha \omega)_{,j} - \nu_{,j} r^\alpha \omega - \nu_{,k}\left(\frac{2\psi_{,k}}{r \sin^\alpha \phi}\right)_{;j} + \nu_{,k}\varepsilon_{3ki}\,\delta_{j3} U_{3;j}\right]_{;j} \end{aligned} \tag{4.17}$$

Selection of a particular coordinate system allows specification of the second term in brackets (modified by the alternating tensor) in terms of the u_3 velocity component. For example, in cylindrical coordinates, with the space spanned by:

$$x_i \equiv \{z, r, \theta\} \tag{4.18}$$

the last term becomes:

$$-\varepsilon_{3ki} u_3 e_{3il}\left(\frac{\psi_{,l}}{r \sin^\alpha \phi}\right)_{;3k} = -\left(\frac{u_3^2}{2r}\right)_{,z} \tag{4.19}$$

Note that u_3 may be a function of independent variables other than x_3.

Determination of u_3, in terms of ω and ψ, is obtained from solution of the x_3 component of the Navier–Stokes equation. From Equations 4.6, 4.12 and 4.14, obtain

$$\begin{aligned} u_{3,t} = & -\varepsilon_{3ij}\left(\frac{\psi_{,j}}{r \sin^\alpha \phi} u_3\right)_{:i} - 2u_3 u_{3:3} - p_{,3} \\ & + \frac{1}{Re}[(\nu u_{3;k})_{;k} + u_{k;3}\nu_{,k}] \end{aligned} \tag{4.20}$$

Note that $u_{3;3}$ does not vanish identically; for example, in cylindrical coordinates, $u_{3:3} = (1/r)u_2$. In Equation 4.20, the derivative of p with respect to x_3 is retained, since it may be non-vanishing for certain flows, e.g. fluid flow through a corkscrew-type device.

The form of Equation 4.7 is particularly well suited for numerical computation for this problem class, since the pressure influence is contained solely within the temporal dependence. Since the pressure, as an explicitly coupled dependent variable, is eliminated from the momentum equations as well, solutions for small Eckert number and steady state problems are possible without coupled determination of pressure. This advantage is noteworthy, since those numerical algorithms for fluid mechanics which contain explicit pressure dependence are specially designed to 'half-step' march (or iterate) between the flow field and the pressure distribution. To cast Equation 4.7 into the desired form, replace the velocity and stress tensor terms by stream-function to obtain:

$$H_{,t} = -\left[\left(\frac{\varepsilon_{3ij}\psi_{,j}}{r\sin^{\alpha}\phi} + u_3\,\delta_{i3}\right)H - \frac{1}{Re}\frac{\nu}{Pr}H_{,i}\right]_{;i}$$
$$+\frac{\nu Ec}{Re}\left\{\left[\left(\frac{\psi_{,k}}{r\sin^{\alpha}\phi}\right)^2_{;i} - \frac{\psi_{,i}}{r\sin^{\alpha}\phi}\left(\frac{\psi_{,j}}{r\sin^{\alpha}\phi}\right)_{;j}\right]\right.$$
$$\left.+\frac{1}{2}\left(u^2_{3;i} + \left(u^2_{3:3} + 2u_{3:j}\frac{\varepsilon_{3jk}\psi_{,k}}{r\sin^{\alpha}\phi}\right)\delta_{i3}\right]\right\}_{;i} + E_c p_{,t} \tag{4.21}$$

The pressure distribution may be recovered by contracting Equation 4.2 with the infinitesimal displacement vector dx_i and integrating. Since the integral of the pressure term is that of a perfect differential, the pressure difference between arbitrary points in the field is determined as:

$$\Delta p = -\int (u_i u_j)_{;j}\,dx_i + \frac{1}{Re}\int \tau_{ij;j}\,dx_i - \frac{\partial}{\partial t}\int u_i\,dx_i \tag{4.22}$$

To obtain the desired computational form, transform the velocity and stress tensor fields into the stream-function–vorticity definition. Performing several integrations by parts, and denoting perfect differential integrations as Δ, obtain:

$$\Delta p = \Delta\left[\left(\frac{\nu}{Re}\frac{\varepsilon_{3jk}\psi_{,k}}{r\sin^{\alpha}\phi}\right)_{;j} - \left(\frac{\psi_{,k}}{r\sin^{\alpha}\phi}\right)^2\right] + \int\left\{\left(\frac{\nu}{Re}\frac{\varepsilon_{3ik}\psi_{,k}}{r\sin^{\alpha}\phi}\right)_{;j}\right.$$
$$\left. - \nu_{,i}\frac{\varepsilon_{3jk}\psi_{,k}}{r\sin^{\alpha}\phi} - \nu_{,j}\frac{\varepsilon_{3ik}\psi_{,k}}{r\sin^{\alpha}\phi} + \frac{\psi_{,i}\psi_{,j}}{r\sin^{\alpha}\phi}\right\}_{;j} dx_i$$
$$+\int\left\{\frac{\nu}{Re}u_{3;3i} - u_{3:3}\frac{\varepsilon_{3ik}\psi_{,k}}{r\sin^{\alpha}\phi} - \frac{\partial}{\partial t}\left(\frac{\varepsilon_{3ik}\psi_{,k}}{r\sin^{\alpha}\phi}\right)\right\} dx_i$$
$$+\Delta x_3\left[\frac{1}{Re}(\nu u_{3;j})_{;j} - u^2_{3;3} - \left(\frac{u_3\varepsilon_{3jk}\psi_{,k}}{r\sin^{\alpha}\phi}\right)_{;j}\right] \tag{4.23}$$

4.3 Finite element solution algorithm

The literature abounds with numerous quasi-variational and integral-solution procedures for problems in mathematical physics. It has been convincingly shown[8] that most of these methods may be viewed in a unifying way as application of specific criteria within the Method of Weighted Residuals (MWR).[1] Included herein are procedures commonly referred to as local potential, collocation, the integral method of strips, Galerkin and Kantorovich methods. The Galerkin criterion within MWR is selected to provide the finite element solution algorithm for the Navier–Stokes problem. The mathematical uniformity pervading the derived governing partial differential equation system allows each to be expressed in the form:

$$\delta Q_{,t} = \kappa\left|K_{jk}Q_{;j}\right|_{;k} - \left|Q\varepsilon_{3ki}\frac{\psi_{,i}}{r\sin^\alpha\phi}\right|_{;k} + f(Q, \omega, \psi, u_3, x_i) \tag{4.24}$$

by the identification of Q, the generalized dependent variable, appropriate for each κ, K_{jk} and f. The left-hand side of Equation 4.24 contains the initial value operator, the first bracket on the right is the generalized elliptic operator, and the remaining two terms involved at most first order derivatives of Q. The delta modifying the initial value operator is unity, except when Q coincides with stream-function, when it vanishes. Table 4.1 lists κ, K_{jk} and f for Q associated with each dependent variable. The correct problem statement is completed by specifying appropriate boundary conditions on the closure $\partial\Omega$ of the domain Ω as well as an initial distribution throughout Ω, for each Q. Their general forms are well known as:

$$a^1 Q(\bar{x}_i, t) + a^2_{jk}Q(\bar{x}_i, t)_{;k}v_j = a^3 \tag{4.25}$$

$$Q(x_i, 0) = Q_0(x_i) \tag{4.26}$$

Table 4.1 Coefficients in generalized elliptic differential (Equation 4.24)

Equation	Q	κ	K_{jk}	f
(4.15)	ψ	1	$\frac{\delta_{jk}}{r\sin^\alpha\phi}$	$r^\alpha\omega + \left[\psi\frac{\varepsilon_{3ki}\psi_{,i}}{r\sin^\alpha\phi}\right]_{;k}$
(4.17)	$r^\alpha\omega$	$\frac{1}{Re}$	$v\delta_{jk}$	$-\left[\frac{v_{,k}}{Re}(r^\alpha\omega) + \frac{2v_{,j}}{Re}\left(\frac{\psi_{,j}}{r\sin^\alpha\phi}\right)_{;k} - v_{,j}\varepsilon_{3ji}\delta_{j3}U_{3;i}\right]_{;k} + \left[\left(u_3\frac{\psi_{,k}}{r\sin^\alpha\phi}\right)_{;3}\right]_{;k}$
(4.20)	u_3	$\frac{1}{Re}$	$v\delta_{jk}$	$[u_{jj3}v_{,j}] - 2u_3u_{3;3} - p_{,3}$
(4.21)	H	$\frac{1}{RePr}$	$v\delta_{jk}$	$-\left[u_3\,\delta_{k3}H + \frac{vEc}{Re}\left\{\left[\left(\frac{\psi_{,i}}{r\sin^\alpha\phi}\right)^2_{;k} - \frac{\psi_{,k}}{r\sin^\alpha\phi}\left(\frac{\psi_{,j}}{r\sin^\alpha\phi}\right)_{;j}\right] + \frac{1}{2}\left[u^2_{3;k} + \left(u^2_{3;3} + 2u_{3;j}\frac{\varepsilon_{3jk}\psi_{,l}}{r\sin^\alpha\phi}\right)\delta_{k3}\right]\right\}\right]_{;k} + Ecp_{,t}$

The superscript bar constrains x_j to $\partial\Omega$, and ν is the outward pointing unit vector, normal to $\partial\Omega$.

The finite element solution algorithm transforms Equations 4.24 to 4.26 into a system of ordinary differential equations written on an arbitrarily selected, discretized variable approximation to Q, that degenerates to coupled algebraic for the steady state solution. Form a series approximation, $Q^*_{m'}$, Equation 4.27 to the dependent variable, Q, constrained to lie within the mth subdomain, Ω_m, defined by $(x_i, t) \in \Omega_m \times |0, \infty)$, and where the union of the M disjoint interior subdomains, Ω_m, forms Ω.

$$Q^*_m(x_i, t) = \{\mathbf{L}(x_i)\}^{\mathrm{T}}\{q(t)\}_m \tag{4.27}$$

Evaluate the weighted residual of Equations 4.25 and 4.26 in each of these finite elements, Ω_m, applying the gradient boundary condition statement to those elements with surfaces coincident with the solution domain closure $\partial\Omega$. Assemble these equations into a global system written on the totality of unconstrained expansion coefficients, $C_{mk}(t)$, by Boolean algebra. The final form of the finite element algorithm for the time dependent distribution of dependent variables of the discretization becomes

$$\mathbf{q}(t)' = \mathbf{A}\mathbf{q}(t) + \mathbf{b} \tag{4.28}$$

Boundary condition specifications are accepted on an element basis by the set of a^j_m, Equation 4.25, for each finite element with nodes occurring on the closure of the global solution domain. For stream-function the left-hand side of Equation 4.28 vanishes identically. The integration algorithm used for Equation 4.28 is an explicit, single-step, multi-stage finite difference procedure with a large region of absolute stability.[7] A banded Cholesky equation solver is used for the algebraic system for stream-function.

4.4 Numerical results

Numerical evaluation of the finite element solution algorithm for the two-dimensional Navier–Stokes equations has been performed using linear natural coordinate approximations[9] over line and triangular elements (Figure 4.1). A problem of particular value for accuracy studies is developing flow of an isothermal fluid in a rectangular duct. This problem retains the full non-linear character of the Navier–Stokes equations, and the outflow boundary conditions of vanishing normal gradient for both stream-function and vorticity are analytically exact for sufficient duct length. Figure 4.2 illustrates the basic 144 finite element discretization used for the duct flow problem. For $Re = 200$, based on duct width, the fully developed velocity flow field should be attained within the duct.[10] Additional discretizations doubled and halved the number of node rows, and doubled the number of node columns to yield 264, 54 and 276 finite element discretizations. Figure

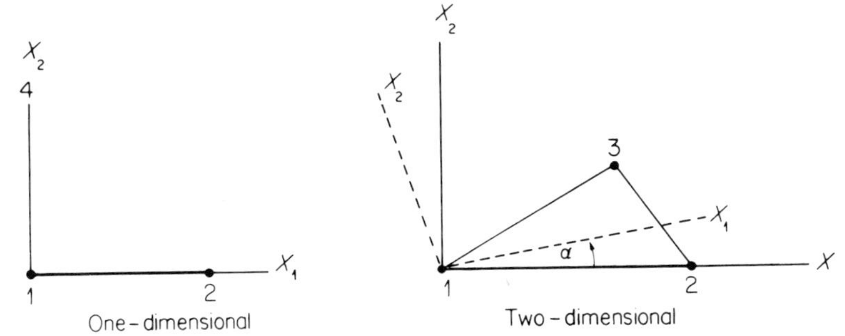

Figure 4.1 Finite element shape for linear approximation functions

4.3 shows the COMOC computed velocity distributions at the terminal node column of the solution domain for the four studied discretizations. Since velocity is a constant within an element for linear stream-function approximation, the velocity contours are correspondingly constructed. For all discretizations, the analytic solution approximately bisects the computed constant velocity within each element, and convergence with discretization is illustrated. Within the constraint of element constant velocity, solution accuracy is good for all discretizations, although refinement of the grid certainly improves the solution for points not coincident with an approximate element centroid. Figure 4.4 presents computed stream-function and vorticity in comparison to the analytic values, and convergence with discretization is illustrated. Figure 4.5 contains computed longitudinal velocity contours at various stations downstream of the duct inlet for the 264 element discretization. The important non-linear effect of velocity overshoot is predicted to occur over the first 4·8 % of the duct length and is maximum at the 1·5 % length station. The magnitude and location of velocity overshoot is in agreement with other solutions of the problem.[9,11]

A crucial analysis for flows governed by the full Navier–Stokes equations is prediction of regions of recirculation imbedded within an arbitrary flow

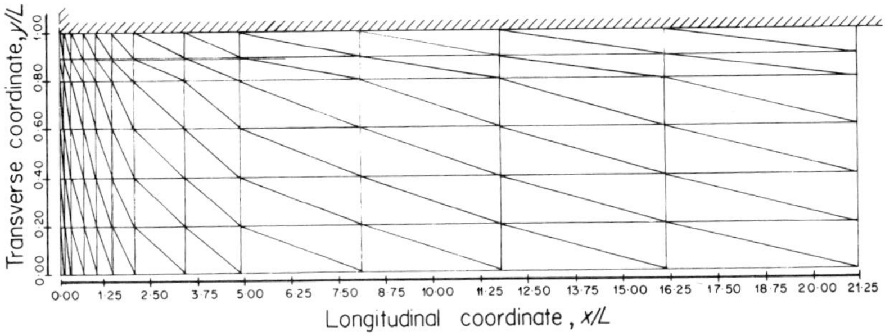

Figure 4.2 Discretization of rectangular duct into 144 triangular finite elements

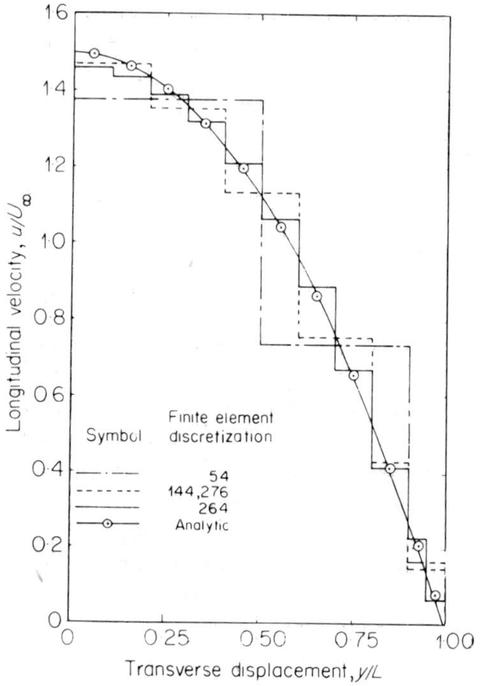

Figure 4.3 Computed fully developed longitudinal velocity distributions, $Re = 200$

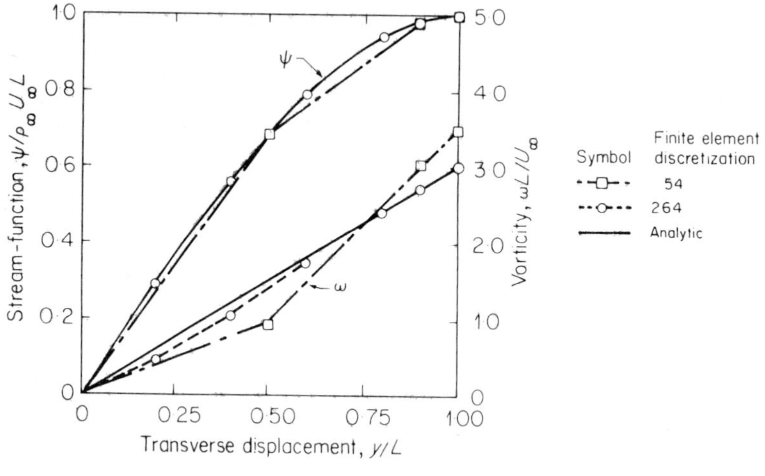

Figure 4.4 Fully developed computed stream-function and vorticity for duct flow, $Re = 200$

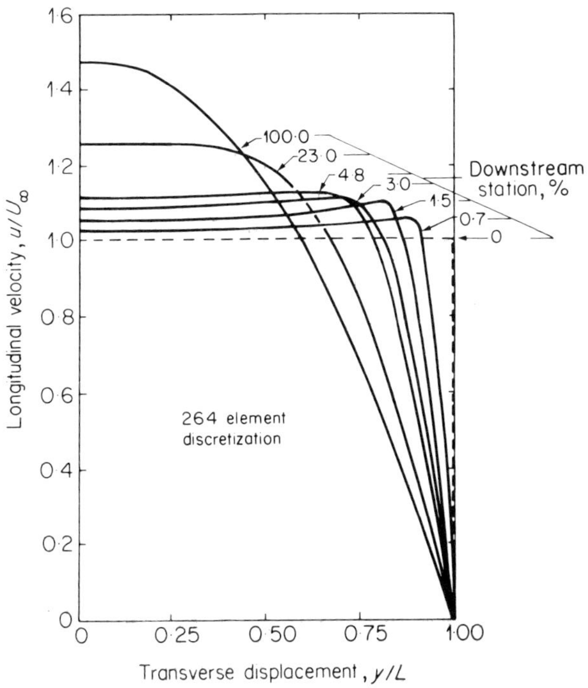

Figure 4.5 Longitudinal velocity distributions for incompressible duct flow, $Re = 200$

field. It is precisely this phenomenon that prevents use of boundary layer theory for regions of separation in otherwise predominantly boundary layer flows. A great deal of laboratory and numerical experimentation has been conducted for the sample geometry illustrated in Figure 4.6, which is flow in a duct with an abrupt enlargement in cross-sectional area. The fluid will not turn the corner, but will instead provide a smooth area transition by self-generation of a recirculation region. Finite difference numerical solution

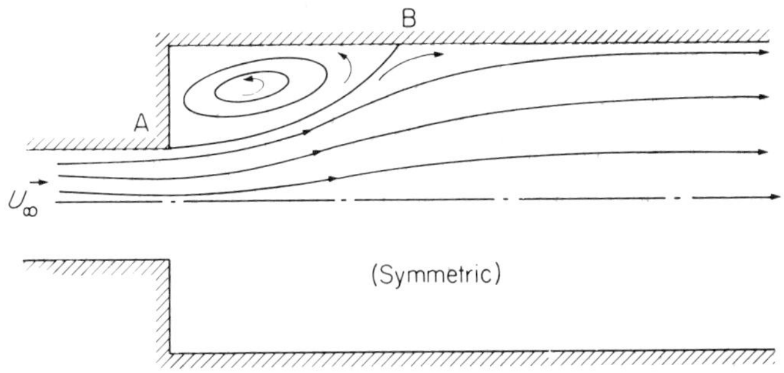

Figure 4.6 Flow over a rearward facing step

procedures for this problem typically require special handling of the corner region to promote generation of the recirculation zone.[11,12] There is no such requirement for the finite element algorithm and all wall-node vorticities are computed in a uniform fashion. Accurate computational representation of this problem is measured by prediction of the points of attachment of the stream line dividing the recirculation region from the main flow, both at the step and at the downstream reattachment point, points *A* and *B* in Figure 4.6. The points of attachment are identified by a vorticity sign change and the stream-function solution must numerically bifurcate to predict recirculation.

The finite element solution algorithm, as embodied in COMOC, predicts the rearward step steady state flow field for $Re = 200$, treating the problem as transient. Shown in Figure 4.7 is the discretization employed for the

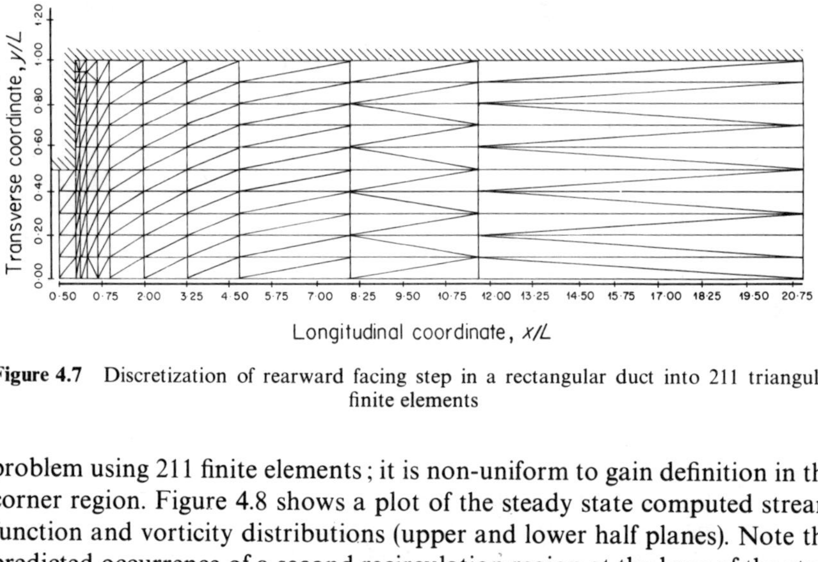

Figure 4.7 Discretization of rearward facing step in a rectangular duct into 211 triangular finite elements

problem using 211 finite elements; it is non-uniform to gain definition in the corner region. Figure 4.8 shows a plot of the steady state computed stream function and vorticity distributions (upper and lower half planes). Note the predicted occurrence of a second recirculation region at the base of the step. Within this region, the flow field corresponds to flow into a forward facing step, and the fluid again shields itself from abrupt changes in cross-sectional area by self-generation of a smooth transition. The point of reattachment of the dividing stream line for steady state is about 21 step-heights downstream.

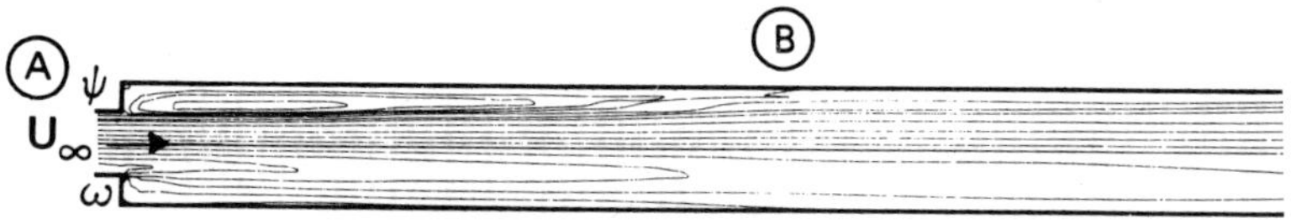

Figure 4.8 COMOC computed stream line and vorticity distributions for flow over a rearward facing step, $Re = 200$. Steady state solution

This prediction agrees within 10% of an extrapolation of data for two-dimensional flows[13] and with the computations for axisymmetric flow.[14] The overall shape of the predicted recirculation zone (Figure 4.8) is in agreement with experimental data[12] obtained for $Re = 130$.

Several additional experiments with the rearward step problem evaluate the versatility of the finite element solution technique. Figure 4.9(a) shows a close-up of the steady-state rearward step solution, which served as the

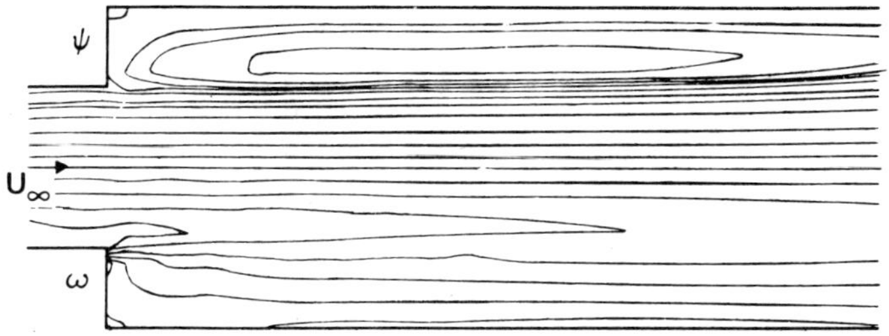

(a) COMOC computed steady state stream-function and vorticity for flow over a rearward step, Re = 200

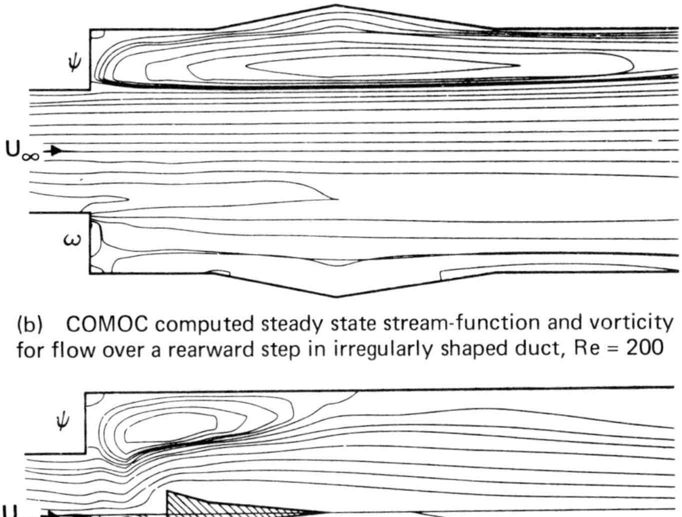

(b) COMOC computed steady state stream-function and vorticity for flow over a rearward step in irregularly shaped duct, Re = 200

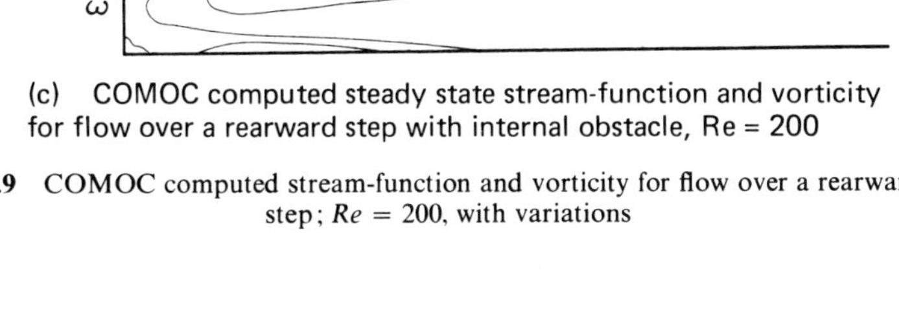

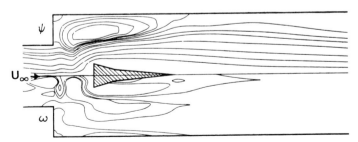

(c) COMOC computed steady state stream-function and vorticity for flow over a rearward step with internal obstacle, Re = 200

Figure 4.9 COMOC computed stream-function and vorticity for flow over a rearward facing step; $Re = 200$, with variations initial condition. Figure 4.9(b) shows the steady state computed solution for an indentation in the wall of 0·4 step heights relative displacement. The recirculation region has undergone a minor readjustment to fill the new interior domain and the downstream point-of-attachment of the dividing stream line was unaffected. However, as shown in Figure 4.9(c), the insertion of a blunt body into the solution domain drastically alters the predicted recirculation region. The inserted obstacle was tapered downstream to preclude the shedding of vortices, hence allow attainment of a steady state solution. The computed vorticity solution is markedly altered from Figure 4.9(a) and the dividing stream line reattachment has moved to within four step-heights downstream. The second recirculation region at the step base is retained in both sample solutions. No independent measures are available to assess the accuracy of either of these solutions, but the computed solutions do appear physically acceptable and amply illustrate the boundary shape variability that is automatically acceptable to the finite element technique.

4.5 Conclusions

A finite element solution algorithm has been derived for two-dimensional hydrodynamical flows. Sample solutions for problems involving general flows, using the COMOC computer program system, have verified accuracy and indicated considerable geometric versatility. Extension to computation of turbulent flows is directly accomplished by interpretation of the kinematic viscosity as an 'eddy' viscosity, since no linearity constraint exists in the derivation. Application to a wide range of realistic problems, including environmental hydrodynamics, is routinely possible.

Acknowledgements

The author wishes to acknowledge the long-term support given this work by Bell Aerospace Division of Textron. The analysis and computational results were supported by NASA-Langley Research Center on Contract NAS1-11809. I particularly wish to acknowledge the significant contributions of my colleagues Messrs Paul Manhardt and Joe Orzechowski.

References

1. B. A. Finlayson and L. E. Scriven, 'The method of weighted residuals—a review', *App. Mech. Rev.*, **19**, 9, 735–748 (1966).
2. A. J. Baker, 'Finite element computational theory for three-dimensional boundary layer flow', *AIAA Paper* 72-108, 1972.
3. F. D. Hains and A. J. Baker, 'Binary diffusion of a jet imbedded in a boundary layer', *AIAA*, **10**, 7, 938–940 (1972).

4. A. J. Baker and S. W. Zelazny, 'A theoretical study of mixing downstream of transverse injection into a supersonic boundary layer', *NASA CR*-112254, 1972.
5. A. J. Baker, 'Finite element solution algorithm for viscous incompressible fluid dynamics', *Int. J. Num. Meth. Engr.*, **6**, 1, 89–101 (1973).
6. A. J. Baker, 'A finite element solution algorithm for the Navier–Stokes equations', *NASA CR*-2391, 1974.
7. A. J. Baker and P. D. Manhardt, 'Finite element solution for energy conservation using a highly stable explicit integration algorithm', *NASA CR*-130149, 1972.
8. B. A. Finlayson and L. E. Scriven, 'On the search for variational principles', *Int. J. Heat Mass Trans.*, **10**, 6, 799–821 (1967).
9. O. C. Zienkiewicz, *The Finite Element Method in Engineering Science*, McGraw-Hill, London, 1971, Ch. 7.
10. H. Schlichting, *Boundary Layer Theory*, McGraw-Hill, New York, 1960.
11. A. D. Gosman, W. M. Pun, A. K. Runchal, D. B. Spalding and M. Wolfshtein, *Heat and Mass Transfer in Recirculating Flows*, Academic Press, London, 1969.
12. T. J. Mueller and R. A. O'Leary, 'Physical and numerical experiments in laminar incompressible separating and reattaching flows', *AIAA Paper 70–763*, 1970.
13. A. A. Dorodnitsyn, 'Review of methods for solving the Navier–Stokes equations', *Proceedings Third Int. Conf. on Num. Mtd. in Fluid Mechanics*, 1973, pp. 1–11.
14. E. O. Macagno and T. K. Hung, 'Computational and experimental study of a captive annular eddy', *J. Flu. Mech.*, **28**, 1, 43–64 (1967).

Chapter 5

Applications of Integral Equations to Fluid Flows in Unbounded Regions

R. D. Milne

5.1 Introduction

A linear partial differential equation subject to boundary values can, in principle, be solved in terms of an integral involving an appropriate Green's function. Since a Green's function cannot generally be found except for simple regions, this method of solution is not often practicable. Instead, a modification may be employed in which a fundamental solution of the differential equation is used in conjunction with a version of Green's theorem to give an integral equation expressing known boundary values in terms of an unknown distribution of such fundamental solutions with singular points lying in the boundary surface. In effect, the solution is expressed as an integral superposition of single and double layers distributed over the boundary: the technique is very familiar and is described in, for example, References 1 and 2. Three points of interest arise in connection with this technique in its application to the numerical solution of technological problems. Firstly, the values of the unknown are often required on the boundary and not in the interior; secondly, the spatial dimension of the problem is reduced by one and, thirdly, unbounded physical domains cause no real difficulty since the fundamental solution with the correct behaviour at infinity can usually be found. Furthermore, although the kernel of the integral equation is singular, the numerical procedures required for an approximate solution are basically those of summation. Indeed, Noble[3] has pointed out that the singular nature of the kernel is beneficial in the computational sense.

The problem of added or virtual mass in a dam-reservoir system is a good example. The pressure in the water consequent upon movement of the dam is needed only on the dam face; the large longitudinal extent of the body of water normally requires an artificial termination of the reservoir and the usual finite element solution must needs find the pressure everywhere by a subdivision of the entire three-dimensional interior.

It is clear that, by using a suitable fundamental solution which can automatically satisfy the free surface condition and perhaps also partly satisfy the conditions on the bed and side walls, the problem can be reduced to a two-dimensional integral equation over the dam face. This latter problem can then be dealt with approximately in the finite element manner by a (two-dimensional) subdivision of the face. Some preliminary calculations of this type are described in Reference 4.

The problem considered in this paper in detail is a very simple one, namely, the determination of the unsteady loading on a thin aerofoil oscillating harmonically in a stream: compressibility is ignored, but its inclusion would not cause any undue difficulty.

The problem has been chosen for four reasons:

(1) Exact solutions are available.[5]
(2) The solution, namely the loading, exhibits an edge singularity of the type $l \sim \delta^{-\frac{1}{2}}$, where δ is distance to the leading edge; it will also exhibit singularities where the derivative of upwash has a jump discontinuity.
(3) The kernel of the associated integral equation is strongly singular (i.e. is not locally integrable).
(4) The kernel is not symmetric and therefore there does not exist an associated variational statement of the problem.

In the linearized theory the aerofoil is represented by its projection onto the interval $(-1, 1)$ of the x axis (Figure 5.1) and the boundary-value problem

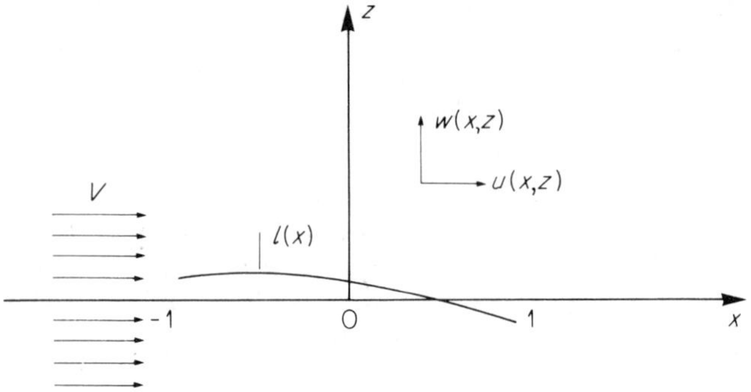

Figure 5.1 Thin aerofoil—coordinate system

is most readily expressed in terms of an acceleration potential $\psi(x, z, t)$ which is effectively a measure of pressure since, in the linearized theory $\psi = -p/\rho_\infty$ where ρ_∞ is the free stream density. The acceleration potential satisfies the differential equation

$$\frac{\partial^2 \psi}{\partial x^2} + \frac{\partial^2 \psi}{\partial z^2} - \frac{1}{a_\infty^2}\left(\frac{\partial}{\partial t} + V\frac{\partial}{\partial x}\right)^2 \psi = 0 \tag{5.1}$$

where a_∞ is the free stream sonic speed and V the free stream speed. We require that:

(1) ψ is continuous everywhere except across $(-1, 1)$ where it should exhibit a discontinuity $\Delta\psi = -\Delta p/\rho_\infty$ representing the loading on the aerofoil.

(2) ψ should represent outgoing waves at infinity.

(3) On the interval $(-1, 1)$, ψ should satisfy the kinematic boundary condition

$$\left(\frac{\partial \psi}{\partial z}\right)_{\substack{z \to 0 \\ -1 \leqslant x \leqslant 1}} = \left(\frac{\partial}{\partial t} + V\frac{\partial}{\partial x}\right) w \tag{5.2}$$

where $w(x, t)$ the upwash velocity in $(-1, 1)$ is given in terms of the known aerofoil camber shape $z_a(x, t)$ by

$$w = \left(\frac{\partial}{\partial t} + V\frac{\partial}{\partial x}\right) z_a \tag{5.3}$$

(4) We also have the auxiliary uniqueness condition (Kutta condition) that the loading vanishes at the trailing edge $x = 1$.

In practice the problem proves too difficult for a general time variation of $z_a(x, t)$ and effectively a Fourier transform procedure is adopted by assuming that the time variation of z_a, w, ψ is simple harmonic with non-dimensional frequency k.

The aerofoil is then represented by a sheet of fundamental solutions of Equation 5.1 which are in effect sinusoidally pulsating doublets, with their axes vertical, whose local strengths are directly proportional to the loading; the upwash on $(-1, 1)$ associated with this sheet is then used in (5.2) to yield an integral equation for the loading of the form

$$w(x) = \frac{1}{\pi}\int_{-1}^{1} l(\xi)K(x - \xi, k)\,\mathrm{d}\xi \tag{5.4}$$

The details can be found in References 5 or 7 and it is sufficient here to quote the final result for the kernel function K for the case of incompressible flow ($M = V/a_\infty \to 0$) viz.

$$K(x, k) = k\left\{i\,e^{-ikx}\left[C_i(k|x|) + i\left(\frac{\pi}{2} + S_i(kx)\right)\right] - \frac{1}{kx}\right\} \tag{5.5}$$

where C_i, S_i are the cosine and sine integral functions. When the aerofoil is steady in the flow ($k \to 0$) the kernel reduces to the simple form

$$K(x) = -\frac{1}{x} \tag{5.6}$$

It can be readily shown that, in general, the loading exhibits an inverse half-power singularity at the leading edge of the aerofoil.

5.2 Projection solution of the integral equation

For a general description of the projection method it is convenient to write the singular integral equation (5.4) in the form

$$w = T(l) \tag{5.7}$$

where

$$T(\,.\,) \equiv \frac{1}{\pi}\int_{-1}^{1} K(x-\xi, k)\,.\,\mathrm{d}\xi$$

is a linear operator mapping the elements of a loading space $\mathscr{L}$ to a space $\mathscr{W}$ of upwash functions. Because of the singular behaviour of the loading at the edge $x = -1$, the loading functions are not square integrable in $[-1, 1]$; consequently T cannot be interpreted as an operator in the Hilbert space L_2. Instead, we distinguish between the spaces $\mathscr{L}$ and $\mathscr{W}$ and, in addition, introduce the conjugate spaces $\mathscr{V}$ and $\mathscr{M}$, the spaces of linear functionals on $\mathscr{L}$ and $\mathscr{W}$ respectively.†

We shall take $\mathscr{L}$ to be a closed subspace of $Lp_1[-1, 1]$[5] with $1 < p_1 < 2$ such that every $l \in \mathscr{L}$ vanishes at $x = 1$ and satisfies a Lipschitz condition there: in this way the Kutta condition is automatically satisfied and $T\colon \mathscr{L} \to \mathscr{W}$ is one-to-one.[8]

We shall take $\mathscr{W}$ to be a subspace of $Lp_2[-1, 1]$ with $p_2 > 2$ but shall not otherwise specify the functions w. Although in practice a given upwash function will be bounded in $[-1, 1]$, for the purposes of the projection method we need to allow for unbounded functions in $\mathscr{W}$.

The conjugate space $\mathscr{M}$ is the space of bounded linear functions on $\mathscr{W}$: that is, if $m \in \mathscr{M}$ then for every $w \in \mathscr{W}$ the linear functional

$$m(w) = \int_{-1}^{1} mw\,\mathrm{d}x \tag{5.8}$$

is bounded where the interval of integration is understood to be $(-1-0, 1+0)$; $\mathscr{M}$ is consequently a subspace of $Lp_1[-1, 1]$.

Similarly, we introduce the space $\mathscr{V}$ conjugate to $\mathscr{L}$ such that if $v \in \mathscr{V}$ then for every $l \in \mathscr{L}$

$$v(l) = \int_{-1}^{1} vl\,\mathrm{d}x$$

is bounded; $\mathscr{V}$ is clearly a subspace of $Lp_2[-1, 1]$ and is, in fact, indistinguishable from $\mathscr{W}$ for $x \in (-1, 1)$.

It is readily seen that, in the neighbourhood of $x = -1$, the functions m must behave no worse than δ^{α} with $\alpha > -\frac{1}{2}$: in fact, we shall impose the more stringent condition that the functions m vanish at $x = -1$ and satisfy a Lipschitz condition there. This allows us to define a conjugate operator $T^*\colon \mathscr{M} \to \mathscr{V}$ which is one-to-one.

Since the spaces $\mathscr{W}$ and $\mathscr{M}$ are reflexive, $\mathscr{W}$ could equally well be regarded as the conjugate of $\mathscr{M}$: in view of this it is convenient to represent the func-

† The usual notation for the conjugate space would be L^* and W^* respectively, but this does not prove to be a convenient notation in the subsequent analysis.

tional $m(w)$ by the symmetric bracket notation

$$m(w) = [w, m] = [m, w] = w(m), \tag{5.9}$$

with the same notation applying to $\mathscr{L}$ and $\mathscr{V}$.†

The linear operator $T:\mathscr{L} \to \mathscr{W}$ induces a conjugate operator[6] $T^*:\mathscr{M} \to \mathscr{V}$ defined by

$$[T(l), m] = [l, T^*(m)] \tag{5.10}$$

for all $l \in \mathscr{L}$, $m \in \mathscr{M}$. The conjugate operator is easily seen to be

$$T^*(\,.\,) \equiv \frac{1}{\pi}\int_{-1}^{1} K(\xi - x, k)\,.\,\mathrm{d}\xi \tag{5.11}$$

which may be interpreted physically as a reversal of the stream direction. Equation 5.10 is a statement of Flax's Reverse Flow Theorem[9] in which both operators T, T^* are rendered one-to-one by imposing the Kutta condition at appropriate 'trailing edges'.‡

It is now natural to refer to $\mathscr{L}$ and $\mathscr{W}$ as spaces of forward loading and upwash functions with $\mathscr{M}$ and $\mathscr{V}$ as spaces of reverse loading and upwash functions; the position is summarized in diagrammatic form in Figure 5.2.

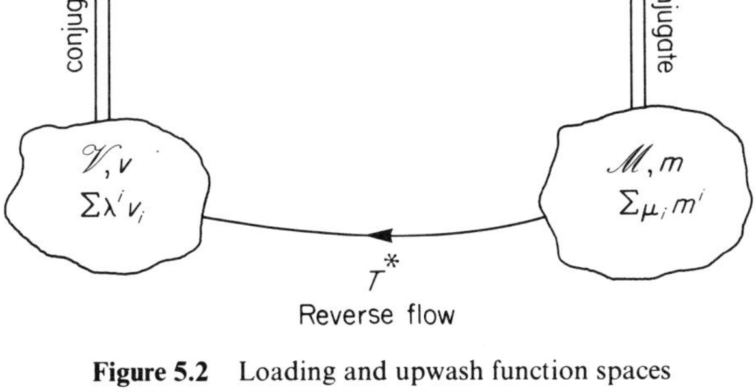

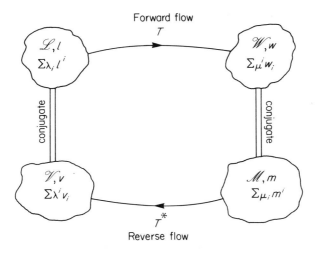

Figure 5.2 Loading and upwash function spaces

† The use of the bracket notation suggests a similarity with an inner product and indeed their properties are similar except that an interchange of elements in the linear functional does not generate the complex conjugate as for the inner product. Weighted inner products could be introduced on $\mathscr{L}$ and $\mathscr{M}$ but it is felt that the distinction between the various spaces introduced above is more satisfying from a physical point of view; furthermore, in computing generalized forces in the unsteady case the hermitian property of the inner product is an embarrassment.

‡ The result (5.10) remains true for less stringent conditions.

Let $\mathscr{W}_n$ be an n dimensional subspace of $\mathscr{W}$ with basis w_i, $i = 1 \ldots n$; then every element of $\mathscr{W}$ can be represented by

$$w = \sum_{i=1}^{n} \mu^i w_i + w' \tag{5.12}$$

for some set of scalars $\{\mu^i\}$ and some $w' \in \mathscr{W}'_n$, a direct complement of $\mathscr{W}_n$ in $\mathscr{W}$.

The decomposition (5.12) induces a similar decomposition on $\mathscr{M}$ such that

$$\mathscr{M} = \mathscr{M}_n \oplus \mathscr{M}'_n$$

where $\mathscr{M}_n$, the subspace reciprocal to $\mathscr{W}_n$, is an n dimensional subspace with basis vectors m^i defined by

$$[m^j, w_i] = \delta^j_i \tag{5.13}$$

Hence, every $m \in \mathscr{M}$ can be represented by

$$m = \sum_{i=1}^{n} \mu_i m^i + m'$$

for some set of scalars $\{\mu_i\}$ and some $m' \in \mathscr{M}'_n$.

Further, $\mathscr{M}'_n$ is an annihilator of $\mathscr{W}_n$ and $\mathscr{W}'_n$ is an annihilator of $\mathscr{M}_n$, so that, for any w, m,

$$[w, m] = \sum_{i=1}^{n} \mu^i \mu_i + [w', m'] \tag{5.14}$$

Taking $m = m^j$ in (5.14) gives

$$\mu^j = [w, m^j] \tag{5.15}$$

and similarly,

$$\mu_i = [m, w_i] \tag{5.16}$$

We call the function

$$w_{(n)} = \sum_{i=1}^{n} [w, m^i] w_i \tag{5.17}$$

the projection of w onto $\mathscr{W}_n$ and

$$m^{(n)} = \sum_{i=1}^{n} [m, w_i] m^i \tag{5.18}$$

the projection of m onto $\mathscr{M}_n$.

Let $\tilde{w} \in \mathscr{W}$ be a given upwash function. Let $l^{(n)} \in \mathscr{L}_n \subset \mathscr{L}$ and $m^{(n)} \in \mathscr{M}_n \subset \mathscr{M}$.

Then we shall call $\tilde{l}^{(n)}$ a projection solution of class n of the operator equation

$$T(l) - \tilde{w} = 0$$

if

$$[T(\tilde{l}^{(n)}) - \tilde{w}, m^{(n)}] = 0 \tag{5.19}$$

for all $m^{(n)} \in \mathscr{M}_n$. Here we regard $T(\tilde{l}^{(n)}) - \tilde{w} = \varepsilon$ as an error function and demand that the projection of ε onto $\mathscr{M}_n$ vanishes.

If $\{l_i\}$ is a basis set for $\mathscr{L}_n$ then

$$\tilde{l}^{(n)} = \sum_{i=1}^{n} \lambda_i l^i$$

where the n scalars $\{\lambda_i\}$ are the solution values of the simultaneous linear equations,

$$\sum_{i=1}^{n} \lambda_i[T(l^i), m^j] = [w, m^j], \qquad j = 1, 2, \ldots, n \tag{5.20}$$

Let $\{\mathscr{L}_n\}$, $\{\mathscr{M}_n\}$ be a sequence of subspaces of equal denumerable dimension such that $\mathscr{L}_m \subset \mathscr{L}_n$ and $\mathscr{M}_m \subset \mathscr{M}_n$ if $m < n$; then we call

$$\tilde{l} = \lim_{n \to \infty} \tilde{l}^{(n)}$$

a weak solution of $T(l) - \tilde{w} = 0$.

This weak solution will not in general converge pointwise or in the norm of Lp_1 or Lp_2 to the true solution.† However, the main quantities of interest in the applications are generalized forces associated with the loading l. These are scalar quantities (generally complex) defined by weighted integrals of the loading over $[-1, 1]$. Thus, for example, for the weighting function $\zeta(x)$, $x \in [-1, 1]$ and the given upwash $\tilde{w}$ the generalized force $Q(\zeta; \tilde{w})$ is given by

$$Q(\zeta; \tilde{w}) = \int_{-1}^{1} \zeta \tilde{l} \, \mathrm{d}x \tag{5.21}$$

where $T(\tilde{l}) = \tilde{w}$. Since, necessarily, $\zeta \in \mathscr{V}$ the generalized force $Q(\zeta; \tilde{w})$ can be regarded as the value of the linear functional $\zeta(\tilde{l})$. For a chosen subspace $\mathscr{L}_n$ let $\mathscr{V}_n \subset \mathscr{V}$ be the corresponding reciprocal subspace with basis $\{v_i\}$ and let $\tilde{l}^{(n)}$ be a projection solution of class n. Then

$$\begin{aligned} Q_n(\zeta; \tilde{w}) &= [\zeta, \tilde{l}^{(n)}] \\ &= [\zeta_{(n)}, \tilde{l}^{(n)}] \end{aligned} \tag{5.22}$$

is an approximation of class n to the generalized force, where $\zeta_{(n)}$ is the projection of ζ onto $\mathscr{V}_n$.

† Note that we have imposed no requirement that the subspaces $\mathscr{L}_n$ are in the predomain of $\tilde{w}$ by T.

Let $\tilde{m}^{(n)}$ be a projection solution of class n for the operator equation

$$T^*(\tilde{m}) - \zeta = 0$$

that is,

$$[T^*(\tilde{m}^{(n)}) - \zeta, l^{(n)}] = 0 \tag{5.23}$$

for all $l^{(n)} \in \mathscr{L}_n$.

Then, using (5.10) and (5.22) we have an alternative form for Q_n, namely

$$\begin{aligned} Q_n(\zeta; \tilde{w}) &= [\tilde{m}^{(n)}, \tilde{w}] \\ &= [\tilde{m}^{(n)}, \tilde{w}_{(n)}] \end{aligned} \tag{5.24}$$

where $\tilde{w}_{(n)}$ is the projection of $\tilde{w}$ onto $\mathscr{W}_n$.

We shall now show that, provided ζ is an element of $R(T^*)$, the range space of T^*, then convergence of $\tilde{l}_{(n)}$ to a weak solution $\tilde{l}$ implies ordinary convergence of the corresponding generalized force.

Suppose

$$\lim_{n \to \infty} Q_n(\zeta; \tilde{w}) = Q(\zeta; \tilde{w})$$

that is,

$$\lim_{n \to \infty} [\tilde{l}^{(n)} - \tilde{l}, \zeta] = 0$$

then, since $\zeta \in R(T^*)$,

$$\lim_{n \to \infty} [\tilde{l}^{(n)} - \tilde{l}, T^*(m)] = 0$$

for some $m \in \mathscr{M}$. But using (5.10) we have

$$\lim_{n \to \infty} [T(\tilde{l}^{(n)}) - \tilde{w}, m] = 0$$

which is satisfied for any $m \in \mathscr{M}$ if

$$\lim_{n \to \infty} \tilde{l}^{(n)} = \tilde{l}$$

This result, together with Equation 5.19, is equivalent to an application of Flax's Variational Principle.[10]

The quality and character of a projection solution of the integral equation is of course dependent on the choice of basis sets in $\mathscr{L}_n$ and $\mathscr{M}_n$, together with the corresponding reciprocal sets. The conditioning of the resulting algebraic equations can generally be expected to be good because of the singular nature of the kernel function.[3]

Section 5.3 develops the method for a particular basis set. Section 5.4 develops an alternative projection method which is based on an extension of the operator T to the interval $(-\infty, \infty)$ giving a projection method which is, in a sense, reciprocal to the one just described.

5.3 A finite element approach

Basis function sets defined over the whole interval $(-1, 1)$ suffer the disadvantage that local singular behaviour of the loading tends to degrade the approximation elsewhere. The case of flap deflection is a good illustration wherein the exact loading exhibits a logarithmic singularity at the hinge line.† While special functions can be introduced to deal locally with such singularities, simplicity and flexibility are lost.

An alternative approach is to adopt what might be termed the finite element philosophy and represent the loading in terms of a set of basis functions having localized support; preferably, the functions should be of a very simple type. Functions of this type can be expected to accommodate local singular behaviour; indeed, one can go further and make no explicit attempt to allow for singularities in the loading, whether at the leading edge or at flap hinge lines. Since these singularities are integrable, an approximation to the values of the generalized forces in terms of an everywhere-bounded loading is quite feasible. Naturally, in those regions where the exact loading exhibits singularities, the approximate loading can be expected to peak. This viewpoint is naive, and yet entirely consistent with the type of approximation to be expected from a projection method.

Let the aerofoil chord $(-1, 1)$ be subdivided into N intervals,

$$J_r, r = 1 \ldots N$$

In each interval define a set of basis functions

$${}_r\psi_k, k = 1 \ldots M$$

which vanish outside J_r. The set of $n + 1 = MN$ functions is taken as a basis for $\mathscr{V}_{n+1}$.

We take as a basis for $\mathscr{L}_{n+1}$ the conjugate basis $\{{}_r\psi^k\}$, consisting of the union of conjugate basis sets on each interval J_r.

The Kutta condition is satisfied by imposing the constraint that, in the interval J_N containing the trailing edge, the net loading should vanish at the edge; the approximating loading space is then of dimension $n = MN - 1$, and

$$l_{(n)} = \sum_{r=1}^{N} \chi_r \sum_{i=1}^{M} {}_r\lambda_i\, {}_r\psi^l \tag{5.25}$$

† It is often stated that, for the case of flap deflection, the lift and moments etc. can be readily calculated by using the reverse flow theorem Equation 5.10. Thus, for the lift L, we have

$$L = [l, 1] = [l, T^*(m_\alpha)] = [T(l), m_\alpha]$$
$$= [w_{\text{flap}}, m_\alpha]$$

where m_α is the loading, in reverse flow, for an aerofoil at incidence and this can be found exactly in terms of Jacobi polynomials. But $m_\alpha \in \mathscr{M}_n$ so that the piecewise constant function w_{flap} will effectively be represented only by its smooth polynomial projection in $\mathscr{W}_n$.

with the constraint that $\tilde{l}_{(n)}(1) = 0$, where χ_r is the characteristic function for the rth interval.

The choice of basis sets in $\mathscr{W}_{n+1}$ and $\mathscr{M}_{n+1}$ is not necessarily directly related to those chosen for $\mathscr{V}_{n+1}$ and $\mathscr{L}_{n+1}$; however, in practice it would appear sensible to adopt similar local basis sets based on the same subdivision of the chord. The Kutta condition is applied at the leading edge to give the reverse loading space $\mathscr{M}_n$.

While not essential, it is natural to impose continuity of loading, and perhaps of its slope, at element boundaries; this immediately fixes the minimum number of basis functions in each sub-interval. For example, if continuity is prescribed, we require a minimum of two basis functions per element. It is convenient then to introduce global nodal loading coordinates, as in a typical structural analysis, numbered or referenced with respect to the whole interval $(-1, 1)$.

Continuity is imposed by means of an assembly matrix or more directly by an implicit computing procedure. In this way, one constructs effectively a continuous basis for $\mathscr{L}_n$; the same continuity conditions must of course be applied to $\mathscr{M}_n$ for application of the projection equation (5.19). The dimension of the final matrix equation when continuity is imposed will clearly be $N(M-1)$ so that with $M = 2$ the solution is in terms of N nodal loading coordinates only.

In the unsteady case, the coordinates of the basis functions (and the global coordinates) are complex numbers and the final complex matrix equation is converted to a real equation having twice the dimension; the deails of the calculations are given fully in Reference 7. The basic quantities required in the steady case, for example, are the numbers

$$ {}_{rs}l^{ij} = -\frac{1}{\pi}\int_{J_r} {}_r\phi^i \, \mathrm{d}x \int_{J_s} \frac{{}_s\psi^j}{x-\xi}\,\mathrm{d}\xi \tag{5.26} $$

which can be formed into $M \times M$ local influence matrices and which are independent of the absolute location of any element whenever sub-intervals of equal length are used; they can consequently be computed once and for all.

For sufficiently simple basis functions the numbers ${}_{rs}l^{ij}$ can be calculated analytically both for the steady and unsteady cases.

5.3.1 Some numerical results

The following results are based on equal intervals with only two basis functions per interval, namely the constant and linear functions

$$ \begin{aligned} {}_r\phi_1 &= 1 \\ {}_r\phi_2 &= \sqrt{3}(1-2\eta), \qquad 0 \leqslant \eta \leqslant 1 \end{aligned} \tag{5.27a} $$

where η is a local, interval coordinate. For the conjugate functions we simply take:

$$\begin{aligned} {}_r\phi^1 &= 1 \\ {}_r\phi^2 &= \sqrt{3}(1 - 2\eta) \end{aligned} \tag{5.27b}$$

If continuity is not imposed, the projection equation (5.19) can be given a simple physical interpretation in this case: the mean value and first moment of the upwash error is required to vanish in each interval with the exception of the leading edge interval where only the first moment about the leading edge is required to vanish.

For the continuous case linear combinations of these requirements are imposed by the projection: over two adjacent intervals the linearly weighted error should vanish. When the loading is approximated continuously, an alternative viewpoint in this case is to consider the basis of $\mathscr{L}_n$ as consisting of rooftop or triangle functions which are non-zero over pairs of adjacent elements.

Owing to the local nature of the kernel the final matrix has elements which decrease outside the triple diagonal band. For the steady case the matrix takes the partitioned form,

$$\left[\begin{array}{c|ccccc} c_1 & & & & & \\ c_2 & & & & & \\ \vdots & & & S & & \\ c_{N-2} & & & & & \\ c_{N-1} & & & & & \\ \hline c_N & c_{N-1} & c_{N-2} & \cdots & c_2 & c_1 \end{array}\right]$$

where $c_r > c_s$ if $r > s$ and S is a skew-symmetric matrix whose elements decrease in magnitude beyond the upper and lower diagonals.

Extensive numerical results are given in Reference 7: we give here only one or two examples. Exact solutions are available in References 5 and 10.

Steady case—incidence and flap deflection Results for these cases are shown in Figures 5.3 and 5.4 and in Tables 5.1 and 5.2.

Moments are calculated about the leading edge which tends to minimize error, but since the moment about a point distant d from the leading edge is given by

$$M_d = M_0 - Ld$$

the maximum error is effectively reflected in the lift.

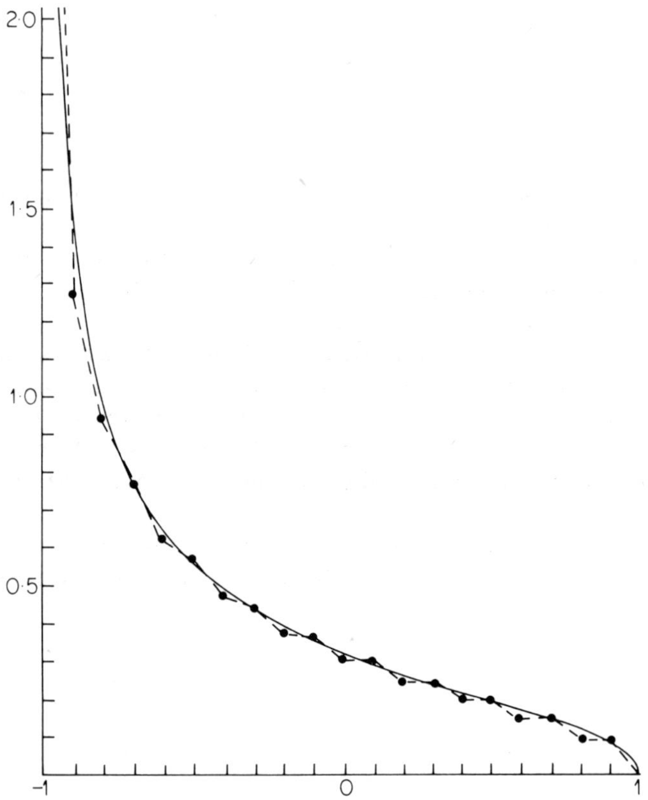

Figure 5.3 Loading for aerofoil at incidence—20 elements

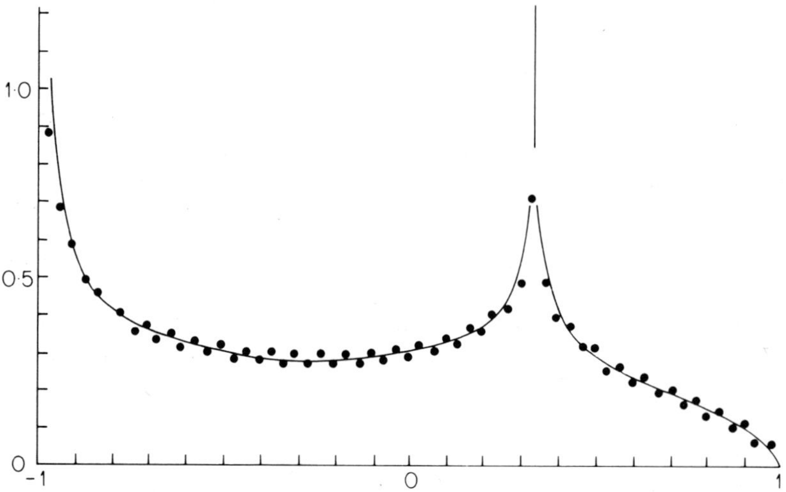

Figure 5.4 Loading for flap deflection—60 elements

Table 5.1 Aerofoil at incidence

No. of elements	Lift	First moment	Second moment
5	0·4864	0·1254	0·0605
10	0·4929	0·1251	0·0616
15	0·4953	0·1250	0·0619
20	0·4964	0·1250	0·0620
Exact	0·5000	0·1250	0·0625

Table 5.2 Aerofoil with flap

No. of elements	Lift	First moment	Hinge moment
12	0·3410	0·1353	0·0756
24	0·3435	0·1359	0·0767
36	0·3443	0·1361	0·0771
48	0·3447	0·1362	0·0773
60	0·3450	0·1362	0·0775
Exact	0·3460	0·1365	0·0780

Unsteady case—linear upwash field Results for a frequency parameter (based on half-chord) of 0·4 are given in Figure 5.5 and in Table 5.3.

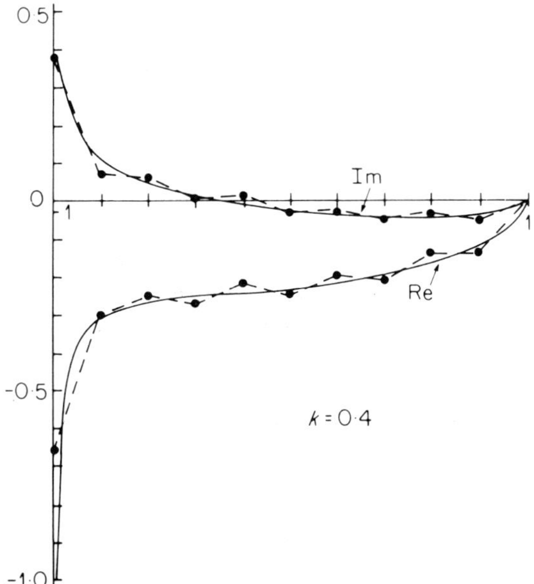

Figure 5.5 Loading for harmonic linear upwash—10 elements

Table 5.3 Harmonic linear upwash, $k = 0{\cdot}4$

No. of elements	Lift	First moment
5	$-0{\cdot}2266 + i0{\cdot}0129$	$-0{\cdot}0866 - i0{\cdot}0107$
8	$-0{\cdot}2295 + i0{\cdot}0127$	$-0{\cdot}0880 - i0{\cdot}0115$
10	$-0{\cdot}2305 + i0{\cdot}0125$	$-0{\cdot}0884 - i0{\cdot}0117$
Exact	$-0{\cdot}2343 + i0{\cdot}0119$	$-0{\cdot}0898 - i0{\cdot}0126$

5.4 A reciprocal projection for the steady case

By extending the interval of integration to $(-\infty, \infty)$ in Equation 5.4 with the kernel (5.6), a type of projection which is, in a sense, reciprocal to that of Section 5.2 can be deduced. Since T is not a symmetric operator and is therefore not associated with a functional, there is no question of obtaining upper and lower bounds for a specific scalar quantity. However, there are resemblances with the classical quadratic reciprocal variational problem[11] in that the projection to be described now requires a special or restricted solution of the operator equation (compare equilibrium solutions in applications of the Complementary Virtual Work Principle); furthermore, the projection can be considered as being, in a sense, 'orthogonal'.

If it is understood implicitly that the loading l vanishes outside $(-1, 1)$ then Equation 5.4 with 5.6 can be written

$$w(x) = -\frac{1}{\pi}\int_{-\infty}^{\infty} \frac{l(\xi)}{x - \xi}\,d\xi \tag{5.28}$$

Equation 5.28 is a convolution and hence the inverse may be written down immediately as

$$l(x) = \frac{1}{\pi}\int_{-\infty}^{\infty} \frac{w(\xi)}{x - \xi}\,d\xi \tag{5.29}$$

But of course we have a knowledge of $w(x)$ only in $(-1, 1)$; outside this interval $w(x)$ is not known. However, by approximating to the upwash outside $(-1, 1)$ in terms of the elements of an upwash function space we can use a projection to ensure that, in a projective sense, the loading off aerofoil vanishes.

We proceed in the following way. Firstly, we may ensure that l always vanishes in $(-\infty, -1)$ by dealing with the operator equation

$$w = S(l)$$

where

$$S(\,.\,) = -\frac{1}{\pi}\int_{-1}^{\infty} \frac{\cdot}{x - \xi}\,d\xi \tag{5.30}$$

instead of Equation 5.28. The inverse of S is almost as easy to find as for the complete convolution operator (5.28). In fact, we have

$$l = S^{-1}(w), \qquad l = 0 \quad \text{in } (-\infty, -1) \tag{5.31}$$

where

$$S^{-1}(\,.\,) = \frac{1}{\pi}\int_{-1}^{\infty}\left(\frac{\xi+1}{x+1}\right)^{\frac{1}{2}}\frac{\cdot}{x-\xi}\,d\xi$$

Let l_0 be any loading function defined in $(-1, \infty)$ such that

$$l_0 = S^{-1}(w_0)$$

where w_0 coincides with the given upwash $\tilde{w}$ in the interval $(-1, 1)$ and $w_0 = \mathbf{0}(1/x)$ as $x \to \infty$† but is otherwise arbitrary.

Let $\mathscr{W}''$ be a linear space of (upwash) functions vanishing outside $(1, \infty)$ and such that, if $w'' \in \mathscr{W}''$,

$$\begin{aligned} &\text{(a)}\ w''(1) = 0; \\ &\text{(b)}\ w''(x) = \mathbf{0}(1/x) \quad \text{as } x \to \infty. \end{aligned} \tag{5.32}$$

The net upwash in $(-1, \infty)$ is then given by

$$w = S(l_0) + w'' \tag{5.33}$$

and the nett loading in $(-1, \infty)$ by

$$l = l_0 + S^{-1}(w'') \tag{5.34}$$

The nett loading should vanish in $(1, \infty)$; a projection principle for this is embodied in the statement,

$$\int_1^{\infty}(l_0 + S^{-1}(\tilde{w}''))w''\,dx = 0 \qquad \forall w'' \in \mathscr{W}'' \tag{5.35}$$

where $\tilde{w}''$ is the true upwash function in $(1, \infty)$. Approximating w'' in an n-dimensional subspace $\mathscr{W}''_n$ with basis w''_i, $i = 1 \ldots n$, we obtain the projection approximation

$$\tilde{w}''_{(n)} = S(l_0) + \sum_{i=1}^{n} v^i w''_i$$

where $\{v^i\}$ is the solution vector of

$$\int_1^{\infty}\left(l_0 + \sum_{i=1}^{n} v^i S^{-1}(w''_i)\right)w''_j\,dx = 0, \qquad j = 1 \ldots n, \tag{5.36}$$

† We know that, sufficiently far downstream the lifting aerofoil will appear as a single vortex and hence that the upwash should behave as $\mathbf{0}(1/x)$.

the approximate loading being given by

$$\tilde{l}_{(n)} = l_0 + \sum_{i=1}^{n} v^i S^{-1}(w_i'')$$

The Kutta condition is not satisfied at the trailing edge in this projection.

An interesting example of this type of projection is described in Reference 12, although the author does not frame the analysis in such terms; rather, he presents an approximate solution of a pair of dual integral equations. Reference 12 is concerned with the loading on a rectangular wing in a sinusoidal gust field. The author having found a method of solution for the infinite span wing[13] (i.e. an approximate equivalent of S^{-1}) then finds the loading on the finite wing by ensuring that 'a weighted chordwise smoothing of the loading is made zero to increasingly high orders for all spanwise stations outside the wing tips'. His first approximation merely makes the lift vanish beyond the wing tips.

A numerical example For the aerofoil at incidence we take $l_0 = S^{-1}(w_0)$ where

$$\begin{aligned} w_0 &= 1, & -1 \leqslant x \leqslant 1 \\ &= 1/x, & 1 \leqslant x \leqslant \infty \end{aligned}$$

and as a basis for $\mathscr{W}_n''$ the functions,

$$w_i'' = \frac{(x-1)}{x^i}, \qquad i = 2 \ldots (n+1)$$

The required scalars

$$-\frac{1}{\pi}\int_1^\infty w_j''(x)\,\mathrm{d}x \int_1^\infty \left(\frac{\xi+1}{x+1}\right)^{\frac{1}{2}} \frac{w_i''(\xi)}{x-\xi}\,\mathrm{d}\xi$$

can be computed analytically for these simple basis functions.

The approximate load distribution is shown in Figure 5.6; the main error arises through lack of an explicit Kutta condition. The approximate upwash in $(1, \infty)$ is virtually indistinguishable from its exact counterpart. Resulting values of lift and moment are shown in Table 5.4. Lift is consistently over-

Table 5.4 Aerofoil at incidence—results of reciprocal projection

No. of polynomials	Lift	First moment
4	0·5125	0·1312
8	0·5130	0·1315
10	0·5135	0·1318

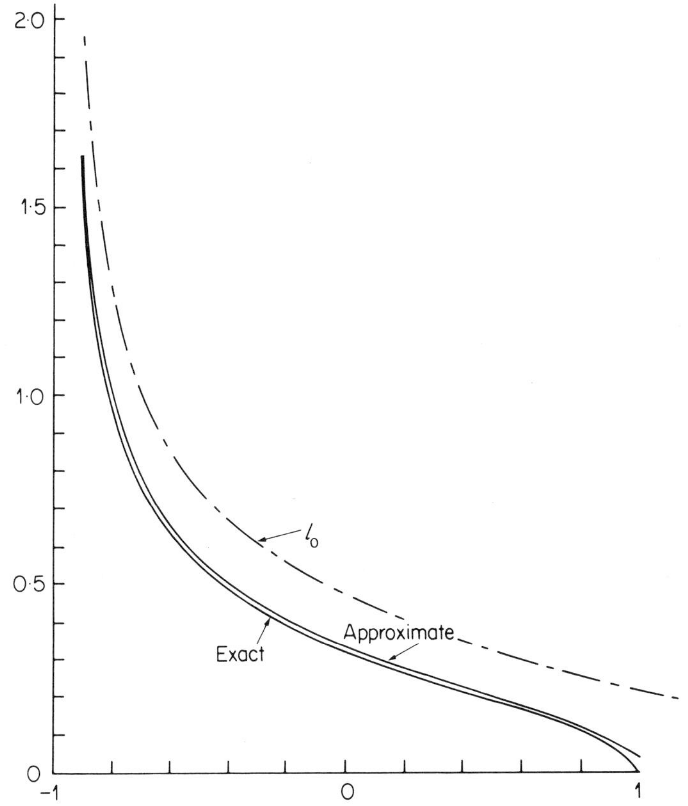

Figure 5.6 Incidence loading—7 wake upwash functions

estimated, whereas the results of the projection of Section 5.3 give underestimates (Table 5.1 and 5.2). The author does not believe there is a rigorous theoretical justification for this result.

Acknowledgements

The author is grateful to Mr. G. D. Padfield for computing the numerical results presented in Section 5.4; he is also grateful to Mr. G. Clapworthy for the results shown in Figure 5.4 which were computed by a more efficient version of the program of Reference 7.

References

1. C. Lanczos, *Linear Differential Operators*, Van Nostrand, New York, 1961, p. 476.
2. C. D. Green, *Integral Equation Methods*, Nelson, London, 1969, Chapter 6.

3. B. Noble, 'Some applications of the numerical solution of integral equations to boundary value problems', Conference on the Applications of Numerical Analysis, Springer-Verlag, *Lecture Notes in Mathematics Series*, **228** (1971).
4. R. J. Astley, *The Dynamic Interaction of Dams and Reservoirs*, Ph.D. thesis, University of Bristol, 1972.
5. R. L. Bisplinghoff *et al.*, *Aeroelasticity*, Addison-Wesley, Reading, 1955, pp. 251–280, 325.
6. G. Bachman and L. Narici, *Functional Analysis*, Academic Press, New York, 1966, Chapters 12 and 14.
7. Niaz A. Sheikh, *The Application of Projection Methods to Linearised Unsteady Lifting Surface Theory*, Ph.D. thesis, London University, 1970.
8. F. G. Tricomi, *Integral Equations*, Interscience, New York, 1967, pp. 173–183.
9. A. H. Flax, 'Reverse flow and variational theorems for lifting surfaces in non-stationary compressible flow', *Journ. Aero. Sci.*, **20**, 2, 120 (1953).
10. D. E. Davies, 'An Application of Flax's Variational Principle to lifting surface theory', *Brit. Aero. Res. Counc., R & M No.* 3564, 1969.
11. R. Courant and D. Hilbert, *Methods of Mathematical Physics*, Vol. 1, Interscience, New York, 1953, pp. 252–257.
12. J. M. R. Graham, 'A lifting surface theory for the rectangular wing in non-stationary flow', *Aero. Quart.*, **22**, 1, 83–100 (1971).
13. J. M. R. Graham, 'Lifting surface theory for the problem of an arbitrarily yawed sinusoidal gust incident on a thin aerofoil in incompressible flow, *Aero. Quart.*, **21**, 2, 182–198 (1970).

Chapter 6

Numerical Solution of Steady State Diffusion Problems Containing Singularities

J. R. Whiteman

6.1 Introduction and physical problem

In the study of steady state diffusion in heterogeneous mixtures some property of the mixture, for example permeability or thermal conductivity, has to be determined. This property is itself related to a characteristic of the physical situation. Thus permeability is related to concentration of diffusing substance and thermal conductivity to temperature. In this paper, the mixtures under consideration are two-component composites made up of isolated islands of impermeable material in a continuum.

We consider a two-dimensional model situation of the type described above in which the function $u(x, y)$ represents either concentration of diffusing substance or temperature. In the composite it is assumed for simplicity that rectangular impermeable blocks are present on a regular lattice in an isotropic substance, see Figure 6.1, and that u is held constant on the sides of the composite. Advantage can be taken of the symmetry so that only a section of the problem in an L-shaped subregion, such as $\overline{\text{OBCDEFO}} \equiv \Omega$ with boundary $\partial\Omega$ as in Figure 6.1, need be considered,[5] where $\overline{\text{OB}} = \overline{\text{BC}} = \overline{\text{EF}} = \overline{\text{FO}} = \frac{1}{2}$, $\overline{\text{CD}} = \overline{\text{DE}} = 1$.

The problem therefore becomes that of finding $u(x, y)$, the solution of the harmonic mixed boundary value problem

$$-\Delta[u(x, y)] = 0, \qquad (x, y) \in \Omega, \tag{6.1}$$

$$u(x, y) = 1, \qquad (x, y) \in \overline{\text{CD}}, \tag{6.2}$$

$$u(x, y) = 0, \qquad (x, y) \in \overline{\text{EF}}, \tag{6.3}$$

$$\frac{\partial u(x, y)}{\partial x} = 0, \qquad (x, y) \in \overline{\text{OB}}, \tag{6.4}$$

$$\frac{\partial u(x, y)}{\partial y} = 0, \qquad (x, y) \in \overline{\text{BC}} \cup \overline{\text{DE}} \cup \overline{\text{FO}}, \tag{6.5}$$

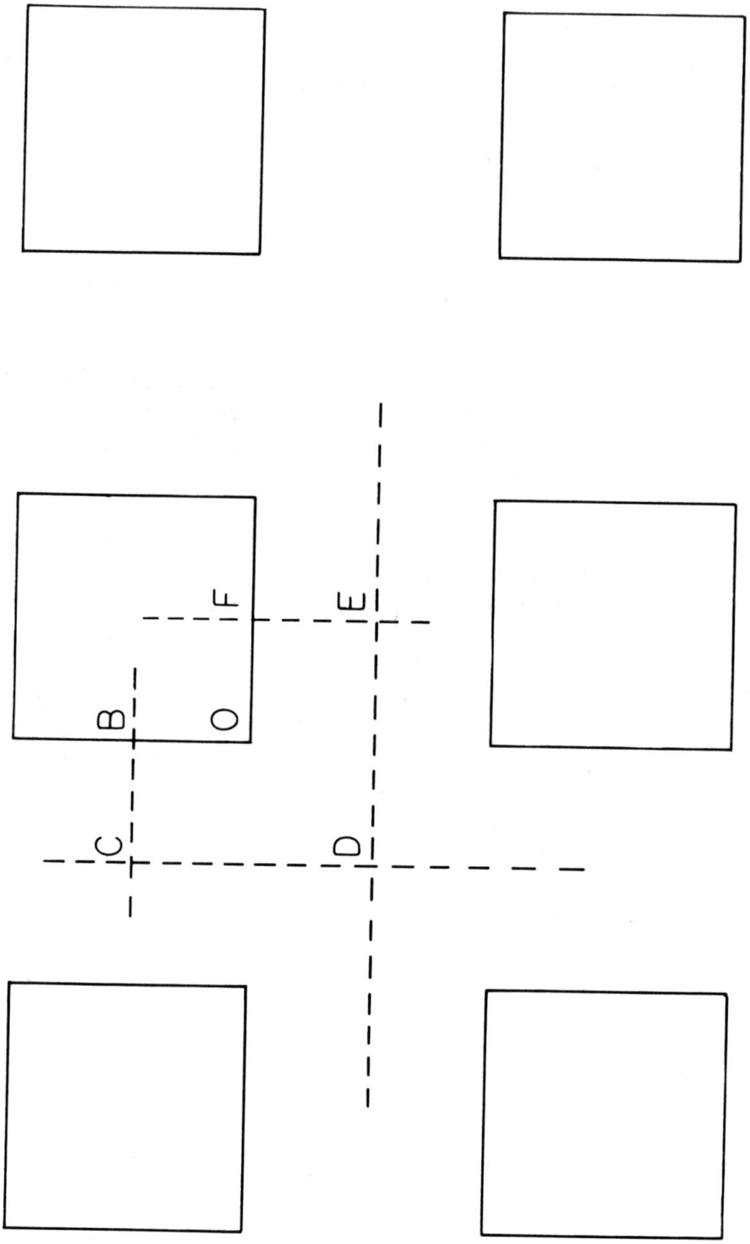

Figure 6.1

where the Dirichlet boundary conditions are taken for convenience as zero and unity and can be scaled as necessary. We define the two disjoint parts of the boundary $\partial\Omega_1 \equiv \overline{CD} \cup \overline{EF}$ and $\partial\Omega_2 \equiv \overline{OB} \cup \overline{BC} \cup \overline{DE} \cup \overline{FO}$, and let $\bar{\Omega} \equiv \Omega \cup \partial\Omega$.

The region $\bar{\Omega}$ contains a *re-entrant* corner at O with internal angle $3\pi/2$, and as a result u possesses a boundary singularity in that it has derivatives which are unbounded at the corner. This is illustrated by the use of an asymptotic expansion of the solution due to Wasow[22] which was later adapted by Lehman.[14] In terms of local polar coordinates (r, θ), with origin at the corner and zero angle along one of the arms of the corner, the asymptotic form of u is

$$\begin{aligned} u(r, \theta) &= \sum_i a_i \phi_i(r, \theta) \\ &= a_1 r^{\frac{2}{3}} \cos\frac{2\theta}{3} + a_2 r^{\frac{4}{3}} \cos\frac{4\theta}{3} + \cdots \end{aligned} \tag{6.6}$$

From (6.6) it is clear that $\partial u/\partial r$ is unbounded at $r = 0$.

A problem of this type containing a slit has recently been considered by Rosser and Papamichael.[18] For this the re-entrant angle at the tip of the slit is 2π. They show that the series expansion about the tip of the slit, corresponding to (6.6), is convergent throughout the region of the problem so that in their case the exact rather than asymptotic form of u is given.

6.2 Weak problem and Galerkin method

Let $W_2^1(\Omega)$ be the Sobolev space of functions which together with their generalized derivatives of order one are in $L_2(\Omega)$. The subspace of functions in $W_2^1(\Omega)$ which satisfy a homogeneous boundary condition on $\partial\Omega_1$ is written $W_2^1(\Omega) \cap (\partial\Omega_1)_0$; that is for $v \in W_2^1(\Omega) \cap (\partial\Omega_1)_0$, $v \in W_2^1(\Omega)$ and $v = 0$ on $\partial\Omega_1$.

The *weak problem* corresponding to (6.1) to (6.5) is:
find $u \in \psi + W_2^1(\Omega)$ such that

$$a(u, v) = 0 \qquad \forall v \in W_2^1(\Omega) \cap (\partial\Omega_1)_0 \tag{6.7}$$

where $\psi \in W_2^1(\bar{\Omega})$ with $\psi = 1$ on $\overline{CD}$ and $\psi = 0$ on $\overline{EF}$. The notation $u \in \psi + W_2^1(\Omega)$ means that $u = \psi + v$, where $v \in W_2^1(\Omega) \cap (\partial\Omega_1)_0$. In (6.7) the bilinear functional $a(u, v)$ is defined as

$$a(u, v) = \iint_\Omega \left(\frac{\partial u}{\partial x}\frac{\partial v}{\partial x} + \frac{\partial u}{\partial y}\frac{\partial v}{\partial y} \right) dx\, dy \qquad \forall u, v \in W_2^1(\Omega)$$

The *energy norm* $\|v\|_E$ is defined by

$$\|v\|_E = (a(v, v))^{\frac{1}{2}} \tag{6.8}$$

The region Ω is discretized into elements, either entirely into triangles or entirely into rectangles so that there are m interior mesh points (nodes), n mesh points on $\partial\Omega_1$ and p mesh points on $\partial\Omega_2$. The elements have generic length h. In similar manner to that in Barnhill and Whiteman[3] $\{B_i(x, y)\}_{i=1}^{m+p}$ and $\{C_j(x, y)\}_{j=1}^{n}$ are defined to be two sets of functions which are biorthonormal (see Reference 10) with respect to point evaluations at the nodes. The B_i and C_j are *basis functions* for our approximation. They are here taken so that, with triangular elements each is linear in x and y, whilst with rectangular elements each is bilinear in x and y.

The set $S^h \in \psi^h + W_2^1(\Omega)$, (i.e. elements of S^h are in $W_2^1(\Omega)$ and take value unity at nodes in $\overline{CD}$ and zero at nodes in $\overline{EF}$), is defined to be $m + p$ dimensional set elements of which can be written in the form

$$s(x, y) = \sum_{i=1}^{m+p} A_i B_i(x, y) + \sum_{j=1}^{n} \psi_j^h C_j(x, y) \tag{6.9}$$

In (6.9) in B_i, ψ_y^h and C_j are known and the A_i are to be found. The $m + p$ dimensional space spanned by the B_i is written S_0^h. Note that elements of S_0^h take on zero value on $\partial\Omega_1$ and that $S_0^h \subset W_2^1(\Omega) \cap (\partial\Omega_1)_0$.

In the Galerkin method used here we seek

$$U \in S^h \text{ such that}$$

$$a(U, B_k) = 0, \qquad k = 1, 2, \ldots, m + p \tag{6.10}$$

The technique (6.10) for determining U is a special case of the more general method in which the test functions are all $V \in S_0^h$. Substitution of (6.9) for U in (6.10) leads to the system of linear equations for the A_i,

$$\sum_{i=1}^{m+p} A_i a(B_i, B_k) = -\sum_{j=1}^{n} \psi_j^h a(C_j, B_k), \qquad k = 1, 2, \ldots, m + p \tag{6.11}$$

A best approximation property for the Galerkin solution is proved in the Lemma of Reference 3 for the case $\partial\Omega_2 = \varnothing$. When that Lemma is applied here, it is found that the Galerkin solution U to (6.10) is the best approximation to the weak solution u of (6.7) from S^h in the energy norm (6.8).

Thus

$$\|u - U\|_E \leqslant \|u - w\|_E \qquad \forall w \in S^h \tag{6.12}$$

Taking for w in (6.12) the function $\tilde{u} \in S^h$ which interpolates u at the $m + n$ nodes in $\Omega \cup \partial\Omega_1$, we obtain

$$\|u - U\|_E \leqslant \|u - \tilde{u}\|_E \tag{6.13}$$

Many bounds for interpolation errors using rectangular and triangular elements have been given: see, for example, References 7 and 9. Under assumptions on the shape of both the triangular and rectangular elements, for the trial functions mentioned above, bounds of the form

$$\|u - \tilde{u}\|_E \leqslant Kh|u|_2 \tag{6.14}$$

can be found, where K is a constant and

$$|u|_2 = \left\{ \left\| \frac{\partial^2 u}{\partial x^2} \right\|^2_{L_2(\Omega)} \left\| \frac{\partial^2 u}{\partial x \partial y} \right\|^2_{L_2(\Omega)} \left\| \frac{\partial^2 u}{\partial y^2} \right\|^2_{L_2(\Omega)} \right\}^{\frac{1}{2}}$$

Galerkin solutions to the weak problem (6.7) have been calculated using (6.10) with partitions of Ω consisting either entirely of rectangular or entirely of right triangular elements, as in Figure 6.2. Numerical results are given in

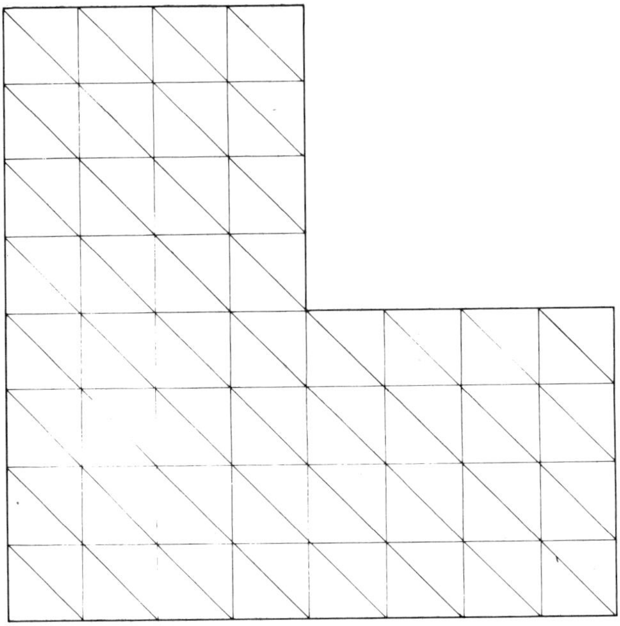

Figure 6.2

Figure 6.3; from these it is seen that the numerical solutions are inaccurate in the neighbourhood of O so that there they are being adversely affected by the boundary singularity. In addition it is clear from (6.6) that $u \in W_2^1(\Omega) - W_2^2(\Omega)$ so that for problem (6.7) Equations (6.13) and (6.14) do not together constitute a meaningful bound for the Galerkin error.

In order that the shortcomings of inaccuracy of the Galerkin approximation and inapplicability of the error analysis for the standard Galerkin

1·0000	0·9698	0·9425	0·9202	0·9055	0·9005					
	0·9687	0·9401	0·9166	0·9011	0·8955					
	0·9669	0·9366	0·9120	0·8958	0·8902					
1·0000	0·9686	0·9398	0·9163	0·9007	0·8953					
	0·9674	0·9374	0·9127	0·8960	0·8900					
	0·9655	0·9337	0·9077	0·8903	0·8843					
1·0000	0·9647	0·9319	0·9044	0·8856	0·8787					
	0·9636	0·9296	0·9007	0·8803	0·8725					
	0·9614	0·9253	0·8944	0·8731	0·8653					
1·0000	0·9584	0·9188	0·8839	0·8581	0·8482					
	0·9573	0·9166	0·8802	0·8521	0·8394					
	0·9551	0·9119	0·8725	0·8409	0·8296					
1·0000	0·9502	0·9013	0·8548	0·8146	0·7948					
	0·9472	0·8992	0·8513	0·8085	0·7808					
	0·9472	0·8950	0·8440	0·7953	0·7928					
1·0000	0·9411	0·8816	0·8207	0·7560	0·6663	0·4884	0·3580	0·2371	0·1170	0·0000
	0·9402	0·8797	0·8173	0·7498	0·6667	0·5003	0·3653	0·2405	0·1196	
	0·9389	0·8771	0·8135	0·7456	0·6635	0·5118	0·3674	0·2418	0·1201	
1·0000	0·9324	0·8632	0·7897	0·7066	0·6026	0·4788	0·3555	0·2355	0·1174	0·0000
	0·9319	0·8622	0·7885	0·7068	0·6096	0·4847	0·3602	0·2386	0·1189	
	0·9311	0·8606	0·7864	0·7043	0·6048	0·4891	0·3637	0·2401	0·1195	
1·0000	0·9254	0·8487	0·7672	0·6774	0·5760	0·4646	0·3490	0·2326	0·1162	0·0000
	0·9253	0·8486	0·7676	0·6794	0·5803	0·4685	0·3522	0·2348	0·1174	
	0·9247	0·8474	0·7658	0·6768	0·5780	0·4693	0·3541	0·2362	0·1180	
1·0000	0·9204	0·8388	0·7528	0·6605	0·5606	0·4539	0·3428	0·2294	0·1149	0·0000
	0·9206	0·8393	0·7540	0·6627	0·5637	0·4569	0·3453	0·2311	0·1158	
	0·9200	0·8382	0·7523	0·6607	0·5621	0·4567	0·3460	0·2319	0·1163	
1·0000	0·9175	0·8331	0·7450	0·6516	0·5524	0·4477	0·3387	0·2271	0·1139	0·0000
	0·9179	0·8340	0·7465	0·6538	0·5550	0·4502	0·3408	0·2286	0·1147	
	0·9173	0·8328	0·7449	0·6520	0·5536	0·4496	0·3410	0·2290	0·1149	
1·0000	0·9163	0·8315	0·7426	0·6486	0·5500	0·4452	0·3376	0·2257	0·1140	0·0000
	0·9170	0·8323	0·7440	0·6510	0·5522	0·4480	0·3393	0·2277	0·1142	
	0·9164	0·8311	0·7425	0·6493	0·5509	0·4473	0·3393	0·2279	0·1145	

At each mesh point P the numbers have the significance:

	CTM[17] See Section 6.4
P	Finite elements with isosceles right triangles, $h = 0{\cdot}1$. Finite elements with squares, $h = 0{\cdot}1$.

Figure 6.3

method in this context may be overcome, two modifications are now proposed. These are

(1) Local mesh refinement in the neighbourhood of O.
(2) Inclusion of singular terms from (6.6) in the trial functions in elements near O.

6.2.1 *Local mesh refinement*

Functions contained in $W_2^1(\Omega)$ possess certain continuity properties which, since $S^h \subset W_2^1(\Omega)$, also have to be exhibited by functions in S^h. The continuity condition which must be satisfied is the *conforming condition*, and in this case it is that $S^h \subset C^0(\Omega)$. Thus to be conforming the global approximating functions of Section 6.2 for both triangular and rectangular meshes must be continuous over Ω. For a standard triangular mesh with linear interpolation to the function values at the vertices of each triangle, or a standard rectangular mesh with bilinear interpolation to the function values at the four corners of each rectangle, the global approximating functions are in $C^0(\Omega)$. Thus the results of Section 6.2 have been derived with conforming functions.

Refinement of the mesh will not make the above error analysis applicable to singular problems. However, as would be expected, refinement does improve the accuracy of the numerical approximations, particularly in the

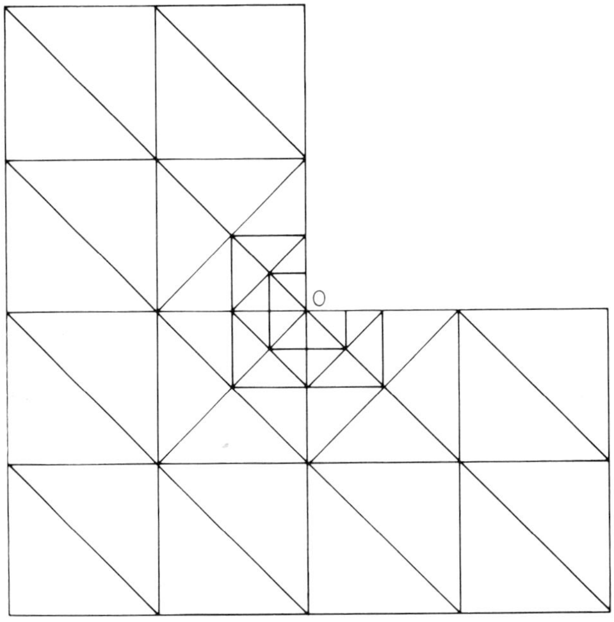

Figure 6.4

neighbourhood of a singularity. The shortcoming of refining over the whole of the region Ω is that many mesh points remote from the singularity are introduced needlessly so that the resulting stiffness systems are unnecessarily large. In order to keep the total number of points in the discretization as small as possible, we refine only in the neighbourhood of O as indicated in Figures 6.4 and 6.5. With right triangular elements as many levels of refinement as required can be performed (in Figure 6.4 two levels are shown) and the elements are all triangles with nodes at the vertices. The linear trial function interpolating the function value at each node is used in

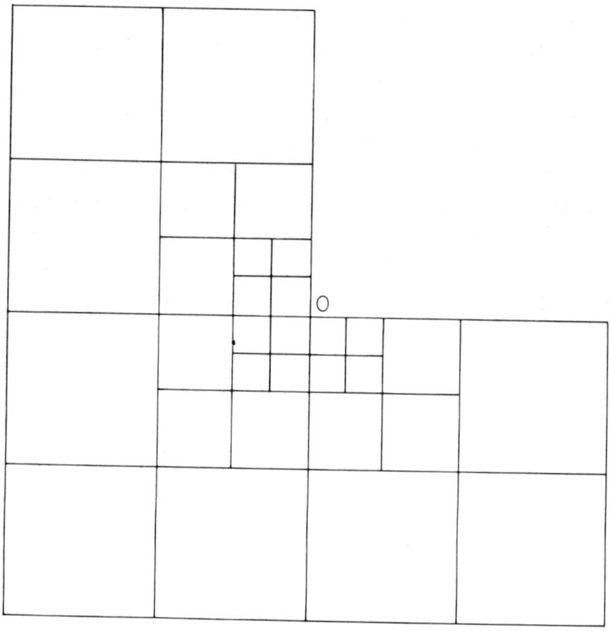

Figure 6.5

each element. A piecewise linear trial function is again obtained which satisfies the conforming condition. For local refinement with rectangular elements the scheme of Figure 6.5 introduces *mid-side nodes* and special interpolants must be used in the *five-node rectangles* so that the global approximating function may again be in C^0. Five node interpolants of this type have been derived by Gregory and Whiteman[12,13] so that merely the general form is given here. This is stated for a square element of unit side length, but can be scaled for a square of side h or for a rectangle.

Consider the five-node square element, Figure 6.6, with nodes at the points $a \equiv (0, 0)$, $b \equiv (\frac{1}{2}, 0)$, $c \equiv (1, 0)$, $d \equiv (1, 1)$ and $e = (0, 1)$. From Reference 13 the trial function which is continuous throughout the element, which

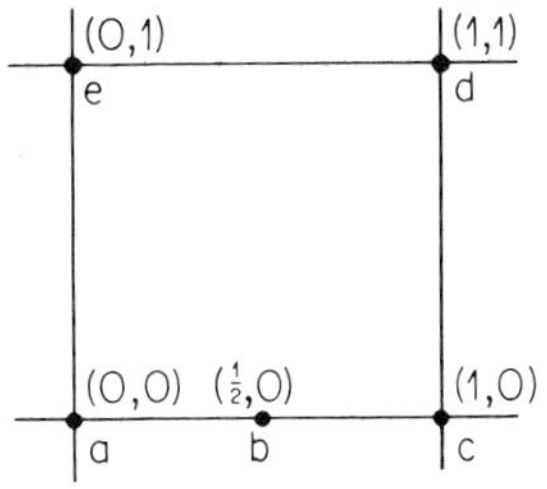

Figure 6.6

is bilinear for $0 \leqslant x \leqslant \frac{1}{2}$ and for $\frac{1}{2} \leqslant x \leqslant 1$, which interpolates the nodal values and which is linear along $\overline{ab}$, $\overline{bc}$, $\overline{cd}$, $\overline{de}$ and $\overline{ea}$, is

$$U(x, y) = (1 - x)yU(0, 1) + xyU(1, 1)$$
$$+ \begin{cases} (1 - 2x)(1 - y)U(0, 0) + 2x(1 - y)U(\frac{1}{2}, 0), & 0 \leqslant x \leqslant \frac{1}{2} \\ 2(1 - x)(1 - y)U(\frac{1}{2}, 0) + (2x - 1)(1 - y)U(1, 0), & \frac{1}{2} \leqslant x \leqslant 1 \end{cases} \tag{6.15}$$

Incorporation of trial functions of the type (6.15) into the space S^h of piecewise bilinear functions will produce a C^0 global approximating function. Local rectangular mesh refinement can therefore be carried out in the manner of Figure 6.5, where two levels of refinement are shown.

The mid-side nodes are equivalent to the *hanging points* familiar in finite differences. We note that, if the equation corresponding to a hanging point in the global stiffness system is considered as a difference equation, the resulting difference approximation at the point is *inconsistent*.

When refinement is done with triangular elements, no special procedure has to be adopted because the elements are all of the same form—isosceles right-angled triangles. In our numerical experiments the mesh has been generated using an automatic mesh generation scheme due to A. Bykat,[8] which demands as data only the coordinates of the boundary nodes. These nodes are thus concentrated near the re-entrant corner and the refined mesh so produced is in effect a graded mesh. This is one advantage over the method of refinement with rectangles where a *problem defined peripheral node ordering scheme* as in Reference 13 has been used.

Results using local refinement schemes are given in Figure 6.7. We note that at the point of singularity itself the solution appears to have converged. This phenomenon occurs repeatedly with local mesh refinement, where the effect of the singularity on the numerical solution has been removed by the refinement. However, it should be remembered that the approximation contains a discretization error which at points remote from the singularity is related to the original mesh length. Thus the apparent convergence will in all probability be to the wrong number.

1·0000	0·9698	0·9425	0·9202	0·9055	0·9005					
	0·9693	0·9414	0·9184	0·9034	0·8981					
	0·9693	0·9412	0·9182	0·9031	0·8977					
1·0000	0·9686	0·9398	0·9163	0·9007	0·8953					
	0·9681	0·9388	0·9146	0·8985	0·8927					
	0·9680	0·9386	0·9144	0·8981	0·8924					
1·0000	0·9647	0·9319	0·9044	0·8856	0·8787					
	0·9643	0·9310	0·9028	0·8832	0·8759					
	0·9641	0·9308	0·9025	0·8827	0·8755					
1·0000	0·9584	0·9188	0·8839	0·8581	0·8482					
	0·9580	0·9180	0·8824	0·8553	0·8448					
	0·9579	0·9178	0·8820	0·8549	0·8442					
1·0000	0·9502	0·9013	0·8548	0·8146	0·7948					
	0·9499	0·9006	0·8535	0·8112	0·7923					
	0·9497	0·9004	0·8531	0·8106	0·7913					
1·0000	0·9411	0·8816	0·8207	0·7560	0·6663	0·4884	0·3580	0·2371	0·1170	0·0000
	0·9408	0·8810	0·8199	0·7553	0·6667	0·4894	0·3603	0·2379	0·1184	
	0·9407	0·8809	0·8196	0·7548	0·6667	0·4904	0·3608	0·2382	0·1185	
1·0000	0·9324	0·8632	0·7897	0·7066	0·6026	0·4788	0·3555	0·2355	0·1174	0·0000
	0·9323	0·8629	0·7895	0·7077	0·6039	0·4811	0·3569	0·2365	0·1179	
	0·9322	0·8628	0·7895	0·7077	0·6043	0·4817	0·3574	0·2376	0·1179	
1·0000	0·9254	0·8487	0·7672	0·6774	0·5760	0·4646	0·3490	0·2326	0·1162	0·0000
	0·9254	0·8488	0·7677	0·6787	0·5776	0·4659	0·3499	0·2312	0·1165	
	0·9254	0·8488	0·7676	0·6788	0·5779	0·4663	0·3502	0·2334	0·1167	
1·0000	0·9204	0·8388	0·7528	0·6605	0·5606	0·4539	0·3428	0·2294	0·1149	0·0000
	0·9206	0·8392	0·7536	0·6618	0·5620	0·4550	0·3436	0·2298	0·1151	
	0·9206	0·8392	0·7537	0·6619	0·5622	0·4553	0·3438	0·2300	0·1152	
1·0000	0·9175	0·8331	0·7450	0·6516	0·5524	0·4477	0·3387	0·2271	0·1139	0·0000
	0·9178	0·8337	0·7459	0·6528	0·5536	0·4487	0·3395	0·2275	0·1141	
	0·9178	0·8338	0·7460	0·6529	0·5537	0·4489	0·3396	0·2277	0·1142	
1·0000	0·9163	0·8315	0·7426	0·6486	0·5500	0·4452	0·3376	0·2257	0·1140	0·0000
	0·9169	0·8319	0·7434	0·6500	0·5510	0·4466	0·3381	0·2267	0·1137	
	0·9169	0·8319	0·7435	0·6502	0·5514	0·4468	0·3382	0·2268	0·1138	

	CTM[17]
P	Finite elements with isosceles right triangles, $h = 0{\cdot}1$ plus 5 levels of local refinement.
	Finite elements with isosceles right triangles, $h = 0{\cdot}1$ plus 2 levels of local refinement.

Figure 6.7

6.2.2 *Inclusion of singular terms*

The idea of augmenting the trial function spaces (S^h of Section 6.2) with terms having the form of the singularity was proposed by Fix[11] who used rectangular elements to solve a second order self-adjoint elliptic problem in a rectangular region with homogeneous Dirichlet boundary conditions. The technique has been extended to triangular elements by Barnhill and Whiteman.[2,3] This modification of the standard finite element method is undertaken with two aims in mind: the adaptation of the error analysis of Section 6.2 and the increase of accuracy of the numerical solution.

Let the neighbourhoods $N(r_i) \subset \bar{\Omega}$ of the corner be defined as

$$N(r_i) \equiv \{(r, \theta); \quad 0 \leqslant r \leqslant r_i, \quad 0 \leqslant \theta \leqslant 3\pi/2\}, \quad i = 0, 1$$

where $r_1 > 0$ is some fixed number and $r_0 = qr_1$, $0 < q < 1$; (r, θ) as in Section 6.1; (note that $N(r_0) \subset N(r_1)$). The functions $w_i(r, \theta)$, $i = 1, 2, \ldots, N$ are constricted in $N(r_1)$, where the $w_i(r, \theta)$ have the form of the singularity in $N(r_0)$, are equal to zero in $(\bar{\Omega} - N(r_1))$ and are in $W_2^2(\bar{\Omega} - N(r_0))$. These last smoothness properties can be achieved by taking in $N(r_1) - N(r_0)$ the functions $w_i(r, \theta)$ as the product of a cubic Hermite polynomial in r and a suitable function of θ; for details see Reference 1.

The function

$$w = u - \sum_{i=1}^{N} a_i w_i(r, \theta)$$

can now be formed, such that w would be in $W_2^2(\Omega)$ if the a_i were known exactly. Thus w could then be approximated with the Galerkin solution $U \in S^h$ and clearly, if the a_i were known, the error bound formed by combining (6.13) and (6.14) would then apply.

However, the a_i cannot be found exactly and in practice approximations $\hat{a}_i$ are calculated by the method of augmenting with singular functions from (6.6) the trial functions spaces S^h. These augmented spaces are denoted by Aug S^h and, in each element, the trial functions of Aug S^h have the form:

for triangular elements:

$$a + bx + cy + \sum_{i=1}^{N} c_i w_i(r, \theta)$$

for rectangular elements:

$$a + bx + cy + dxy + \sum_{i=1}^{N} c_i w_i(r, \theta)$$

Galerkin approximations $\hat{U} \in \text{Aug } S^h$ to u are calculated, and the best approximation Lemma of Reference 3 is now applied to Aug S^h. Thus it follows immediately that

$$\|u - \hat{U}\|_E \leqslant \|u - v\|_E \qquad \forall v \in \text{Aug } S^h \tag{6.16}$$

In particular let $\tilde{u} \in S^h$ interpolate to $u - \sum_{i=1}^{N} a_i w_i$ at the $m + p$ nodes in $\Omega \cup \partial\Omega_1$ and take $v \in \text{Aug } S^h$ as

$$v = \sum_{i=1}^{N} a_i w_i(r, \theta) + \tilde{u}$$

Then (6.16) gives

$$\|u - \hat{U}\|_E \leqslant \left\| u - \left(\sum_{i=1}^{N} a_i w_i + \tilde{u} \right) \right\|_E = \left\| \left(u - \sum_{i=1}^{N} a_i w_i \right) - \tilde{u} \right\|_E \tag{6.17}$$

The Ciarlet–Raviart theorem can be applied so that

$$\left\| \left(u - \sum_{i=1}^{N} a_i w_i \right) - \tilde{u} \right\|_E \leqslant Kh \left| u - \sum_{i=1}^{N} a_i w_i \right|_2 \tag{6.18}$$

As the a_i are the correct values of, and not approximations to, the constants in the expansion (6.6), the function $u - \sum_{i=1}^{N} a_i w_i$ is in $W_2^2(\Omega)$ so that (6.17) and (6.18) combined give an $O(h)$ bound on the Galerkin error.

Galerkin approximations have been calculated with the inclusion of singular functions in the trial function spaces for both triangular and rectangular elements. Augmention with one singular function ($N = 1$) in each case causes considerable improvement in accuracy in the neighbourhood of O. It is found that the inclusion of more singular terms does not appreciably further improve the accuracy. This conclusion is in agreement with that of Wait and Mitchell[21] who use a combination of mesh refinement and singular function augmentation with rectangular elements to solve a harmonic problem in a region containing a slit.

6.3 Finite difference methods

The systems of linear equations which result from the use of finite element methods, as considered in the previous sections, can be thought of as systems of *difference equations*. Many well known finite difference schemes can thus be produced by the use of the finite element method with appropriate trial functions on meshes of squares and right-angled triangles. Clearly in such cases error analyses as in Section 6.2 are applicable.

In particular the use of linear trial functions interpolating function values at the element vertices with a mesh of isosceles right triangles having short sides of length h, as in Figure 6.2, produces at a mesh point (x, y) the standard five-point finite difference replacement for Laplace's equation

$$4U(x, y) - U(x + h, y) - U(x, y + h) - U(x - h, y) - U(x, y - h) = 0 \tag{6.19}$$

Thus the relevant results given in Figure 6.3 are exactly those obtained with this scheme for the problem (6.1) to (6.5) with the mesh length h as shown.

If bilinear trial functions are used in the same way with the square mesh of side h as in Figure 6.2, these lead to the *non-standard* nine-point finite difference replacement for Laplace's equation, see Reference 7,

$$\begin{aligned}8U(x, y) &- \{U(x + h, y) + U(x, y + h) + U(x - h, y) + U(x, y - h)\} \\ &- \{U(x + h, y + h) + U(x - h, y + h) + U(x - h, y - h) \\ &+ U(x + h, y - h)\} = 0\end{aligned} \tag{6.20}$$

Use of the replacement (6.20) therefore produces again the relevant results of Figure 6.3.

The *standard* nine-point replacement for Laplace's equation,

$$\begin{aligned}20U(x, y) &- 4\{U(x + h, y) + U(x, y + h) + U(x - h, y) + U(x, y - h)\} \\ &- \{U(x + h, y + h) + U(x - h, y + h) + U(x - h, y - h) \\ &+ U(x + h, y - h)\}\end{aligned} \tag{6.21}$$

has also been used to produce numerical solutions to the problem (6.1) to (6.5). For completeness results obtained using (6.21) with a square mesh of length $h = 0{\cdot}1$ are given in Figure 6.8. It is seen that these results are less accurate in the neighbourhood of the singularity than those obtained with (6.19) or (6.20). This might be expected since the truncation error in (6.21) involves the sixth derivatives of u, the solution of (6.1) to (6.5). Thus use of (6.21) implicitly assumes the continuity and boundedness of higher order derivatives of u than those presupposed by the use of (6.14) and (6.20). It was shown in Section 6.1 that the solution u does not possess these properties.

Modifications for improving accuracy of finite difference solutions can be performed as in Sections 6.2.1 and 6.2.2. Experiments using local and general mesh refinement have been performed and accuracy is increased. The technique of incorporating singular functions into finite difference methods has also been much used: see References 6, 16, 23, 25.

For non-modified difference methods uniform convergence with decreasing mesh size of the finite difference solution to the exact solution of the problem defined in a rectangle containing a slit considered by Rosser and Papamichael[18] is proved by Whiteman and Webb.[24] The latter exploit the symmetry of the problem and reflect the region across those parts of the resulting boundary which meet at the singular point. Such reflecting cannot be used for the problem in the L-shaped region.

	CTM[17]
P	Standard 9-point finite difference replacement, $h = 0{\cdot}1$.

1·0000	0·9698	0·9425	0·9202	0·9055	0·9005					
	0·9670	0·9367	0·9121	0·8159	0·8902					
1·0000	0·9686	0·9398	0·9163	0·9007	0·8953					
	0·9656	0·9339	0·9079	0·8905	0·8843					
1·0000	0·9647	0·9319	0·9044	0·8856	0·8787					
	0·9615	0·9255	0·8950	0·8736	0·8656					
1·0000	0·9584	0·9188	0·8839	0·8581	0·8482					
	0·9548	0·9118	0·8732	0·8431	0·8299					
1·0000	0·9502	0·9013	0·8548	0·8146	0·7948					
	0·9460	0·8940	0·8438	0·7969	0·7624					
1·0000	0·9411	0·8816	0·8207	0·7560	0·6663	0·4884	0·3580	0·2371	0·1170	0·0000
	0·9344	0·8748	0·8115	0·7427	0·6559	0·5013	0·3639	0·2394	0·1190	
1·0000	0·9324	0·8632	0·7897	0·7066	0·6026	0·4788	0·3555	0·2355	0·1174	0·0000
	0·9293	0·8584	0·7839	0·7011	0·6023	0·4830	0·3591	0·2377	0·1184	
1·0000	0·9254	0·8487	0·7672	0·6774	0·5760	0·4646	0·3490	0·2326	0·1162	0·0000
	0·9237	0·8457	0·7638	0·6747	0·5755	0·4660	0·3508	0·2339	0·1169	
1·0000	0·9204	0·8388	0·7528	0·6605	0·5606	0·4539	0·3428	0·2294	0·1149	0·0000
	0·9194	0·8370	0·7508	0·6590	0·5601	0·4543	0·3436	0·2301	0·1153	
1·0000	0·9175	0·8331	0·7450	0·6516	0·5524	0·4477	0·3387	0·2271	0·1139	0·0000
	0·9168	0·8319	0·7436	0·6505	0·5518	0·4476	0·3391	0·2275	0·1141	
1·0000	0·9163	0·8315	0·7426	0·6486	0·5500	0·4452	0·3376	0·2257	0·1140	0·0000
	0·9160	0·8303	0·7413	0·6479	0·5492	0·4455	0·3376	0·2266	0·1137	

Figure 6.8

6.4 Numerical conformal transformation method

As no closed form solution is available for the problem (6.1) to (6.5), an accurate approximation, obtained by transforming the problem into a simple problem which can be solved by inspection, has been used throughout this paper for comparing the accuracy of the finite element and finite difference solutions. This method, the Numerical Conformal Transformation Method (NCTM), is given in Papamichael and Whiteman.[17] The NCTM consists of four successive conformal mappings, the first of which is performed numerically using a technique due to Symm[19] which involves the numerical solution of a Fredholm integral equation of the first kind with a logarithmic kernel. The results for problem (6.1) to (6.5) obtained in Reference 17 have been displayed in each of Figures 6.3, 6.7 and 6.8. They are also quoted by Bell and Crank.[5] Symm[20] uses an integral equation technique modified to deal with the singularity to solve the boundary value problem (6.1) to (6.5) numerically. His results are given in Figure 6.9 together again with those obtained with the NCTM.

6.5 Discussion

The results of the previous sections show the shortcomings of the finite element and finite difference methods when boundary singularities are present. The success of the modifications to the finite element method in improving the accuracy of the numerical solutions is evident. Indeed with continued local mesh refinement the stage has been reached by Gregory and Whiteman[13] at which the Galerkin solutions are more accurate near the singularity than they are at points in Ω remote from O. The effect of the singularity on the numerical solution has thus been neutralized by the refinement. For problems of this type and magnitude there seems to be little to choose between the finite element and finite difference methods when comparing accuracy of solutions for a certain amount of computation. The NCTM produces accurate approximations in a fraction of the computation time taken by the other methods. However, it is a much less general method and the range of problems to which it can be applied is limited.

We have found that for these problems where there is considerable regularity of mesh the programming involved to produce the finite element solutions can be much simplified. The technique is to divide the totality of mesh points into several different classes, where each mesh point of a class has the same pattern of neighbouring mesh points. The linear equation derived with the finite element method for a particular point is then treated as a *difference equation*, so one type of difference equation is associated with each class. This removes the need to generate local stiffness matrices, a fact that will be hotly disputed by some on the grounds that it removes

At each mesh point **P** the numbers represent:

	CTM[17]
P	Integral equation method[20]

1·0000	0·9698	0·9425	0·9202	0·9055	0·9005					
0·9999	0·9700	0·9427	0·9205	0·9060	0·9009					
1·0000	0·9686	0·9398	0·9163	0·9007	0·8953					
1·0000	0·9687	0·9400	0·9166	0·9012	0·8957					
1·0000	0·9647	0·9319	0·9044	0·8856	0·8787					
1·0000	0·9648	0·9322	0·9048	0·8860	0·8793					
1·0000	0·9584	0·9188	0·8839	0·8581	0·8482					
1·0000	0·9584	0·9191	0·8843	0·8587	0·8487					
1·0000	0·9502	0·9013	0·8548	0·8146	0·7948					
1·0000	0·9503	0·9015	0·8553	0·8154	0·7961					
1·0000	0·9411	0·8816	0·8207	0·7560	0·6663	0·4884	0·3580	0·2371	0·1170	0·0000
1·0000	0·9412	0·8818	0·8210	0·7565	0·6667	0·4869	0·3579	0·2364	0·1177	0·0000
1·0000	0·9324	0·8632	0·7897	0·7066	0·6026	0·4788	0·3555	0·2355	0·1174	0·0000
1·0000	0·9325	0·8633	0·7898	0·7066	0·6019	0·4780	0·3549	0·2352	0·1172	0·0000
1·0000	0·9254	0·8487	0·7672	0·6774	0·5760	0·4646	0·3490	0·2326	0·1162	0·0000
1·0000	0·9254	0·8487	0·7671	0·6772	0·5756	0·4642	0·3486	0·2323	0·1161	0·0000
1·0000	0·9204	0·8388	0·7528	0·6605	0·5606	0·4539	0·3428	0·2294	0·1149	0·0000
1·0000	0·9204	0·8387	0·7527	0·6603	0·5604	0·4536	0·3425	0·2291	0·1147	0·0000
1·0000	0·9175	0·8331	0·7450	0·6516	0·5524	0·4477	0·3387	0·2271	0·1139	0·0000
1·0000	0·9175	0·8331	0·7449	0·6515	0·5521	0·4474	0·3385	0·2269	0·1138	0·0000
1·0000	0·9163	0·8315	0·7426	0·6486	0·5500	0·4452	0·3376	0·2257	0·1140	0·0000
1·0001	0·9166	0·8313	0·7424	0·6487	0·5495	0·4453	0·3371	0·2261	0·1134	0·0000

Figure 6.9

one of the main advantages of versatility possessed by the finite element method.

In this paper we started by considering a specific class of composites. We now return to the physical situation and note that an important application of composites of this type is in the production of materials that have 'effective diffusion properties' which are equivalent to those of isotropic materials. In this way the isotropic material may be simulated. The pattern of flow in the composite is in general dependent on the arrangement of the impermeable islands in the continuum. The effect of different arrangements is discussed by Barrer.[4] In the study of the flow rates through a composite of the type under consideration here, the total flow across a section (x = constant) of the L-shaped region will be sought. This total flow is

$$\int_0^\alpha \frac{\partial u(x, y)}{\partial x} \mathrm{d}y$$

where

$$\alpha = \begin{cases} 1 & \text{for } 0 \leqslant x < \frac{1}{2} \\ \frac{1}{2} & \text{for } \frac{1}{2} \leqslant x \leqslant 1 \end{cases}$$

and can easily be approximated with a quadrature formula from the numerical results for a particular value of x.

We have concentrated on trying to eliminate the effect of the singularity on the solution of a particular diffusion problem. Another important field is that of stress and deformation of this type of composite. Leissa, Claussen and Agrawal[15] have attempted an analysis of this using point matching techniques. In the stress situation the governing differential equation is biharmonic and, when the finite element method is used to solve this, the conforming condition is that the relevant finite dimensional subspace be contained in $C^1(\Omega)$. Gregory and Whiteman[13] have derived a C^1 element for the local mesh refinement scheme with rectangles as in Figure 6.4. Thus this local refinement scheme may again be used.

Acknowledgements

It is clear from the references made throughout this paper that much of this work has been done in collaboration with colleagues. I acknowledge with great pleasure the contributions of R. E. Barnhill, J. A. Gregory, N. Papamichael and J. Barkley Rosser, and also the programming assistance of A. Bykat and P. Theodorou. Many of the numerical results will appear in the dissertation of Theodorou for the degree of Master of Technology at Brunel University.

References

1. R. E. Barnhill and J. R. Whiteman, 'Error analysis of finite element methods with triangles for elliptic boundary value problems', in J. R. Whiteman (ed.), *The Mathematics of Finite Elements and Applications*, Academic Press, London, 1973, pp. 83–112.
2. R. E. Barnhill and J. R. Whiteman, 'Singularities due to re-entrant boundaries in elliptic problems', in L. Collatz (ed.), *ISNM* 19, Birkhauser-Verlag, Basel, 1974, 29–45.
3. R. E. Barnhill and J. R. Whiteman, 'Error analysis of Galerkin methods for Dirichlet problems containing boundary singularities', *Technical Report TR/19*, Department of Mathematics, Brunel University, 1973.
4. R. M. Barrer, 'Diffusion and permeation in heterogeneous media', in J. Crank and G. S. Park (eds.), *Diffusion in Polymers*, Academic Press, London, 1968, pp. 165–217.
5. G. E. Bell and J. Crank, 'Diffusion in composite media. Part I: Steady state', *Technical Report TR/23*, Department of Mathematics, Brunel University, 1973.
6. M. J. M. Bernal and J. R. Whiteman, 'Numerical treatment of biharmonic boundary value problems with re-entrant boundaries', *Computer Journal*, **13**, 87–91 (1970).
7. G. Birkhoff, M. H. Schultz and R. S. Varga, 'Piecewise Hermite interpolation in one and two variables with applications to partial differential equations', *Numer. Math.*, **11**, 232–256 (1968).
8. A. Bykat, 'Automatic triangulation of two-dimensional regions' *Report ICSI* 420, University of London, Institute of Computer Science, 1972.
9. P. G. Ciarlet and P. G. Raviart, 'General Lagrange and Hermite interpolation in R^n with applications to finite element methods', *Arch. Rat. Mech. Anal.*, **46**, 177–199 (1972).
10. P. J. Davies, *Interpolation and Approximation*, Blaisdell, Waltham, Mass., 1963.
11. G. Fix, 'Higher-order Rayleigh-Ritz approximations', *J. Math. Mech.*, **18**, 645–657 (1969).
12. J. A. Gregory, *Piecewise Interpolation Theory and Finite Element Analysis*, Ph.D. Thesis, Department of Mathematics, Brunel University, (forthcoming).
13. J. A. Gregory and J. R. Whiteman, 'Local mesh refinement with finite elements for elliptic problems', *Technical Report TR/24*, Department of Mathematics, Brunel University, 1973.
14. R. S. Lehman, 'Development at an analytic corner of solutions of elliptic partial differential equations', *J. Math. Mech.*, **8**, 727–760 (1959).
15. A. W. Leissa, W. E. Clausen and G. K. Agrawal, 'Stress and deformation analysis of fibrous composite materials by point matching', *Int. J. Numer. Meth. Eng.*, **3**, 89–101 (1971).
16. H. Motz, 'Treatment of singularities in partial differential equations by relaxation methods', *Q. Appl. Math.*, **4**, 371–377 (1946).
17. N. Papamichael and J. R. Whiteman, 'A numerical conformal transformation method for harmonic mixed boundary value problems in polygonal domains', *Z. angew. Math. Phys.*, **24**, 304–316 (1973).
18. J. B. Rosser and N. Papamichael, 'A power series solution for a harmonic mixed boundary value problem', *Technical Report TR/35*, Department of Mathematics, Brunel University, 1973.
19. G. T. Symm, 'An integral equation method in conformal mapping', *Numer. Math.*, **9**, 250–258 (1966).
20. G. T. Symm, 'Treatment of singularities in the solution of Laplace's equation by an integral equation method', *Report NAC31*, National Physical Laboratory, 1973.

21. R. Wait and A. R. Mitchell, 'Corner singularities in elliptic problems by finite element methods', *J. Comp. Phys.*, **8**, 45–52 (1971).
22. W. Wasow, 'Asymptotic development of the solution of Dirichlet's problem at analytic corners', *Duke Math. J.*, **24**, 47–56 (1957).
23. J. R. Whiteman, 'Singularities due to re-entrant corners in harmonic boundary value problems', *Technical Report No. 829*, Mathematics Research Center, University of Wisconsin, Madison, 1967.
24. J. R. Whiteman and J. C. Webb, 'Convergence of finite difference techniques for a harmonic mixed boundary value problem', *B.I.T.*, **10**, 366–374 (1970).
25. L. C. Woods, 'The relaxation treatment of singular points in Poisson's equation', *Q. J. Mech. Appl. Math.*, **6**, 163–189 (1953).

Chapter 7

Numerical Analysis of Transient and Non-Linear Waves

Y. Allouard and J. F. Coudert

7.1. Introduction

Recent research in pure and applied mathematics has produced new tools for solving systems of partial differential equations coupled with differential equations. The aim of this paper is to illustrate how to use some of those tools to solve problems of practical interest. More specifically we are interested in calculating both the forces resulting from wave action and the resulting motion of floating bodies.

Without taking up the general three-dimensional case, we use a simple model to present two recent numerical methods, one using a direct approximation in a space of spline functions and the other based on a formulation in terms of optimal control. After explaining the notations we set up the equations of the problem with some standard assumptions (perfect fluid, irrotational motion) in transient and non-linear cases.

We choose the case of a laboratory flume with a piston-type wave maker at one end and with a freely moving vertical wall with known mechanical features (mass, damping, rigidity) at the other end. This test enables us to face all the difficulties involved in wave generation, propagation and reflection as well as in the calculation of the motion of a body. In addition, it can easily be modified to handle the case of any shape of generator or body in the presence of an irregular bottom.

We then go on to describe the functional properties of the mathematical tools used to justify the methods and see that calculations are valid. Lastly, we take up two examples and compare the results obtained with theoretical results.

We feel that these methods will help improve the design of offshore structures for which, because of recent developments in offshore operations, allowances are increasingly strict and operating conditions increasingly tough.

7.2. Nomenclature

The plane is referred to a system of orthonormal axes $(0, x, z)$. The origin O and the x axis or abscissa are on the free surface at rest. This axis is directed from left to right. The z axis of ordinates is directed upward.

The following notation is used:

$\tilde{D}$ = bounded domain occupied by the fluid at time t.
$\tilde{\Gamma}$ = boundary of $\tilde{D}$, $\tilde{\Gamma} = \tilde{\Gamma}_1 \cup \tilde{\Gamma}_f \cup \tilde{\Gamma}_m \cup \tilde{\Gamma}'_m$ at time t.
$\tilde{\Gamma}_f$ = part of $\tilde{\Gamma}$ formed by a material surface with an unknown motion.
$\tilde{\Gamma}'_m$ = part of $\tilde{\Gamma}$ formed by a material surface with an imposed motion.
D', Γ, Γ_1, Γ_f, Γ_m designate the same domain and boundaries when the fluid is at rest, at the initial time $t = 0$. D is a rectangle.
$D = \{x, z \mid x \in)0, L(\ , z \in)-h, 0(\}$
$Q =)0, T(\times D$
u, w = respective velocity components along the x axis and the z axis.
$\varphi(x, z)$ = velocity potential in the case where time is separable (steady state)
$\Phi(x, z, t) = \varphi(x, z) f(t)$
$\Delta D = \Delta x \times \Delta z$
$\Delta Q = \Delta x \times \Delta z \times \Delta t$
φ_{ijk} = value of the potential at node (x_i, z_j, t_k) belonging to $\Delta x \times \Delta z \times \Delta t$.
$\eta(x, t)$ = ordinate of the free surface.
h = water depth.
g = gravitational acceleration.
$V(t)$ = velocity of left wall (wave maker).
$X(t)$ = velocity of right wall.
$S^d_{\Delta x}$ = space of spline functions of degree d defined on Δx.
$S^d_{\Delta z}$ = space of spline functions of degree d defined on Δz.
$\sigma_i(x)$, $i = 1, \ldots, I$ = ith spline function of the base of the space $S^d_{\Delta x}$.
$\tau_j(x)$, $j = 1, \ldots, J$ = jth spline function of the base of the space $S^d_{\Delta z}$.
ε = perturbation parameter.
τ = period.
ρ = specific mass.
m = mass of the right wall.
α = damping of the right wall.
K = rigidity of the right wall.
n = the normal to $\tilde{\Gamma}$ directed toward the inside of the liquid.
$\zeta(t)$, $\xi(t)$ = respective x and z coordinates of a fluid particle.
$\tilde{X}(\lambda, t)$, $\tilde{Z}(\lambda, t)$ = parametric equation of $\tilde{\Gamma}$ surfaces.
$\Gamma(x, z, t)$ = intrinsic equation of $\tilde{\Gamma}$ surfaces.
ψ, p, q, Y, U = auxiliary variables.

7.3 Physical formulation of the problem

7.3.1 Hypothesis

Let us consider gravity waves in a fluid which is supposed to be incompressible and inviscid. The material surfaces Γ_f, Γ_m and Γ'_m are assumed to be impermeable to the fluid, and pressure P is assumed to be constant on free surface Γ_1. Surface tension is neglected. With these assumptions, there is a velocity potential, as proved in Reference 2, p. 10, $\Phi(x, z, t)$ such that

$$u = \frac{\partial \Phi}{\partial x}, \qquad w = \frac{\partial \Phi}{\partial z}$$

This reduces the number of unknown physical variables to two, namely, $\Phi(x, z, t)$ and $\tilde{P}(x, z, t)$.

Two equations are then sufficient to describe the mechanics of the fluid in the time dependent domain $\tilde{D}$, i.e. the mass balance

$$\Delta \Phi = 0$$

and an integral equation describing the conservation of momentum throughout the entire fluid (Bernoulli):

$$\rho \frac{\partial \Phi}{\partial t} + \tfrac{1}{2}\rho\left[\left(\frac{\partial \Phi}{\partial x}\right)^2 + \left(\frac{\partial \Phi}{\partial z}\right)^2\right] + \rho g z + \tilde{P} = C(t)$$

Without loss of generality, the Bernoulli equation can be written

$$\rho \frac{\partial \Phi}{\partial t} + \tfrac{1}{2}\rho\left[\left(\frac{\partial \Phi}{\partial x}\right)^2 + \left(\frac{\partial \Phi}{\partial z}\right)^2\right] + \rho g z + P = 0$$

The problem is completely defined by the boundary conditions and the initial conditions. On $\tilde{\Gamma}_f$, $\tilde{\Gamma}_m$ and $\tilde{\Gamma}'_m$, the adherence with slippage[1] of the fluid on the material surface can be written

$$\operatorname{grad} \Phi \, . \, n = v_n \tag{7.1}$$

where v_n represents the normal component of the velocity of the moving surface (we know that v_n is an intrinsic concept).[3] On the free surface $\tilde{\Gamma}_1$ we write[1] a dynamic condition,

$$p = 0$$

and a condition of material invariance. This condition reflects that all the particles located on the free surface make up a material system. Basic mechanics[3] then require that during movement this system must always be made up of the same particles.

Let us write the equation for the free surface in its standard form

$$\tilde{\Gamma}_1(x, z, t) = z - \eta(x, t) = 0$$

Let $\xi(t)$ and $\zeta(t)$ be the coordinates of a particle at time t. The condition of material invariance is expressed by

$$\xi(t) = \eta(\zeta(t), t) \; \forall \; t$$

By differentiation with respect to time, we obtain

$$\frac{d\xi}{dt} = \frac{\partial \eta}{\partial x}\frac{d\zeta}{dt} + \frac{\partial \eta}{\partial t} \qquad w = \frac{\partial \eta}{\partial x}u + \frac{\partial \eta}{\partial t} \qquad \frac{\partial \Phi}{\partial z} = \frac{\partial \eta}{\partial x}\frac{\partial \Phi}{\partial x} + \frac{\partial \eta}{\partial t}$$

We choose the state of rest as the initial condition:

$$\Phi(x, z, 0) = 0$$
$$\eta(x, 0) = 0$$

The method of perturbation[1] allows us to replace the problem of solving this set of non-linear equations in a time dependent domain by the simpler problem of solving several sets of linear equations in the domain at rest.

7.3.2 *Presentation of the cases handled*

7.3.2.1 Description of the model The subject of this study is not the actual calculation of how complex floating structures behave. We have limited ourselves to presenting various numerical methods that could make this calculation possible in a subsequent phase.

We shall concentrate on a test example. Consider a channel in which both ends are made up of mobile vertical walls (Figure 7.1). One of these walls

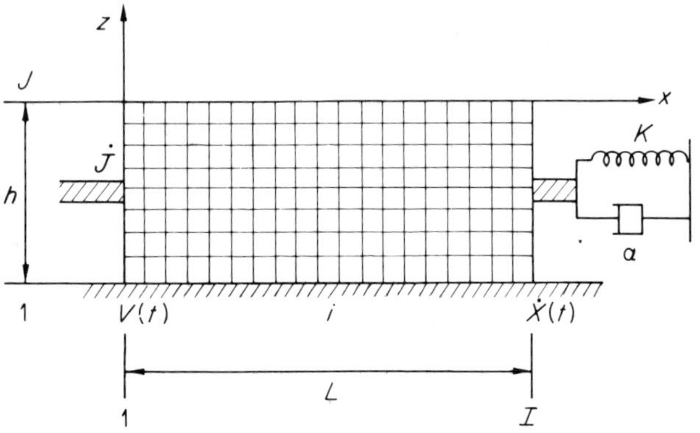

Figure 7.1 Problem studied

is set in motion to generate waves. The other wall is thus subjected to the wave force. The equation for its motion is similar to the one for the motion of a floating body, with the damping effect taking into consideration the effects resulting from the viscosity in the perfect-fluid solution.

7.3.2.2 Equations for the model in transient conditions at the first order

$$\Delta\Phi = 0$$

$$\left(\frac{\partial\Phi}{\partial x}\right)_{x=0} = V(t), \qquad \left(\frac{\partial\Phi}{\partial x}\right)_{x=1} = X(t), \qquad \left(\frac{\partial\Phi}{\partial z}\right)_{z=-h} = 0 \tag{7.2}$$

$$\left(\frac{\partial\Phi}{\partial z} + \frac{1}{g}\frac{\partial^2\Phi}{\partial t^2}\right)_{z=0} = 0 \tag{7.3}$$

$$\Phi_{t=0} = \left(\frac{\partial\Phi}{\partial t}\right)_{t=0} = 0$$

$$m\ddot{X} + \alpha\dot{X} + KX = -\rho\int_{-h}^{0}\left(\frac{\partial\Phi}{\partial t}\right)_{x=L}\mathrm{d}z \tag{7.4}$$

$$X(0) = \dot{X}(0) = 0$$

It is assumed that the force of the spring at time $t = 0$ balances the hydrostatic force.

7.3.2.3 Equations for the model in steady state conditions In the model studied, the motion as described using first and second order perturbation is described as follows:

$$V(t) = Re\,\{\varepsilon v_1\,\mathrm{e}^{-i\omega t} + \varepsilon^2(v_{2,1}\,\mathrm{e}^{-i\omega t} + v_{2,2}\,\mathrm{e}^{-2i\omega t})\}$$

where we assume $\omega = \omega_0 + \varepsilon\omega_1 + \varepsilon^2\omega_2$. We then obtain

$$\Phi(x, z, t) = Re\,\{\varepsilon\varphi_1\,\mathrm{e}^{-i\omega t} + \varepsilon^2(\varphi_{2,1}\,\mathrm{e}^{-i\omega t} + \varphi_{2,2}\,\mathrm{e}^{-2i\omega t})\}$$

$$X(t) = Re\,\{\varepsilon x_1\,\mathrm{e}^{-i\omega t} + \varepsilon^2(x_{2,1}\,\mathrm{e}^{-i\omega t} + x_{2,2}\,\mathrm{e}^{-2i\omega t})\}$$

The equations for the model, to the first order, are found in Reference 4 while those of order (2, 1) are

$$\Delta\varphi_{2,1} = \mathrm{O}\left(\frac{\partial\varphi_{2,1}}{\partial x}\right)_{x=0} = v_{2,1}\left(\frac{\partial\varphi_{2,1}}{\partial x}\right)_{x=L} = x_{2,1}\left(\frac{\partial\varphi_{2,1}}{\partial z}\right)_{z=-h} = 0$$

$$\left[\frac{\partial\varphi_{2,1}}{\partial z} - \frac{\omega_0^2}{g}\varphi_{2,1}\right]_{z=0} = 2\omega_0\omega_1\varphi_1(x, 0) \tag{7.5}$$

$$(m\omega_0^2 + i\alpha\omega_0 - k)x_{2,1} = -(2m\omega_0\omega_1 + i\alpha\omega_1)x_1$$

$$- i\rho\omega_0\int_{-h}^{0}\varphi_{2,1}(L, z)\,\mathrm{d}z - i\rho\omega_1\int_{-h}^{0}\varphi_1(L, z)\,\mathrm{d}z$$

Likewise, those of order (2, 2) are:

$$\Delta\varphi_{2,2} = \mathrm{O}\left(\frac{\partial\varphi_{2,2}}{\partial x}\right)_{x=0} = v_{2,2}\left(\frac{\partial\varphi_{2,2}}{\partial x}\right)_{x=L} = \chi_{2,2}\left(\frac{\partial\varphi_{2,2}}{\partial z}\right)_{z=-h} = 0$$

$$\left[\left(\frac{\partial\varphi_{2,2}}{\partial z} - 4\frac{\omega_0^2}{g}\varphi_{2,2}\right)\right]_{z=0} = \left\{-g\eta_1\frac{\partial^2\varphi_1}{\partial z^2} + g\frac{\partial\eta_1}{\partial x}\frac{\partial\varphi_1}{\partial x} + 2\omega_0^2\eta_1\frac{\partial\varphi_1}{\partial z}\right.$$

$$\left. + i\omega_0\left[\left(\frac{\partial\varphi_1}{\partial x}\right)^2 + \left(\frac{\partial\varphi_1}{\partial z}\right)^2\right]\right\}_{z=0} \tag{7.6}$$

$$(4m\omega_0^2 + 2i\alpha\omega_0 - k)\,.\,\chi_{2,2} = -2i\omega_0\rho\int_{-h}^{0}\varphi_{2,2}(L, z)\,\mathrm{d}z$$

$$+ \tfrac{1}{2}\rho\int_{-h}^{0}\left[\left(\frac{\partial\varphi_1}{\partial x}\right)^2 + \left(\frac{\partial\varphi_1}{\partial z}\right)^2\right]_{x=L}\mathrm{d}z$$

7.4 Functional properties

7.4.1 Definition of spaces

The following spaces are defined:[5–8]

$H^1(D)$ = the Sobolev space of real-valued functions defined on D. Boundary Γ of D is assumed to be suitably smoothed near the corners of D. Functions $gH^1(D)$ have generalized derivatives up to the second order, and the following norm: $\|g\|^2_{H^1(D)} = \int_D [g^2 + (\text{grad } g)^2]\,\mathrm{d}x\,\mathrm{d}z$.

$H^{\frac{1}{2}}(\Gamma)$ = the Sobolev space of order $\frac{1}{2}$ of the real-valued functions defined on $H^{\frac{1}{2}}(\Gamma)$ is the trace of $H^1(D)$.

$H^{-\frac{1}{2}}(\Gamma)$ = the dual of $H^{\frac{1}{2}}(\Gamma)$.

$L_0^2[0, T; H^1(D)]$ = the space of real-valued functions h of $)0, T($ in $H^1(D)$ measurable on $)0, T($ so that:

$$\|h\|^2_{L_0^2} = \int_0^T \|h\|^2_{H^1(D)}\,\mathrm{d}t < \infty\,; \qquad h_{t=0} = 0$$

Likewise, we define the following spaces:

$$L_0^2[0, T; H^{\frac{1}{2}}(\Gamma)], \qquad L_0^2[0, T; H^{-\frac{1}{2}}(\Gamma)]$$

$S^d_{\Delta x}$ = the space of the spline functions $\sigma(x)$ of degree d on Δx. Let (a', b') be an interval containing $(0, L)$ and $H^d(a', b')$ be the space of the real-valued functions defined on (a', b') whose $d - 1$th derivative is integrable. $H^d(a', b')$

is a Hilbert space with the following scalar product:

$$\langle \sigma_1, \sigma_2 \rangle = \sum_{i=0}^{d} \int_{a'}^{b'} \sigma_1^{(i)}(x)\sigma_2^{(i)}(x)\,\mathrm{d}x$$

in which $\sigma^{(i)}$ designates the ith derivates of σ ($\sigma^{(0)} \equiv \sigma$). Let I be f_i values defined on Δx. σ is a polynomial of degree $2d-1$ on (x_i, x_{i+1}) and a polynomial degree d on $(a', x_1 = 0)$, $(x_I = L, b')$. Let

$$J_f = \{h \in H^d(a', b') \mid h_i = f_i\}$$

σ must satisfy

$$\int_{a'}^{b'} \sigma^{(d)}\,\mathrm{d}x = \min_{h \in J_f} \int_{a'}^{b'} h^{(d)}\,\mathrm{d}x$$

We build a base $\sigma_i(x)$ for space $S^d_{\Delta x}$ by stating

$$\sigma_{i'}(x_i) = \delta_{i'i} \qquad i' = 1', \ldots, I; \quad i = 1, \ldots, I$$

We then get for any function $\sigma(x)$,

$$\sigma(x) = \sum_{i'=1}^{I} \alpha_{i'}\sigma_{i'}(x) \qquad \forall \sigma \in S^d_{\Delta x}$$

It is clear that σ is continuous up to the $2d-2$ order on $(0, L)$. If $I = d$, then $S^d_{\Delta x}$ is the space of the polynomials of degree $d-1$ on $(0, L)$, whereas if $I > d$, then $S^d_{\Delta x}$ is a vectorial subspace with dimension I. We also introduce $S^d_{\Delta D}$, the space of the spline functions on ΔD defined by[9,10]

$$S^d_{\Delta D} = S^d_{\Delta x} \otimes S^d_{\Delta z}$$

The space $\bigcup_{I,J} S^d_{\Delta D}$ is dense in $H^1(D)$.[10,11]

7.5 Direct approximation in S^d_D of the solution to the problems studied

7.5.1 *First-order transient model*

Using meshing in time, we seek the solution of the problem as

$$\Phi(x, z, t_k) = \sum_{i'=1}^{I} \sum_{j'=1}^{J} \varphi_{i'j'k}\sigma_{i'}(x)\tau_{j'}(z)$$

$$X(t_k) = X_k, \qquad V(t_k) = V_k$$

Including this in Equations 7.2, 7.3 and 7.4, we obtain

$$\left.\begin{array}{l}\displaystyle\sum_{i'=1}^{I}\varphi_{i'jk}\sigma''_{i'}(x_i)+\sum_{j'=1}^{J}\varphi_{ij'k}\tau''_{j'}(z_j)=0\end{array}\right\}\begin{array}{l}i=2,\ldots,I-1\\ j=2,\ldots,J-1\\ k=2,\ldots,K\end{array}$$

$$\left.\begin{array}{l}\displaystyle\sum_{i'=1}^{I}\varphi_{i'jk}\sigma'_{i'}(x_1)=V_k\\ \displaystyle\sum_{i'=1}^{I}\varphi_{i,jk}\sigma'_{i'}(x_I)=\dot{X}_k\end{array}\right\}\begin{array}{l}j=2,\ldots,J-1\\ k=2,\ldots,K\end{array}\qquad(7.7)$$

$$\left.\begin{array}{l}\displaystyle\sum_{j'=1}^{I}\varphi_{ij'k}\tau'_{j'}(z_J)+\frac{1}{g}\ddot{\varphi}_{iJk}=0\\ \displaystyle\sum_{j'=1}^{J}\varphi_{ij'k}\tau'_{j'}(z_1)=0\end{array}\right\}\begin{array}{l}i=2,\ldots,I-1\\ k=2,\ldots,K\end{array}$$

$$m\ddot{X}_k+\alpha\dot{X}_k+KX_k=-\rho\sum_{j'=1}^{J}\dot{\varphi}_{Ij'k}\int_{-h}^{0}\tau_{j'}(z)\,\mathrm{d}z$$

In practice, auxiliary variables such as $Y_k=\dot{X}_k$, etc., are introduced and $\dot{\varphi}_{ijk}, \dot{X}_k, \dot{Y}_k$, etc. are obtained numerically by an explicit fourth-order method of the Runge–Kutta type (and by finite differences for $k=2$). If $\psi_{ijk}=\dot{\varphi}_{ijk}$ and

$$v_k=\{\varphi_{ijk},\psi_{ijk},X_k,Y_k;i=1,\ldots,I;j=1,\ldots,J;(i,j)\neq(1,1),(I,1),(I,J),(1,J)\}$$

the preceding system of equations can be written in matrix form, $Av_2=v_1$, or $Av_k=Bv_{k-1}+Cv_{k-2}$ $(k\geqslant 3)$. The values are then calculated at nodes $(1,1),(I,1),(I,J)$ and $(J,1)$ so as to satisfy an approximate continuity equation in the adjacent mesh (extrapolation of the solution).

7.5.2 Steady state model

We write

$$\varphi_1=\sum_{i'=1}^{I}\sum_{j'=1}^{J}\varphi^1_{i'j'}\sigma_{i'}(x)\tau_{j'}(z)$$

$$\varphi_{2,1}=\sum_{i'=1}^{I}\sum_{j'=1}^{J}\varphi^{2,1}_{i'j'}\sigma_{i'}(x)\tau_{j'}(z)$$

$$\varphi_{2,2}=\sum_{i'=1}^{I}\sum_{j'=1}^{J}\varphi^{2,2}_{i'j'}\sigma_{i'}(x)\tau_{j'}(z)$$

in which φ^1_{ij}, $\varphi^{2,1}_{ij}$ and $\varphi^{2,2}_{ij}$ are complex numbers. The equations are written in the same way as in Equation 7.7 and the vectors are defined

$$v_1 = \{\varphi_{ij}^1, x_1\,;\, i = 1, \ldots, I\,;\, j = 1, \ldots, J\,;\, (i, j) \neq (1, 1), (I, 1), (1, J), (I, J)\}$$

$$v_{1,2} = \{\varphi_{ij}^{1,2}, x_{1,2}\,;\, i = 1, \ldots, I\,;\, j = 1, \ldots, J\,;\, (i, j) \neq (1, 1), (I, 1), (1, J), (I, J)\}$$

$$v_{2,2} = \{\varphi_{ij}^{2,2}, x_{2,2}\,;\, i = 1, \ldots, I\,;\, j = 1, \ldots, J\,;\, (i, j) \neq (1, 1), (I, 1), (1, J), (I, J)\}$$

Recurrence formulae are again written in matrix form and values are calculated at nodes (1, 1), $(I, 1)$, (I, J) and $(1, J)$ as in the preceding section.

7.6 Formulation A in terms of optimal control

Let us consider the transient problem at the first order. Since we know its solution, we have seen that the problem at the second order can be solved in the same way. Systems of Equations 7.2 and 7.3 can actually be divided up into two separate systems coupled by a condition on one boundary, i.e. system (7.2) of differential equations in Φ, describing the motion of the fluid, systems (7.3) and (7.4) of differential equations in X, describing the motion of the centre of mass of the body (here the right wall). It is then possible to use the optimal control formulation,[14,15] in seeking to solve one of the systems as the minimum of a functional with constraint (checking the other systems).

In what follows, we will give a formulation in terms of boundary control. Let $\psi(z, t)$ be the value of the potential on the right wall. We write for the fluid

$$\Delta\Phi = 0 \quad \text{in } D$$

$$\left(\frac{\partial\Phi}{\partial x}\right)_{x=0} = V(t), \qquad \left(\frac{\partial\Phi}{\partial x}\right)_{z=-h} = 0, \qquad \left[\frac{\partial\Phi}{\partial z} + \frac{1}{g}\frac{\partial^2\Phi}{\partial t^2}\right]_{z=0} = 0 \tag{7.8}$$

$$\Phi_{x=L} = \psi$$

$$\Phi_{t=0} = \psi_{t=0} = \left(\frac{\partial\Phi}{\partial t}\right)_{t=0} = \left(\frac{\partial\psi}{\partial t}\right)_{t=0} = 0$$

and for the wall (in a canonical form)

$$Y = \dot{X}$$

$$m\dot{Y} + \alpha Y + KX = -\rho\int_{-h}^{0} \frac{\partial\psi}{\partial t}\,\mathrm{d}z \tag{7.9}$$

$$X(0) = Y(0) = 0$$

Here Φ, $\varphi \in L_0^2[0, T; H^1(D)]$; $\psi, \dot{\psi} \in L_0^2[0, T; H^{\frac{1}{2}})-h, 0(]$ and X, $Y \in L_0^2(0, T)$. Let Φ_0 be the solution to Equation 7.8 when ψ is zero and write $V = \Phi - \Phi_0$

in (7.8). The problem is to find ψ so that $(\partial\Phi/\partial x)_{x=L} = \dot{X}$. Thus, we write

$$Z(z, t) = -\left(\frac{\partial\Phi_0}{\partial x}\right)_{x=L} = Z(t) \qquad \forall z \in (-h, 0)$$

and

$$Z \in L_0^2[0, T; H^{-\frac{1}{2}})-h, 0(]$$

$$\left(\frac{\partial U}{\partial x}\right)_{x=L} = \dot{X} + Z$$

An operator $\mathscr{A}$ form of $H^{\frac{1}{2}}(-h, 0)$ in $H^{-\frac{1}{2}}(-h, 0)$ is defined by $\mathscr{A}\psi = (\partial V/\partial x)_{x=L}$. With the integrals on $(0, T(x)-h, 0)$ being designated by $\langle , \rangle$ we write

$$J(\psi) = \int_0^T \int_{-h}^0 (\mathscr{A}\psi - \dot{X} - Z)^2 \, \mathrm{d}z \, \mathrm{d}t$$

$$= \langle \mathscr{A}\psi - Y - Z, \mathscr{A}\psi - Y - Z \rangle$$

We then try to find ψ which satisfies

$$J(\tilde{\psi}) = \min_{\psi \in H^{1/2}(\Gamma)} J(\psi)$$

Since Y is linked to ψ by the homogeneous linear systems (7.9), the application $\psi \to Y$ is linear. Therefore, we can show that $J(\psi) = J(\psi, Y(\psi))$ is convex with respect to ψ.

The gradient is obviously

$$\delta J = 2\langle \mathscr{A}\psi - Y - Z, \delta(\mathscr{A}\psi)\rangle - 2\langle \mathscr{A}\psi - Y - Z, \delta Y\rangle$$

Equation 7.9 can be used to express the second term of δJ as a function of $\delta\psi$. Let us consider the differential system

$$\dot{p} = q$$

$$m\dot{q} - \alpha q + Kp = -\int_{-h}^0 (\mathscr{A}\psi - Y - Z) \, \mathrm{d}z \qquad (7.10)$$

where $p(T) = q(T) = 0$ and $p, q \in L^2(0, T)$. System 7.9 can be written in terms of variations

$$\dot{\overline{(\delta X)}} = \delta Y$$

$$m\dot{\overline{(\delta Y)}} + \alpha \delta Y + K \delta X = -\rho \int_{-h}^0 \frac{\partial}{\partial t}(\delta\psi) \, \mathrm{d}z$$

$$\delta X(0) = \delta Y(0) = 0$$

Using this system we can calculate

$$I = \langle \mathscr{A}\psi - Y - Z, \delta Y \rangle = \int_0^T [m\dot{q} - \alpha q + Kp]\, \delta Y \, dt$$

$$= \int_0^T q \int_{-h}^0 \frac{\partial}{\partial t}(\delta\psi)\, dz\, dt = \int_{-h}^0 \int_0^T q \frac{\partial}{\partial t}(\delta\psi)\, dt\, dz$$

$$= \int_{-h}^0 [q\, \delta\psi]_0^T\, dz - \int_0^T \int_{-h}^0 \dot{q}\, \delta\psi\, dz\, dt = -\langle \dot{q}, \delta\psi \rangle$$

since $(\delta\psi)_{t=0} = 0$. Since system 7.9 is homogeneous except on the right wall, $\mathscr{A}$ is linear, and $\delta(\mathscr{A}\psi) = \mathscr{A}(\delta\psi)$, we introduce the adjoint operator $\mathscr{A}^*$ of $\mathscr{A}$, to get

$$\langle \mathscr{A}\psi - Y - Z, \mathscr{A}(\delta\psi) \rangle = \langle \mathscr{A}^*(\mathscr{A}\psi - Y - Z), \delta\psi \rangle$$

Finally,

$$\delta J = 2\langle \mathscr{A}^*(\mathscr{A}\psi - Y - Z) + \dot{q}, \delta\psi \rangle$$

7.6.2 Application

Let ΔQ be a meshing of Q. A solution to system 7.10 is sought, as in Section 7.7.1. We write

$$\psi_{jk} = \psi_j(t_k); \qquad Y_k = Y(t_k); \qquad X_k = X(t_k); \qquad p_k = p(t_k);$$
$$q_k = q(t_k); \qquad U_{ijk} = U_{ij}(t_k); \qquad W_{ijk} = U_{ijk}$$

and then we replace J by the functional

$$\tilde{J}(\psi) = \sum_{k=1}^{K} |\mathscr{A}(\psi_{jk}) - Y_k - Z|^2$$

We successively try to find $\{\tilde{\psi}_{jk}\}$ by performing

$$|\mathscr{A}(\psi_{jk}) - Y_k - Z|^2 = \min_{\tilde{\psi}_{jk}} |\mathscr{A}(\tilde{\psi}_{jk}) - Y_k - Z|^2 \qquad k = 1, \ldots, K$$

The following vectors are defined:

$$v_k^{(1)} = \{U_{ijk}, W_{ijk}; i = 1, \ldots, I - 1; j = 1, \ldots, J; (i, j) \neq (1, 1), (1, J)\}$$
$$v_k^{(2)} = \{\psi_{jk}; j = 2, \ldots, J - 1\}$$

By integrating, as in Section 7.5.1, the problem can be written in the following matrix form:

$$\begin{aligned} Av_2^{(1)} &= -Bv_2^{(2)} + C_2 \\ Av_k^{(1)} &= -Bv_k^{(2)} + C_k(v_{k-1}^{(1)}, v_{k-2}^{(1)}) \end{aligned} \tag{7.11}$$

Since the problem has a unique solution for a given $v_k^{(2)}$, matrix A has an inverse. Equation 7.11 can be used to express the values of δU_{ijk} and δW_{ijk} as a function of $\delta\psi_{jk}$ only, because the solutions at the preceding times are known (i.e., $\delta v_{k-1}^{(1)}, \delta v_{k-1}^{(2)}, \ldots$, etc. are zero). The operator is then discretized by writing $(\mathscr{A}\psi)_{jk} = (\psi_{jk} - U_{I-1,jk})/\Delta x$. In this way, we do not have to calculate $\mathscr{A}^*$, because we get in a straightforward way.

7.6.3 *Algorithm*

Let

$$\{v_k^{(1)}, v_k^{(2)}; k = 1, \ldots, K\}^{-1}$$

be a set of starting values. We calculate $\{X_k, Y_k; k = 1, \ldots, K\}^1$ and $\{p_k, q_k; k = 1, \ldots, K\}^1$ by solving systems (7.9) and (7.10). We then calculate the gradiant, and use standard minimization methods[15] to calculate the vector

$$\{v_k^{(2)}, k = 1, \ldots, K\}^2$$

This enables us to calculate $\{v_k^{(1)}, k = 1, \ldots, K\}^2$, $\{X_k, Y_k, k = 1, \ldots, K\}^2$, and $\{p_k, q_k, k = 1, \ldots, K\}^2$.

7.7 Numerical results

The methods proposed were programmed so as to test their practical application.

7.7.1 *Direct approximation*

7.7.1.1 Linearized transient model So as to compare the solution with a known solution and to visualize the reflection on the right wall, a solitary wave was generated in the flume. In very shallow water, system (7.2) of equations can have an asymptotic solution called a solitary wave[2] that is characterized by the following velocity:

$$C = \sqrt{g(h + H)}$$

and the following profile:

$$\eta = H\left[\operatorname{sech}\frac{3H}{4h^3}(x - Ct)\right]^2$$

with H designating the amplitude of the wave.

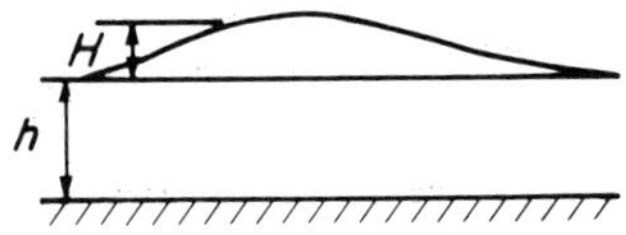

The motion of the wall to generate a solitary wave is given by[16]

$$V(t) = \frac{D\omega}{2}\{1 - th^2[\omega(t + t_0)]\}$$

in which $D^2 = 16Hh(1 + H/h)/3h$ is the square of the total amplitude of the motion and

$$\omega = \sqrt{\frac{3Hg}{4h^2}\left(1 + \frac{H}{h}\right)}$$

We employed the values $h = 0{\cdot}376\,m$; $L = 14$ m; $H = 0{\cdot}1$ m; $D = 0{\cdot}5$ m; $\omega = 2{\cdot}57\,\mathrm{s}^{-1}$. This gives $C = 2{\cdot}16$ m/s and $V(t) = 0{\cdot}647\{1 - th^2[2{\cdot}57(t - 1)]\}$. We also chose $m = 50$ kg; $\alpha = 100\,\mathrm{kg\,.\,s^{-1}}$; $K = 50\,\mathrm{kg\,.\,s^{-2}}$.

7.7.1.2 Results At time $t = 1{\cdot}48$ s, the height H of the calculated wave is 0·0948 m. The amplitude attenuation visible in Figure 7.2 can be explained by the fact that a low-amplitude wave train is created behind the wave. The energy accumulated in these waves is indeed lost by the solitary wave. This phenomenon is found to exist even for very small integration steps (0·3 s). Hence it does not appear to have a numerical nature. In reality, this wave train is always found to exist in the experiments.[17] The solitary wave solution

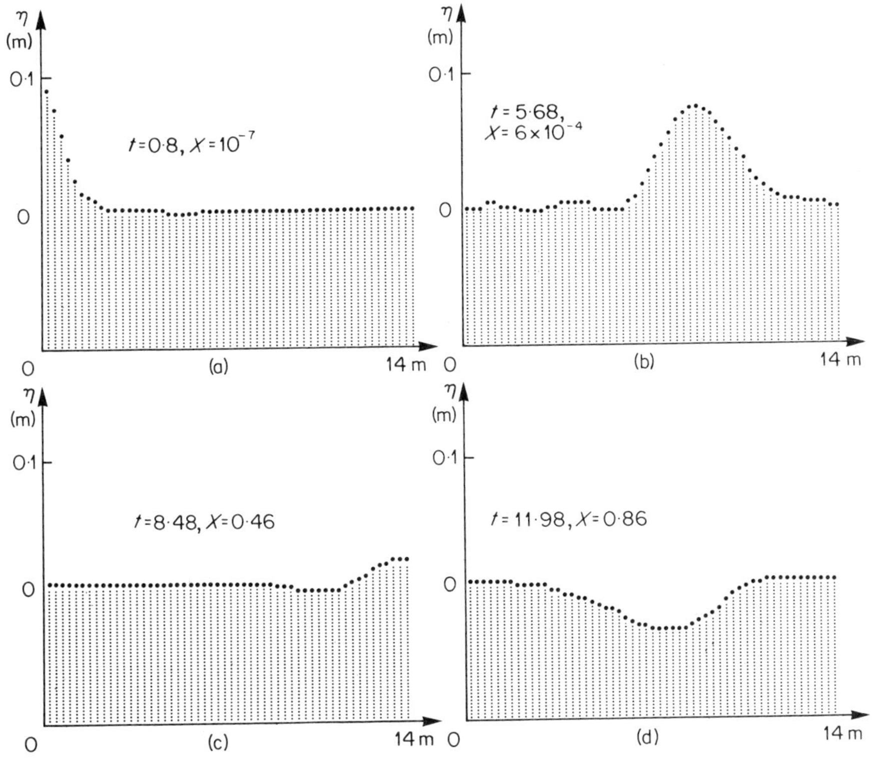

Figure 7.2 Ware propagation and reflection

is an asymptotic one. It thus does not exactly correspond to the solutions of Equation 7.2, even with the assumptions made (perfect fluid, linearization). Because of its characteristics, the right wall moves 0·86 m when the wave reaches it, thus causing a negative-amplitude reflected wave to form.

The calculated wave propagation velocity is $c = 1{\cdot}98$ m/s. Figure 7.3 compares the theoretical and observed profiles for the wave obtained at time $t = 2{\cdot}92$ s. The calculated points are slightly above the theoretical points. This result is always found in the experiments.[17]

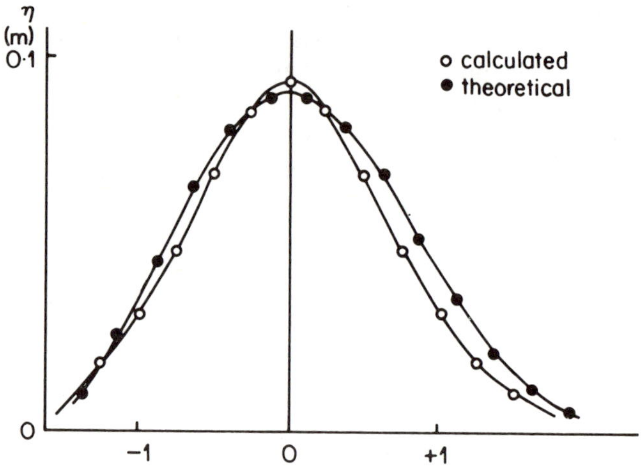

Figure 7.3 Theoretical and observed profiles

7.7.1.3 Second-order non-linear model When the channel length is not a multiple of the wave length, or when the right wall is moving, the shape of the free surface is difficult to interpret on account of the numerous harmonics present in the solution. To visualize a simple non-linear wave, we will take the case of the immobile right wall with the following data: $L = 3{\cdot}026$ m; $h = 0{\cdot}5$ m; $\omega = \omega_0 = 2\,\mathrm{s}^{-1}$ ($\tau_0 = 1$ s); $\omega_1 = 0$; $v_1 = 0{\cdot}14\,\mathrm{m}\,.\,\mathrm{s}^{-1}$; $v_{1,2} = 0$, $v_{2,2} = 0$.

The motion of the left wall is given by $V(t) = 0{\cdot}14\,\mathrm{e}^{-2i\pi t}$. Figure 7.4 shows the free surface obtained at different times. We can see that a second harmonic is caused by the term $\mathrm{e}^{-2i\pi t}$ in the second-order solution. We can also see the variations in the mean level as the result of the motion of the left beater. The very sharp crest and the flat trough that are characteristic of non-linear waves appear quite clearly in Figure 7.4(d). The wave in this figure has an amplitude of $H = 0{\cdot}28$ m. It is thus at the breaking limit H_d, according to standard criteria given in Reference 2, p. 106, i.e. $h/\tau^2 = 0{\cdot}50\,\mathrm{m/s}^2$ gives $H_d/\tau^2 = 0{\cdot}27$ $\mathrm{m/s}^2$. The crest has no cusp because of the smoothing of the spline functions. However, we can see that the estimated crest angle measures 114° as shown by

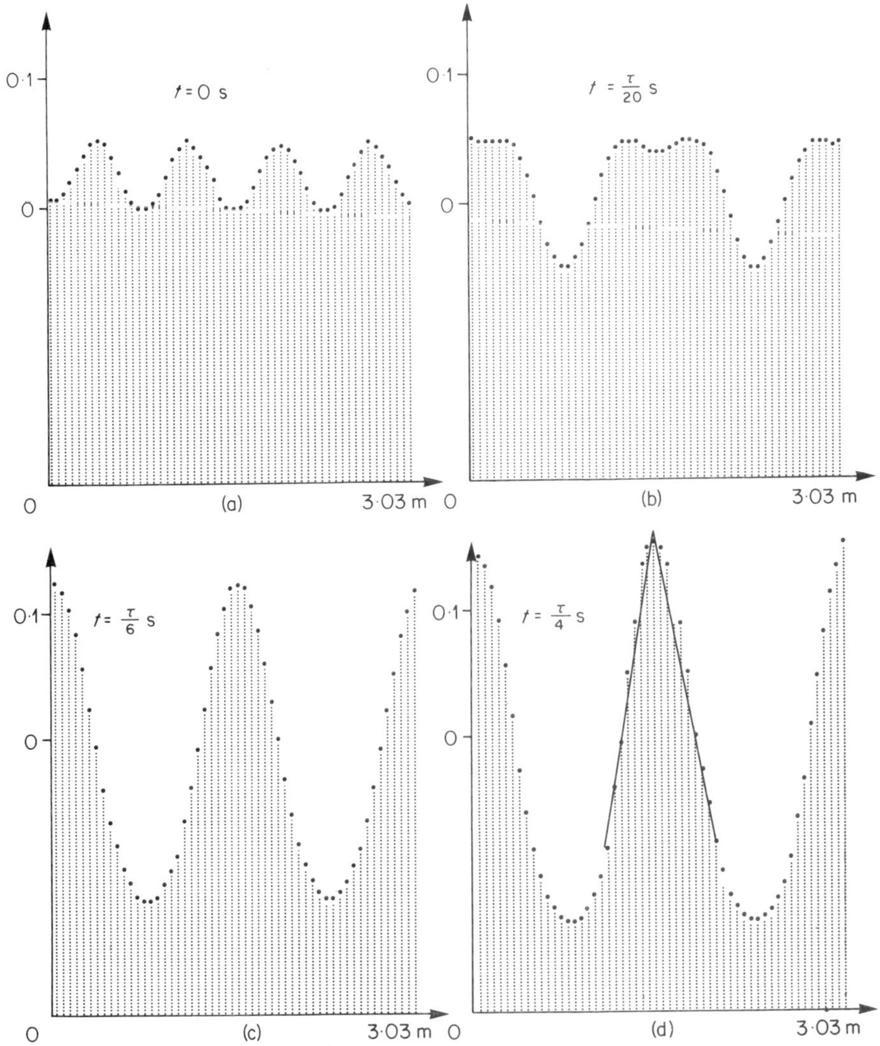

Figure 7.4 Calculated free surface

the two straight lines in Figure 7.4(d). This value is quite near the value of 120° determined by other investigators (Reference 2, p. 14). Figure 7.5 shows the free surface at times $t = 0$ and $t = \tau/4$ when the right wall is mobile with the following characteristics:

$$m = 50\ \text{kg}; \qquad \alpha = 100\ \text{kg}\,.\,\text{s}^{-1}; \qquad K = 50\ \text{kg}\,.\,\text{s}^{-2}$$

for the same motion of the left wall. The calculated response of the right wall is then

$$X = \tfrac{1}{100}(-0{\cdot}15 + 2{\cdot}8i)\,\text{e}^{-2i\pi t} + (-0.25 + 0.024i)\,\text{e}^{-4i\pi t}$$

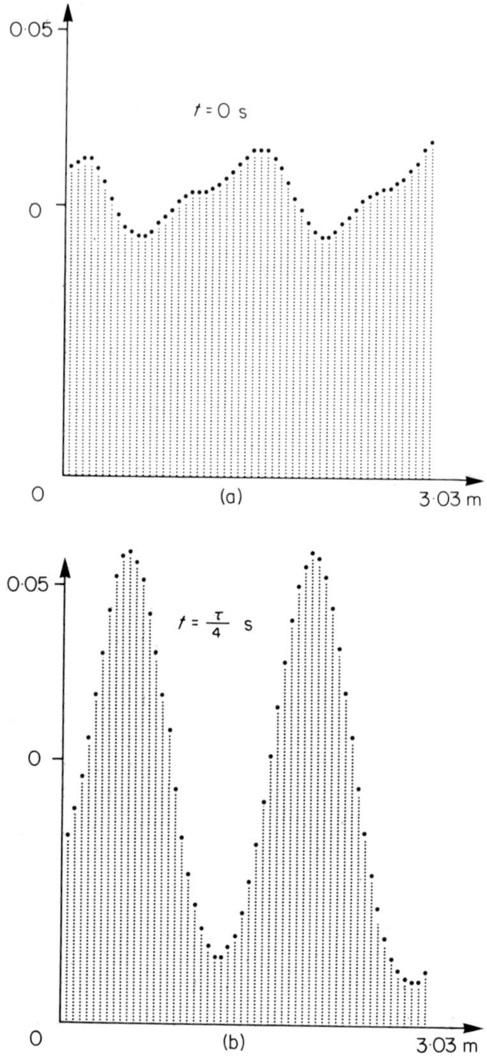

Figure 7.5 Stationary wave with moving right wall

7.7.2 Optimal control solution

A comparison was made of the solutions obtained by control and by direct approximation in the following case:

$$V(t) = 0{\cdot}07 \sin(2t) \text{ m/s}$$

$$h = 0{\cdot}5 \text{ m}; \qquad L = 1{\cdot}513 \text{ m}$$

Control was performed by using both the finite differences (Program I) in the fluid and the spline functions (Program II). In both cases, the number of iterations required to reach the minimum (zero) of J was less than 20, and relative accuracy was 10^{-11}.

The results for Program II were extremely close to those obtained in Program I.

Figures 7.6 and 7.7 show the displacement and the velocity of the right wall, as obtained with Program I and with the direct approximation in splines. There is excellent agreement between the velocities. The difference in the displacement is greater, but the curves behave in exactly the same manner. The result using the control seems to be much better than the result

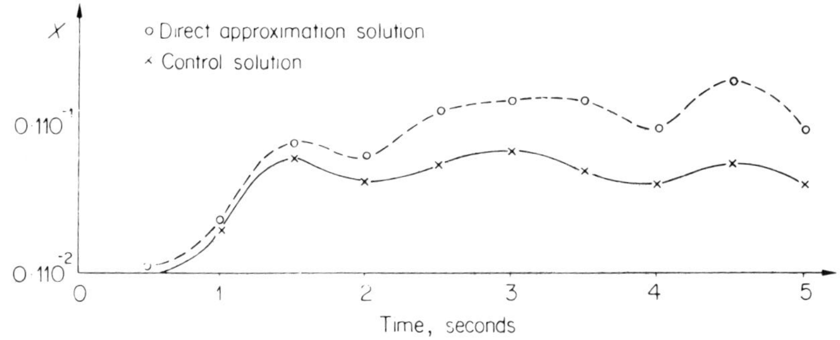

Figure 7.6 Movement of the right beater

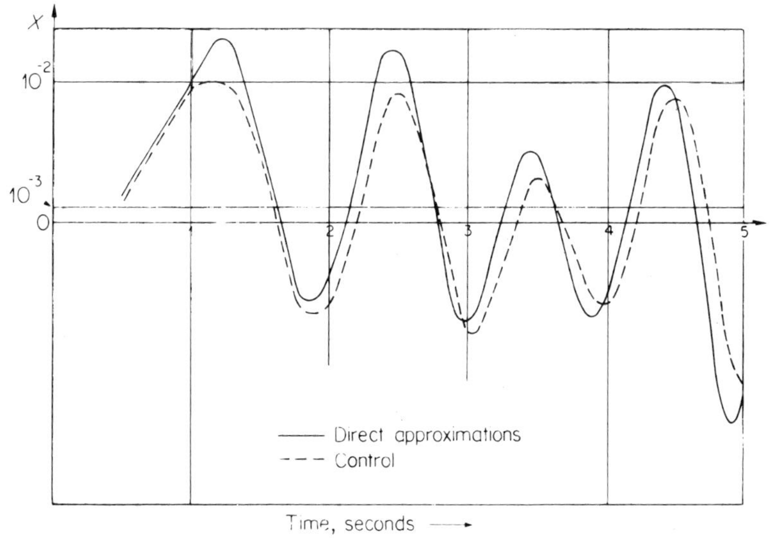

Figure 7.7 Comparison of velocities

obtained by the direct approximation. However, the numerical method (finite differences) used in the control is less accurate than the splines used in the direct approximation. This proves the interest of the last formulation. At the initial time, the left wall is at its maximum negative amplitude (because of the choice of $V(t)$). This explains the initial translation of the right beater toward its median position.

7.8 Conclusion

After stating the physical problem of the motion of a body in water waves, we used two recent mathematical methods to solve it. The first one is a direct approximation in a space of spline functions which is dense in the space of the solutions. This method is an easy one to use and has the advantage over conventional methods of giving a continuous and derivable analytic solution up to an arbitrarily chosen order.

In addition, we found that this smoothness of the solution enabled considerably larger meshing to be used than with other methods (finite differences, finite elements), for the same accuracy (recent research[17] confirms this finding).

Likewise, for a given meshing, the accuracy increases very fast with the degree d of the spline functions, yet without causing an appreciable increase in computing time. In this respect, degree 4 gives much better results than degree 2 (bicubic) which is sometimes used.

The method can easily be extended to domains of any shape by defining different spline function spaces on each row of the mesh.[18,19]

The second method uses a formulation in terms of optimal control. It is more a mathematical formulation than a numerical technique. We built a process which reduces the problem of solving two coupled systems of partial differential equations and differential equations to the solution of each system independently. This process converges to the solution of the coupled problem. It enables numerical methods that are appropriate to each problem to be used. At the same time, the fulfilment of the boundary condition on the overall surface of the body rather than by points guarantees the quality of the solution in the area of particular interest to the user. Lastly, most of the assumptions made in this article to illustrate these methods are not required for this formulation, thus allowing for the possibility of tackling extremely general cases, for example, without going via the perturbation technique.

This project enabled us to study the generation, propagation and reflection of irregular and non-linear wave patterns in laboratory flumes. In addition to the significance of the response of the right wall to these wave patterns, we succeeded in simulating the actual damping of waves in the laboratory flume (beaches, glass wool, etc.) by the damping caused by the moving wall. In this way we created a tool that is well suited to be compared with experiments for making kinematic or pressure-distribution analyses. Apart from infinite-space models now being built, we will extend such laboratory flume models to the three-dimensional case.

Acknowledgements

We would like to thank the Institut Français du Pétrole, its Division des Techniques de Production et de Forages, and its Département Action des Eléments Marins, and Société Franlab, which made available the funds required for this project. We would also like to thank Messrs. Philippe Rast, Patrick Legoulven, Francis Pellerin and Jean-Loup Chenot for the help they gave us.

Thanks must also be extended to Madame Girard for her extensive secretarial assistance.

References

1. J. Wehausen and E. Laitone, 'Surface waves', in *Encyclopedia of Physics*, S. Flügge (ed.), Vol. X, Springer Verlag, Berlin, 1960.
2. T. Ippen, *Estuary and Coastline Hydrodynamics*, McGraw-Hill, New York, 1966.
3. C. Truesdell and R. Toupin, 'The classical field theory', *Encylopedia of Physics*, S. Flügge (ed.), Vol. III-1, Springer Verlag, Berlin, 1960.
4. P. Rast, 'Canal à houle—solution stationnaire et non linéaire', in-service report, Institut Français du Pétrole, Division T.P.F., 1973.
5. J. L. Lions and E. Magènes, *Problèmes aux limites non homogènes et applications*, Vol. 1, Dunod, Paris, 1968.
6. R. Temam, *Analyse Numérique*, Presse Universitaire de France, Paris, 1970.
7. J. Laurent, *Approximation et Optimisation*, Herman, Paris, 1972.
8. A. Sard and S. Weintraub, *A Book of Splines*, Wiley, New York, 1971.
9. M. Schultz, 'Multivariate spline functions and elliptic problems', *Approximation with Special Emphasis on Spline Functions*, I. J. Schoenberg (ed.), Academic Press, 1969.
10. M. Schultz, ''Approximation theory of multivariate spline functions in Sobolev spaces', *SIAM J. Numer. Anal.*, **6**, 4 (1969).
11. M. Schultz, 'Rayleigh–Ritz–Galerkin methods for multidimensional problems', *SIAM J. Numer. Anal.*, **6**, 4 (1969).
12. A. Friedman and M. Shinbrot, 'The initial value problem for the linearized equation of water waves', *Journal of Mathematics and Mechanics*, **17**, 2 (1967).
13. R. M. Garipov, 'On the linear theory of gravity waves: the theorem of existence and uniqueness', *Archive for Rational Mechanics and Analysis*, **24**, 5 (1967).
14. J. L. Lions, *Contrôle Optimal des Systèmes Gouvernés par des Equations aux dérivées Partielles*, Dunod, Paris, 1968.
15. J. Céa, *Optimisation—Théorie et Algorithmes*, Dunod, Paris, 1971.
16. J. French, 'Wave uplift pressures on horizontal platforms', thesis, California Institute of Technology, Pasadena, *Report N. KH-R-19*, 1969.
17. M. Caprilli, A. Cella and G. Gheri, 'Spline interpolation techniques for variational methods', *Int. Journal for Num. Meth. in Eng.*, **6**, 565–576, 1973.

18. Y. Allouard, 'Interpolation et lissage par splines généralisées', in-service report, Franlab, 1969.
19. P. Legoulven, Y. Allouard, J. F. Coudert and Y. Lamy, 'Diffraction de la houle par un corps de révolution à méridienne quelconque', *S.P.E. European Spring Meeting*, Amsterdam, 1974, (in press).

General Bibliography

L. M. Milne Thomson, *Theoretical Hydrodynamics*, Macmillan, 1972.
K. Yosida, *Functional Analysis*, Springer Verlag, Berlin, 1968.
O. C. Zienkiewicz, *The Finite Element in Engineering Science*, McGraw-Hill, London, 1971.
S. G. Mikhlin and K. L. Smolitskiy, *Approximate Methods for Solution of Differential and Integral Equations*, Elsevier, New York, 1967.
J. H. Ahlberg, E. N. Nilson and J. L. Walsh, *The Theory of Splines and Their Applications*, Academic Press, New York, 1967.

Chapter 8

Computation of Flows in Turbomachines

H. J. Perkins and J. H. Horlock

8.1 Introduction

We review in this paper the many and varied methods of calculation that have been, and are, used to predict the flow behaviour in turbomachines.

The flow in the interior of a turbomachine is extremely complex, being in general three-dimensional, time dependent, compressible and subjected to viscous shear at all exposed surfaces. The three-dimensional unsteady boundary layers and wakes are affected by tip clearance and other leakage flows and by irregularities in the machine boundaries. In addition, the working fluid is seldom perfectly incompressible or a perfect compressible gas may be multiphased or a mixture of chemically reacting substances.

For the purposes of estimating the thermodynamic performance of a turbomachine (its output, thermal efficiency, etc.) several simplifying assumptions are normally employed.

(1) Calculation is confined to those regions of the turbomachine that are approximately axisymmetric. The flows to and from the plenum chambers, chimneys, spiral volutes etc., are not normally computed.
(2) All 'duct' flows upstream and downstream of blades are assumed to be steady and axisymmetric.
(3) Flow in blade rows is taken to be circumferentially periodic with a period equal to the blade spacing. Periodic flows of longer wave length, such as those that occur in surging compressors, are excluded from consideration.
(4) In a blade row, time dependence is taken to be negligible when in a coordinate system fixed relative to that blade row.
(5) Boundary layers are dealt with separately and their effects on available flow area, fluid deflection and irreversible losses are incorporated into an otherwise inviscid calculation. The problems encountered in estimating the growth of boundary layers in turbomachines will be discussed fully in a future publication.
(6) Calculations are performed for the flow of a single dominant phase or chemical species whose properties are known.

Having removed the time dependence by using a rotating coordinate system for moving rows we are left within the blade rows with a three-dimensional problem. Full three-dimensional passage flow solutions, with or without boundary layers, have so far been confined to complex duct flows and turbomachines with few blade rows, such as centrifugal compressors and turbines, because of the limitations of computer storage, run times and cost. For multistage axial machines it is normal to solve the flow by matching two-dimensional solutions obtained in intersecting surfaces (Figure 8.1).

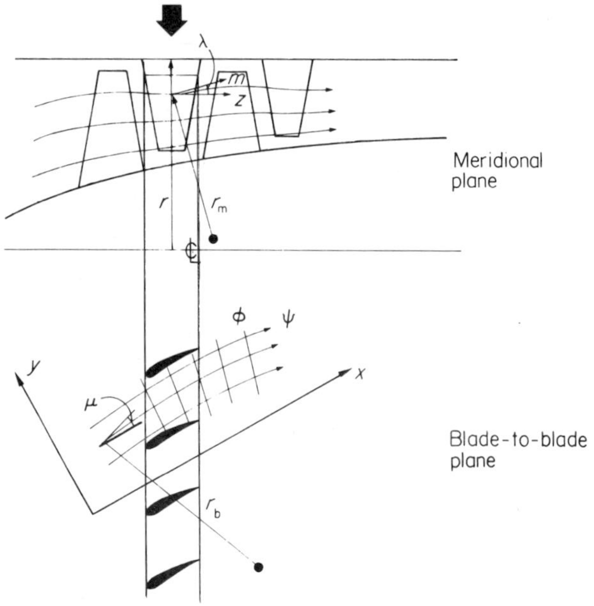

Figure 8.1 The meridional and blade-to-blade planes

Radial planes lying along the axis of the machine form one such family and the 'blade-to-blade' surfaces joining together the stream lines in adjacent radial planes form an intersecting family. By cycling the solution between these two sets of surfaces it is possible, in principle, to generate the complete three-dimensional solution. In practice, a single representative meridional (radial) plane is obtained by pitch averaging the equations of motion across one blade pitch. This process is described fully in Reference 1. The resulting equations describe hypothetical axisymmetric flow, the blade introducing a circumferentially 'smeared' blockage due to its finite thickness, a system of body forces necessary for deflecting the flow and relative total pressure losses due to blade surface boundary layers and the wake. Owing to the inexactness of this procedure it has not proved possible to select a set of self-consistent pitch averaged variables. The surfaces forming the second family are in general twisted in each blade passage making the blade-to-blade solution very difficult. To overcome this problem a blade-to-blade surface of revolution is formed by rotating a stream line from the solution in the representative meridional plane. In assessing the performance of multistage compressors and turbines it is usual to take this simplification further by assuming that the blade-to-blade flow behaviour is known from correlations of two-dimensional cascade data. For well tested traditional blade sections in conservative applications this works very well.

The splitting of the problem into two separate but interacting two-dimensional steady flow problems has been fundamental to the success of turbomachine performance calculations. This review considers firstly those solution procedures applicable to the blade-to-blade plane and then extends the discussion to include the more complex meridional plane.

8.2 Blade to blade plane solution methods

Classical solutions, obtained by conformal transformation, do exist[2,3] for incompressible cascade flow but have not proved widely useful, serving only as test cases with which numerical solutions may be compared.

'Singularity' methods[4] in which the blades are replaced by vortex and sink-source distributions, have been most successful in the study of incompressible flow, but it should be emphasized that their success is strongly dependent on an assumption which must be made about the trailing edge flow—either that the Kutta condition holds at a sharp trailing edge or that the location of the stagnation point on a rounded trailing edge is known. Linearized compressibility corrections have permitted the extension of these methods to locally sonic flow on the blade and more recently account has been taken of cascade sweep and lean. When combined with a detailed boundary layer solution these methods are computationally economical and have gained almost universal acceptance.

For compressible steady flow in the blade-to-blade plane, two powerful 'field' methods have been used. In the first, the stream line curvature method, a differential equation for the gradient of the velocity along the normal, or near normal, to the stream line is written in terms of the (assumed) radius of curvature of the stream line. This equation is integrated across the blade passage, the constant of integration being determined by the continuity equation. The stream lines are located, the radius of curvature is re-determined, and a solution is obtained by iteration. The method appears to give satisfactory answers for isentropic transonic flow, but its validity in a flow with shocks must be open to doubt.

In the second method, an equation for the stream function is derived and rewritten as a finite difference equation based on a particular form of grid. This equation is solved (simultaneously with an equation for local density) by matrix inversion or by relaxation. The method, initially thought to be valid for subsonic flow only, can be extended to deal with isentropic transonic flow.

8.2.1 *Singularity methods*

Singularity methods have been used largely, but not entirely, in the blade-to-blade plane. The essence of these methods is to replace the blade surfaces by singularities (sources, sinks, vortices), to calculate the induced velocity in the field by the (as yet unknown) singularity distribution and to satisfy the boundary conditions of the flow (that the velocity normal to the surface is zero).

The earliest solutions were obtained by Schlichting and his colleagues who distributed vertices $\gamma(x)$ along the blade camber lines and matched the camber surface at three chordwise locations. Separately they placed sinks and sources $q(x)$ along the chord line to simulate the blade thickness. The vortex and source-sink distributions were expressed as Fourier series of the chordwise location. The induced velocity, perpendicular to the mean flow V_x in the x direction, due to these distributions on a single aerofoil (V_{yo}) is known analytically and Schlichting and his colleagues calculated numerically the *extra* induced velocity (V_{y_i}) due to adding other aerofoils in cascade.

Thus the method may be summarized as follows:

$$\left.\begin{aligned} &\text{Camber slope} \quad \frac{dy_c}{dx} = \frac{V_{yo} + V_{y_i}}{V_x} \\ &\text{where } V_{y_i} = f(\gamma, \text{blade geometry}) \\ &\text{Thickness slope} \quad \frac{dy_t}{dx} = \frac{V_{yo} + V_{y_i}}{V_x} \end{aligned}\right\} \tag{8.1}$$

where $V_{y_i} = f(q, \text{blade geometry})$.

Schlichting calculated V_{y_i} and presented his results in tabular form. For n coefficients A_n, B_n in the series for γ and q, satisfying Equation 8.1 at n points yielded the coefficients A_n, B_n.

Clearly the full advantages of numerical computation had not been realized in this elementary method, for singularities were placed along the chord line and the velocities were also matched on the chord line. Developments in the airframe industry followed in which a finite number of finite sources or sinks. This is the basis of the Isay–Martensen[5] method and its specified normal to the aerofoil surface at points between the sources, again on the surface. The internal flow aerodynamicists followed a similar approach but surprisingly replaced the surface by finite vortices rather than finite sources or sinks. This is the basis of the Isay–Martensen[5] method and its more recent development by Wilkinson.[6] Essentially the method involves the solution of an integral equation for the velocity $V(m, n)$ on the surface of the cascade aerofoil (m, n) in the x, y plane. The velocity implies an element of surface vorticity $V(m, n)\,ds$ which induces a velocity elsewhere in the field giving a stream function $\psi(x, y)$ of the form

$$\psi(x, y) = \psi_m + \int_0^l V(m, n)K(x - m, y - n)\,ds \tag{8.2}$$

where ψ_m is the mean stream function based on the mean of the upstream and downstream flows and K is a kernel function.

Differentiation of ψ normal to the aerofoil surface gives the local velocities on the surface at $x = m$, $y = n$. Equation 8.2 is thus an integral equation for surface velocity distribution. The method of solution is to satisfy the equation at N points on each side of the known blade surface and to solve by matrix inversion the $2N$ equations for the velocity.

8.2.2 *Field methods*

In order to simplify the arguments in this section the equations of fluid motion are written in Cartesian coordinates (x, y), the x direction being taken roughly parallel to the mean flow so that $V_x \gg V_y$ (see Figure 8.1).

In steady inviscid flow two equations of motion are available, the statements of flow continuity,

$$\frac{\partial}{\partial x}(\rho V_x) + \frac{\partial}{\partial y}(\rho V_y) = 0 \tag{8.3}$$

and irrotationality,

$$\frac{\partial V_x}{\partial y} - \frac{\partial V_y}{\partial x} = 0 \tag{8.4}$$

A further equation for the density, ρ, follows from the assumption that the flow is esentropic,

$$\frac{\rho}{\rho_0} = \left\{\frac{H - V^2/2}{H}\right\}^{1/\gamma - 1} \tag{8.5}$$

where ρ_0 and H are upstream stagnation values of density and enthalpy,

$$V^2 = V_x^2 + V_y^2,$$

and γ is the isentropic index.

These three equations can be found in many alternative forms in the literature. For example a velocity potential $\phi(x, y)$ may be defined from Equation 8.4,

$$\frac{\partial \phi}{\partial x} = V_x; \qquad \frac{\partial \phi}{\partial y} = V_y$$

and Equation 8.3 becomes

$$\nabla^2\phi = f_1(\phi, \rho) \tag{8.6}$$

Alternatively substituting for ρ in the right-hand side,

$$(1 - M_x^2)\frac{\partial^2\phi}{\partial x^2} + (1 - M_y^2)\frac{\partial^2\phi}{\partial y^2} - 2M_xM_y\frac{\partial^2\phi}{\partial x\,\partial y} = 0 \tag{8.7}$$

This is a non-linear equation involving the flow Mach numbers M_x, M_y which must be solved in conjunction with Equation 8.5 in the form

$$\frac{\rho}{\rho_0} = \left[1 - \frac{1}{2H}\left\{\left(\frac{\partial\phi}{\partial x}\right)^2 + \left(\frac{\partial\phi}{\partial y}\right)^2\right\}\right]^{1/\gamma - 1} \tag{8.8}$$

Gradient boundary conditions on the flow boundaries make potential slightly difficult to use.

Alternatively a stream function may be defined from Equation (8.3)

$$\frac{\partial\psi}{\partial y} = \rho V_x; \qquad \frac{\partial\psi}{\partial x} = \rho V_y$$

Equation 8.4 then becomes either

$$\nabla^2\psi = f_2(\psi, \rho) \tag{8.9}$$

or

$$(1 - M_x^2)\frac{\partial^2\psi}{\partial x^2} + (1 - M_y^2)\frac{\partial^2\psi}{\partial y^2} + 2M_xM_y\frac{\partial^2\psi}{\partial x\,\partial y} = 0 \tag{8.10}$$

which must be solved together with

$$\frac{\rho}{\rho_0} = \left[1 - \frac{1}{2H\rho}\left\{\left(\frac{\partial\psi}{\partial x}\right)^2 + \left(\frac{\partial\psi}{\partial y}\right)^2\right\}\right]^{1/\gamma - 1} \tag{8.11}$$

An alternative statement of Equation 8.7 or 8.10 by Stanitz[7] using the assumption of a tangent gas, leads to a Laplace equation for flow velocity or flow direction but using the operator

$$\frac{\partial^2}{\partial\phi^{*2}} + \frac{\partial^2}{\partial\psi^{*2}}$$

with ϕ^* and ψ^* the non-dimensional potential and stream function. These equations are useful in deriving a blade shape (design mode solution) by specifying values of ϕ^* and ψ^* along the blade boundaries and far upstream and downstream. From straightforward point-by-point relaxation solutions of the Laplace equations for velocity and direction as functions of ϕ^* and ψ^* the blade shape in the physical plane can be recovered. Many other transformed variables have been used in design methods.

Examining Equations 8.7 and 8.10 we see that so long as M_x is small the equation is elliptic and can be solve by techniques such as relaxation or matrix inversion with the dependent variable, or its gradient, specified at the boundaries. When V_x is near sonic the first term may become small and the equation can be integrated directly in y. For supersonic flow the equation becomes hyperbolic and the 'method of characteristics' is appropriate. In the blade–blade plane it is quite likely that all three situations exist simultaneously. However, since supersonic regions are normally terminated by a system of shock waves our initial assumption of inviscid irrotational flow would then be invalid. A more general method of solution must be sought where supersonic flow exists and such a method (time marching) is discussed later. The present arguments are therefore confined to:

(a) incompressible flow,
(b) compressible subsonic flow,
(c) basically subsonic flow with small closed supersonic patches of isentropic flow,
(d) situations where the sonic line passes across the entire blade passage and an isentropic supersonic expansion takes place downstream. (Shocked flows between types (c) and (d) cannot be solved using isentropic equations.)

The stream function (ψ) with its attractively simple boundary conditions for a known mass flow was used by one of the earliest workers in this field, Emmons,[8] who produced hand relaxation solutions of Equation 8.9 in transonic nozzle flows. Emmons's solutions were not obtained without difficulty however, since the implicit equation for the fluid density, Equation 8.11, is double valued for given stream function gradients, giving a subsonic and a supersonic root. The inability of the density equation to choose for itself the appropriate root in transonic flow places a severe limitation on the use of stream function. Emmons overcame this problem by reversing the calculation method in regions of high Mach number. The normal course of the calculation involved setting f_2 in Equation 8.9 from the existing ψ and ρ fields, relaxing for a new ψ distribution, computing a new ρ and then recalculating f_2. In the reversed calculation the stream lines were assumed known (initially set in their incompressible positions). Equation 8.9 was used to revise the density profile by trial and error and Equation 8.11 then integrated for a new ψ distribution given the density. Quite remarkable results were achieved in this way. The modern counterpart using a computer is typified by the blade–blade method of Smith[9] who solves Equation 8.9 by matrix inversion together with an iterative statement of Equation 8.11. The Taylor series finite difference expansion for $\nabla^2\psi$ about a point is written for an arbitrary disposition of the surrounding nodes along parallel grid lines in the y direction. Ten points are used to represent the Laplacian operator leading to a banded coefficient matrix $[M]$ in the equation

$$[M][\psi] = [f_2]$$

Inversion of this equation yields a solution for ψ which can be used (suitably weighted with the previous solution) to recompute f_2. Again the ambiguity in the density equation (8.12) at higher subsonic Mach numbers limits the usefulness of Smith's method.

The family of solution procedures known as 'stream line curvature (SLC) methods' (see, for example, References 10 to 13) attempt a solution directly of Equations 8.3 to 8.5. The essential approach in the SLC method is to integrate a velocity gradient equation along a series of lines drawn roughly normal or truly normal to the stream lines. In the roughly normal (or quasi-orthogonal) method, Equation 8.4 is rewritten as

$$\frac{\partial}{\partial y}V_x^2 - 2V_x\frac{\partial}{\partial x}V_y = \frac{\partial}{\partial y}V_x^2 - 2V_x^2\frac{\partial}{\partial x}\left(\frac{V_y}{V_x}\right) - 2\frac{V_y}{V_x}V_x\frac{\partial V_x}{\partial x} = 0$$

$$\frac{\partial}{\partial y}V_x^2 + 2\frac{V_x^2}{r_b} - 2\tan\mu V_x\frac{\partial}{\partial x}V_x = 0 \qquad (8.12)$$

where r_b is the local radius of curvature of the stream line in the blade-to-blade plane and μ is the angle of inclination of the flow to the direction x (Figure 8.1). By guessing V_x at one boundary (one blade surface) this equation can be integrated step by step across the passage if the stream line slopes and curvatures are known from a previous solution. At each point the fluid density can be obtained without ambiguity from Equation 8.5. On reaching the opposite blade surface the passage mass flow can be computed and if this does not correspond to the required mass flow the guessed V_x can be adjusted and the integration of (8.12) repeated. The resulting velocity and density profiles, and the revised stream line locations along this grid line, will now satisfy the integral continuity equation and the velocity gradient equation but with slopes and curvatures appropriate to the previous solution. Because of the latter feature the change from the old stream line locations to the new ones is heavily damped to avoid large curvature changes between iterations. Wilkinson[12] has discussed the stability and convergence of SLC methods and derived a suitable damping factor which turns out to be a function of the differentiation routine used to establish the stream line slopes and curvatures. He has shown the advantage of high resolution curve fitting methods which require heavy damping but rapidly yield a high solution accuracy and the disadvantages of low resolution spline and least square fits where so much damping is inherent in the estimation of curvature that deceptively little damping is needed in the solution procedure which has an inescapably low accuracy.

In order to compare this procedure with the stream function methods described earlier it is convenient to regard SLC as a line-by-line integration, or line relaxation, of Equation 8.10. With the stream-function method we had, in point-by-point relaxation form

$$(1 - M_x^2)\frac{\partial^2 \psi_o}{\partial x^2} + (1 - M_y^2)\frac{\partial^2 \psi_o}{\partial y^2} + 2M_x M_y \frac{\partial^2 \psi_o}{\partial x\, \partial y} = \Delta(x, y) \qquad (8.13)$$

where subscript letter o refers to the 'old' ψ distribution and Δ is the array of residuals. We now require a new distribution, ψ_n, satisfying,

$$(1 - M_x^2)\frac{\partial^2 \psi_n}{\partial x^2} + (1 - M_y^2)\frac{\partial^2 \psi_n}{\partial y^2} + 2M_x M_y \frac{\partial^2 \psi_n}{\partial x\, \partial y} = 0 \qquad (8.14)$$

Assuming that ψ_o is composed of the true solution, ψ_n, and an error field $\varepsilon(x, y)$, then for $M_y \ll M_x$ the error field approximately satisfies the equation

$$(1 - M_x^2)\frac{\partial^2 \varepsilon}{\partial x^2} + \frac{\partial^2 \varepsilon}{\partial y^2} = \Delta$$

The process of relaxation involves writing

$$\psi_n = \psi_o + \frac{\alpha \Delta}{C_0}$$

C_0 being the coefficient of ψ at central node and α the relaxation or acceleration parameter. It follows that

$$\alpha = -C_0\varepsilon_0 \bigg/ \left[(1 - M_x^2)\frac{\partial^2 \varepsilon}{\partial x^2} + \frac{\partial^2 \varepsilon}{\partial y^2}\right] = -C_0\varepsilon_0 \bigg/ \left[\sum_i C_i\varepsilon_i - C_0\varepsilon_0\right] \qquad (8.15)$$

By choosing a method of differentiation and prescribing an error field, a suitable relaxation factor can be estimated. If, however, we consider a localized error such that ε_i is at all surrounding nodes less than ε_0 then we see that α is of order unity. This indicates that little or no damping is needed in this type of solution since thc change in curvature, $\partial^2\psi/\partial x^2$, implied by a stream line shift is always included in estimating the value of the shift.

Line relaxation, or SLC, on the other hand, presented with the situation described by Equation 8.13 seeks a new ψ variation in y alone, keeping the curvature $\partial^2\psi/\partial x^2$ fixed at its previous value,

$$(1 - M_x^2)\frac{\partial^2 \psi_o}{\partial x^2} + (1 - M_y^2)\frac{\partial^2 \psi_n}{\partial y^2} + 2M_x M_y \frac{\partial^2 \psi_n}{\partial x\, \partial y} = K\Delta$$

or if M_y is small

$$\frac{\partial^2}{\partial y^2}(\psi_n - \psi_o) = \frac{\partial^2 \varepsilon}{\partial y^2} = (K - 1)\Delta \qquad (8.16)$$

When all grid lines have been processed the new ψ field made up of the new 'line distributions' should satisfy Equation 8.14. The problem has therefore been transformed to one of choosing how much of the local residual Δ to remove so that the final field solution contains less or preferably no error, that is, choosing K. We have two statements,

$$(1 - M_x^2)\frac{\partial^2 \varepsilon}{\partial x^2} + \frac{\partial^2 \varepsilon}{\partial y^2} = \Delta; \qquad \frac{\partial^2 \varepsilon}{\partial y^2} = (K - 1)\Delta$$

so that

$$(K - 1) = \frac{\partial^2 \varepsilon}{\partial y^2} \bigg/ \left[(1 - M_x^2)\frac{\partial^2 \varepsilon}{\partial x^2} + \frac{\partial^2 \varepsilon}{\partial y^2}\right] = 1 \bigg/ \left[1 + (1 - M_x^2)\frac{\partial^2 \varepsilon/\partial x^2}{\partial^2 \varepsilon/\partial y^2}\right] \tag{8.17}$$

Again by choosing a differentiation routine and postulating an error field, $(K - 1)$ can be estimated and Equation 8.16 integrated for the new distribution, ψ_n. Equation 8.17 is in agreement with the damping factor derived by Wilkinson. Taking, for example, a point error

$$(K - 1) = 1 \bigg/ \left[1 + (1 - M_x^2)\left(\frac{\Delta y}{\Delta x}\right)^2\right]$$

or with Wilkinson's more restrictive error field having a wave length in x of $2\Delta x$ and in y of twice the passage height, h,

$$(K - 1) = 1 \bigg/ \left[1 + (1 - M_x^2)k\left(\frac{h}{\Delta x}\right)^2\right] \tag{8.18}$$

where k depends on the differentiation formulae (approximately 0·2 for high resolution methods) and $(h/\Delta x)$ is the aspect ratio of the grid. We see that for a grid of aspect ratio 6 and a Mach number of 0·5 the damping factor amounts to about 0·16. If, however, the real error is of much larger wavelength in x (or shorter in y) then a great deal of unnecessary damping has been employed and a great many extra iterations performed.

Small damping factors and a lack of ambiguity in the density equation form the essential features of the SLC methods, when compared with point-by-point relaxation or matrix inversion methods.

8.3 Meridional plane solution methods

In the representative meridional plane the flow equations, which result from averaging local equations across the blade pitch, can be written in the form of a gradient in pitch averaged velocity or pressure, usually along the radial direction in axial flow machines.

$$\frac{1}{\rho}\frac{\partial p}{\partial r} = V_m^2\left[\frac{\cos\lambda}{r_m} - \frac{\sin\lambda}{V_m}\frac{\partial V_m}{\partial m}\right] + \frac{V_\theta^2}{r} - F_\theta \tan\beta$$

where $1/r_m = -\partial\lambda/\partial m$ is the curvature of the stream line in the meridional plane, β is the angle of inclination of the blade surface to the radial direction, F_θ is the tangential blade force, and V_θ is the tangential velocity, or more usually in the form

$$\frac{1}{2}\frac{\partial}{\partial r}V_m^2 = \frac{\partial H}{\partial r} - T\frac{\partial s}{\partial r} - F_\theta \tan\beta - \frac{1}{2r^2}\frac{\partial}{\partial r}(rV_\theta)^2 - \frac{V_m^2}{r_m}\cos\lambda + V_m \sin\lambda\frac{\partial V_m}{\partial m} \tag{8.19}$$

This is referred to as the radial equilibrium equation. Again the stream line curvature method Novak[14] Denton[15] can be used to obtain a solution but in this case radial stagnation enthalpy and entropy gradients arise. Additional parabolic equations for the change of H and s along the meridional stream lines must be simultaneously satisfied. Alternatively, as in the blade-to-blade plane, stream function can be chosen, Wu,[16] to satisfy the continuity equation and the governing equation rephrased in a relaxation format, Perkins,[17] or for matrix inversion, Marsh.[18] Marsh's method was the forerunner of Smith's blade–blade program and was the first attempt, based on Wu's equations, at a detailed field solution of the meridional flow in turbomachines (although the stream function equation used strictly relates to a stream surface passing between the blades). The restrictive coordinate system chosen by Marsh and the use of matrix inversion led to further developments by Perkins who uses a perfectly general non-orthogonal curvilinear coordinate system sympathetic to the form of the duct.

For compressible irrotational flow in an annulus (with no swirl or with free vortex swirl) the equations and the available solution methods differ little from those already described in Section 8.2.2. The additional tangential equation of motion states that angular momentum is conserved along stream lines in duct regions and in blade rows is determined by the geometry of the representative stream surfaces chosen to replace the blade row. When work transfers, losses and rotational flows occur the situation becomes more complex. After a new stream line pattern has been obtained it is necessary to integrate the energy and loss equations along each stream line in turn to find the new stagnation conditions. Any changes in these will affect the balance of the radial equilibrium equation and the subsequent position of the stream lines. It is important therefore not to iterate for stream line positions downstream of a working blade row until the stream line pattern has been established within that blade row; time consuming, inappropriate and often destabilizing stream line adjustments in the downstream duct or blade rows are to be avoided until the correct distribution of stagnation conditions has been established. This means that matrix inversion techniques which consider simultaneously the algebraic equations for all nodes are at a disadvantage in such cases. With SLC methods or by using point-by-point relaxation, the solution can be stopped at any streamwise station (offering a substantial advantage in multi-stage axial flow compressor calculations for example).

In stream function methods the fluid density is again double valued. This is a troublesome feature in the vicinity of sonic relative velocity within blade rows and sonic meridional velocity in duct regions. As in the blade-to-blade plane, the problem can be overcome by using the imbalance of the radial equilibrium equation (computed using the stream function) to compute a new velocity profile in the manner of SLC methods or a new density profile, after Emmons. The appropriate damping factor can be derived in the manner described earlier with the additional effects of radius and stream surface angle included.

In general the flow within and at exit from blade rows is rotational. Although the tangential component of vorticity is correctly reflected in the radial equilibrium equation commonly used, an axisymmetric statement of the axial vorticity component has been employed

$$\frac{1}{r}\frac{\partial}{\partial r}(rV_\theta)$$

This expression ignores the twisting of the true blade-to-blade stream surface on passing through the blade row. If this twisting stops at the blade trailing edge (a free vortex blade) the above expression is correct at exit, although within the blade the error remains. An approximate method for including this effect in meridional calculations is suggested by Hawthorne and Novak.[19]

8.4 Current developments

As indicated earlier, our ability to deal with mixed subsonic/supersonic flows with or without shock waves is somewhat lacking, and it is in this area that much current research is concentrated. The need is a very real one since transonic turbine blading exhausts at modest supersonic velocities; and in highly loaded subsonic impulse sections the flow may become supersonic locally over the blade suction surface. In a high speed compressor rotor tip section the flow at entry will be supersonic and will shock down to subsonic conditions within the blade.

Several numerical methods designed to tackle this problem will be discussed briefly.

(1) Hobson[20] writes the equations of compressible, inviscid, irrotational, isentropic steady flow in terms of a stream function (ψ) in the hodograph plane using M^*, μ (critical Mach number (V/a^*), flow angle) coordinates.

The equation in the subsonic region is written in finite difference form

$$\psi_{ij} + f_1\psi_{i+1,j} + f_2\psi_{i-1,j} + f_3\psi_{i,j+1} + f_4\psi_{i,j-1} = 0 \qquad (8.20)$$

where f_1, f_2, f_3, f_4 are functions of M^*, ΔM^* and $\Delta\mu$.

The equation in supersonic flow is written in terms of the characteristic variables

$$\mathrm{I} = [\mu + \omega(M^*)]/2$$

$$\mathrm{II} = [\mu - \omega(M^*)]/2$$

(where $\omega(M^*)$ is the usual Prandtl–Meyer angle).

$$f_5 \frac{\partial^2 \psi}{\partial \mathrm{I}\, \partial \mathrm{II}} - f_6 \left(\frac{\partial \psi}{\partial \mathrm{I}} - \frac{\partial \psi}{\partial \mathrm{II}} \right) = 0 \tag{8.21}$$

where f_5 and f_6 are functions of M^*, ΔM^* and γ, so that with the stream function known at three corners of a characteristic cell, the stream function can be calculated at the fourth corner.

The difficult part of thc calculation lies in first guessing a stream function distribution along the sonic line, and then in matching $\partial\psi/\partial M^*$ along that line from the two regions. Hobson's method is essentially a design method in which a blade shape is derived from specified M^*, μ boundaries.

(2) For choked turbine blading Wilkinson *et al.*[21] propose a subsonic stream line curvature method matched with a supersonic charateristics method for the downstream flow. A very similar procedure is referred to by Craig and Hobson.[22]

The isentropic SLC calculation is terminated on the first completely supersonic quasi-orthogonal. This is made possible by using backward (upstream) differencing for the stream line slopes and curvatures in supersonic regions. The remainder of the solution is obtained from a field characteristics method started either at the sonic line or on a supersonic quasi-orthogonal. The characteristic equations are derived directly from the hyperbolic wave equations and are not restricted to isentropic irrotational flow. At blade exit the two expansions about the trailing edge are terminated in a triangular base pressure region which in turn is terminated through compression shocks. By setting the base pressure, the extent of the triangular region, and hence the strength and angle of the compression shocks, is fixed giving a unique set of exit conditions.

(3) Many time marching procedures (see References 23 to 27) for solving the transient hyperbolic equations of motion are now available. The space plane flow solution is matched in time until a steady state is reached.

The unsteady two-dimensional equations of gas dynamics may be written in the form

$$\frac{\partial \overline{w}}{\partial t} + \frac{\partial \bar{f}}{\partial x} + \frac{\partial \bar{g}}{\partial y} = 0$$

where

$$\bar{w} = \begin{bmatrix} \rho \\ \rho V_x \\ \rho V_y \\ E \end{bmatrix}, \quad \bar{f} = \begin{bmatrix} \rho V_x \\ \rho V_x + p/\gamma \\ \rho V_x V_y \\ (E + (\gamma - 1)p)V_x \end{bmatrix} \quad \bar{g} = \begin{bmatrix} \rho V_y \\ \rho V_x V_y \\ \rho V_y^2 + p/\gamma \\ (E + (\gamma - 1)p)V_y \end{bmatrix} \tag{8.22}$$

where $E = p + \gamma(\gamma - 1)\rho(V_x^2 + V_y^2)$, V_x and V_y are non-dimensionalized by the stagnation sound speed and p and ρ are non-dimensionalized by their stagnation values.

The equations are written in finite difference form and in the MacCormack method an interior mesh point w_{ij}^n (at location ij at time n) is advanced in two steps

$$\begin{aligned} \bar{\bar{w}}_{ij}^{n+1} &= \bar{w}_{ij}^n + \frac{\Delta t}{\Delta x}(\bar{f}_{i+1,j}^n - \bar{f}_{ij}^n + \bar{g}_{ij+1}^n - g_{ij}^n) \\ \bar{w}_{ij}^{n+1} &= \frac{1}{2}\left(\bar{w}_{ij}^n + \bar{\bar{w}}_{ij}^{n+1} + \frac{\Delta t}{\Delta x}(\bar{f}_{ij}^n - \bar{f}_i^n + \bar{g}_{ij}^n - \bar{g}_{i,j-1}^n)\right) \end{aligned} \tag{8.23}$$

where Δy is taken equal to Δx. The maximum time step for stability is limited by the relation

$$\frac{\Delta t}{\Delta x} \leqslant \min_{ij} \frac{1}{|V_x| + |V_y| + a\sqrt{2}}$$

An alternative time marching scheme developed by MacDonald should be mentioned—a finite area method. MacDonald considers a six-sided elementary control surface and the rate of change of density (ρ) and momentum ($\rho V_x, \rho V_y$) at its central grid point due to fluxes into the element, area ΔA with sides s.

$$\begin{aligned} \frac{\partial \rho}{\partial t} &= -\frac{1}{\Delta A}\oint_s [\bar{i}(\rho V_x) + \bar{j}(\rho V_y)] \,.\, \bar{n}\, \mathrm{d}s \\ \frac{\partial}{\partial t}(\rho V_x) &= -\frac{1}{\Delta A}\oint_s [\bar{i}(p + \rho V_x^2) + \bar{j}(\rho V_x V_y)] \,.\, \bar{n}\, \mathrm{d}s \\ \frac{\partial}{\partial t}(\rho V_y) &= -\frac{1}{\Delta A}\oint_s [\bar{i}(\rho V_x V_y) + \bar{j}(p + \rho V_y^2)] \,.\, \bar{n}\, \mathrm{d}s \end{aligned} \tag{8.24}$$

Knowing the distribution of p, ρ, V_x and V_y at time t, MacDonald calculates mass and momentum and fluxes along each of the six boundaries of the elementary control surface, assuming linear variation between grid points. The change in ρ, ρV_x, ρV_y at the central grid point after time t is obtained from Equation 8.24 and pressure is obtained from isentropic relation $p/\rho^\gamma =$ constant.

8.5 The future

The stream line curvature and stream function solutions for the flow in the meridional and blade-to-blade planes have reached a high degree of sophistication, although there is a severe limitation in both methods to shock free flow. However, we may expect the time marching, characteristics and hodograph methods to develop rapidly over the next few years and to be able to cope with supersonic flow with shocks, and with transonic flows in general.

With these developments taking place, is there a need for more calculation methods, such as for example the finite element method? (Note Thompson's method[28] for low speed cascade flows.) Before we attempt to answer that question we should consider the limitations of the present methods and the assumptions that are implicit in them.

Firstly, all the methods described in this paper assume the flow relative to the blading is steady. In fact the real flow is unsteady, owing to wakes shed from upstream blade rows and to potential flow interaction between adjacent blade rows. Although much progress has been made in the analysis of unsteady flow through cascades, we cannot expect these special solutions to be used in a full three-dimensional design calculation.

Secondly, all the blade-to-blade solutions involve an assumption about the trailing edge flow in order to fix the bound vorticity—either that the Kutta condition applies (that the flow leaves a cusped trailing edge smoothly, with equal velocity (and equal pressure) from the two surfaces) or that the location of the trailing edge stagnation point is known on a rounded trailing edge.

Thirdly, as described in the introduction, it is implied that it is possible to iterate between the meridional and blade–blade planes, the solution from one plane being used in the other. Recently, attention has been directed towards the inclusion of terms previously neglected in the averaging process across the pitch in order to derive a more correct set of equations (8.1). It is clear than when the blade lift is large or varying along the blade span, extra terms previously neglected should be included in the meridional flow equations.

In parallel to the work described in this review fuller treatments of the flow through single blade rows have been made (see, for example References 29 to 31). McCune[29] solves for the potential between the blades, discontinuities in potential representing the shed vorticity. Namba[30] considers pressure dipoles on the blade surfaces, and Nally and Hawthorne[31] consider shear flows through a cascade, solving the equations for potential derived by Honda.[32] These analyses (essentially singularity analyses) are mathematically very complex and the amount of computation involved is substantial. The detailed solutions do, however, show us the true complexity of the flow pattern and serve as a target for the two-dimensional methods reviewed. If all the terms are included in the meridional and blade–blade equations, the two methods of analysis should be in agreement, provided the same pitch (and time) averaging procedure is used in deriving the meriodional equations and in reducing the three-dimensional singularity solution to two dimensions.

The answer to the question posed earlier is therefore that finite elements will find their place in the repertoire of turbomachine flow calculation methods if they, in competition with existing methods whose development seems unceasing, can overcome the limitations listed above. On the evidence so far, and finite element programs are still rare in this field, no particular advantage has been demonstrated. It is hoped that this review will stimulate workers on finite element methods to attempt solutions of the very complex internal flows that occur in turbomachines.

References

1. J. H. Horlock and H. Marsh, 'Flow models in turbomachines', *J. Mech. Eng. Sci.*, **13**, 5, 358–368 (1971).
2. W. Merchant and A. R. Collar, 'Flow of an ideal fluid past a cascade of blades', *Aero. Research Council R. & M. No. 1893* (1941).
3. J. P. Gostelow, 'Potential flow through cascades, extensions to an exact theory,' *Aero. Research Council Current Paper 808* (1964).
4. H. Schlichting and N. Scholz, 'Uber die theoritsche Berechnung der Stromungsverluste eines ebenen Schaufelgitters', *Ing.-Arch.*, Bd. XIX, Heft 1 (1951).
5. E. Martensen, 'Calculation of pressure distribution over profiles in cascade in two dimensional potential flow by means of a Fredholm integral equation', *Arch. for Rat. Mech. and Anal.*, **3**, 3, 325 (1959).
6. D. H. Wilkinson, 'A numerical solution of the analysis and design problems for the flow past one or more aerofoils in cascade', *Aero. Research Council. R. & M. No.* 3545 (1968).
7. J. D. Stanitz, 'Design of two-dimensional channels with prescribed velocity distribution along the channel walls', *N.A.C.A. Tech. Notes 2595 and 2493* (1952).
8. H. W. Emmons, 'The theoretical flow of an ideal frictionless adiabatic perfect gas inside of a two-dimensional hyperbolic nozzle', *N.A.C.A. Tech. Note 1003* (1946).
9. D. J. L. Smith and D. H. Frost, 'Calculation of the flow past turbomachine blades', *Proc. I. Mech. E.*, **184**, 3G, 219–233 (1970).
10. T. Katsanis, 'Use of arbitrary quasi-orthogonals for calculating the flow distribution on a blade-to-blade surface in a turbomachine', *N.A.S.A.*, *T.N. D-2809* (1965).
11. D. H. Wilkinson, 'Stability, convergence and accuracy of two-dimensional streamline curvature methods using quasi-orthogonals', *Iroc. I. Mech. E.*, **184**, 3G, 108–119 (1970).
12. D. H. Wilkinson, 'Calculation of blade-to-blade flow in a turbomachine by streamline curvature', *Aero. Research Council R. & M. No. 3704* (1970).
13. J. P. Bindon and A. D. Carmichael, 'Streamline curvature analysis of compressible and high Mach number cascade flows', *J. Mech. Eng. Sci.*, **13**, 5, 344–357 (1971).
14. R. A. Novak, 'Streamline curvature computation procedures for fluid flow problems', *Am. Soc. Mech. Engrs. J. Eng. for Power*, **89**, A, 478–490 (1967).
15. J. D. Denton, 'A computer program for steam turbine performance prediction', *A.R.C. 34 229—Turbo 248* (1972).

16. Chung Hua Wu, 'A general theory of three-dimensional flow in subsonic and supersonic turbomachines of axial, radial and mixed flow types', *N.A.C.A., T.N. 2604* (1952).
17. H. Perkins, 'The analysis of steady flow through turbomachines', *G.E.C. Power Engineering Internal Report W/M (3C)*, p. 1641 (1970).
18. H. Marsh, 'A digital computer program for the through-flow fluid mechanics in an arbitrary turbomachine using a matrix method', *Aero. Research Council. R. & M. No. 3509* (1968).
19. W. R. Hawthorne and R. A. Novak, 'The aerodynamics of turbomachinery', *Annual Review of Fluid Mechanics*, **1**, 341–366 (1969).
20. D. Hobson, 'A hodograph method for the design of transonic turbine blades', *Aero. Research Council Report No. 34 131*.
21. E. M. Curtis, M. F. Hutton and D. H. Wilkinson, 'Theoretical and experimental work on two-dimensional turbine cascades with supersonic outlet flow', *Proc. I. Mech. E., Warwick Conference* (1973).
22. H. R. M. Craig and G. Hobson, 'Development of long-last-stage turbine blades', *G.E.C. Journal of Sci. and Technology*, **40**, 2 (1973).
23. H. Marsh and H. Merryweather, 'The calculation of subsonic and supersonic flows in nozzles', *Proc. I. Mech. E., Salford Conference on Internal Flow, Paper 22* (1971).
24. P. W. McDonald, 'The computation of transonic flow through two-dimensional gas turbine cascades', *Am. Soc. Mech. Engrs., Paper 71-GT-89* (1971).
25. R. W. MacCormack, 'The effect of viscosity in hypervelocity impact cratering', *A.I.A.A. paper No. 69-345, A.I.A.A. Hypervelocity Impact Conference, Cincinatti, Ohio* (1969).
26. S. Gopalakrishnan and R. Bozzola, 'A numerical technique for the calculation of transonic flows in turbomachinery cascades', *Am. Soc. Mech. Engrs. Paper 71-GT-42* (1971).
27. D. A. Oliver and P. Sparis, 'A computational study of three-dimensional transonic shear flow in turbomachinery cascades', *A.I.A.A. Paper 71-83, A.I.A.A. 9th Aerospace Sciences Meeting* (1971).
28. D. S. Thompson, 'Finite element analysis of the flow through a cascade of aerofoils', *Aero Research Council Report No. 34 412* (1973).
29. J. E. McCune and O. Okurounmu, 'Three-dimensional flow in transonic axial compressor blade rows fluid mechanics and design of turbomachines', *N.A.S.A., S.P. 304* (1973).
30. M. Namba, 'Lifting surface theory for a cascade of blades in subsonic shear flow', *J. Fluid Mech.*, **36**, 735–757 (1969).
31. M. C. Nally and W. R. Hawthorne, 'A numerical solution for shear flow through a cascade of aerofoils', *N.E.L. Report No. 432*, Ministry of Technology (1969).
32. M. Honda, 'Theory of shear flow in a cascade', *Proc. Roy. Soc.*, A 45–69 (1961).

Chapter 9

A Dual Perturbation Expansion and Variational Solution for Compressible Flows Using Finite Elements

G. F. Carey

9.1 Introduction

The recent interest in vertical- or short-take-off-and-landing (V/STOL) aircraft, and in subsonic flows for architectural structures and advanced modes of rapid terrestrial and maritime transportation, has led to a renaissance in subsonic aerodynamics. A need is now evident for more precise determination of flows than has previously been warranted. Computational techniques have become established as an adjunct and possible alternative to the traditional wind-tunnel experiment and the advantages of the computer in its new experimental role are apparent. The concurrent evolution of numerical methods for partial differential equations and computer technology have strengthened the field of numerical fluid dynamics. Since existing finite difference techniques are better-developed and simpler than their finite element counterparts, it is necessary to justify a serious practical interest in the latter. The practical value of the finite element method centres on the ease of treatment of boundary shape, an improved computational speed induced by the structure of the variational problem, and an experienced faculty for flexible general programming.

The fundamental question of the existence and construction of an equivalent variational principle is the first important issue that arises in consideration of general flows. Of course, an equivalent variational problem can always be posed as a least squares residual minimization. However, under appropriate conditions of self-adjointness, the variational statement involves lower-order integrand derivatives than the corresponding Euler equation. In this analysis variational functionals of this type are termed classical or low-order variational functionals. For those problems admitting this convenient variational formulation, and particularly for linear problems, these techniques have, in a relatively short period, essentially superseded existing finite difference approximate methods. Even when a low-order variational functional cannot be constructed, the method of weighted residuals may be extended as Galerkin, least squares and collocation procedures on finite elements. This admits a broader description of the theory and range of applications. The principal disadvantages of these residual techniques are the higher order differentiability requirements on the finite element basis functions, and inferior computational characteristics.

In this chapter a concise discussion of the variational compressible flow situation is first presented to indicate the scope and degree of difficulty of the general problem. The appropriate variational principles for various flow types are examined as energy functionals of Hamilton's principle and, mathematically, using Fréchet derivatives.

Several viable alternatives are possible for the inviscid irrotational flows then considered: weighted residual methods or a non-linear variational functional induce non-linear algebraic problems. Another scheme separates the non-linear and linear operator terms. Then, beginning with incompressible flow (non-linear terms zero), the problem is replaced by an iterative sequence of linear problems with non-linear terms now simply functions of the previously computed solution iterate. A better approach is to introduce perturbation expansions in terms of the non-dimensional Mach number. The non-linear partial differential equation generates a sequence of linear partial differential equations. Mathematically this can be related to the previous linear iteration but produces a successive ordering of correction terms and exhibits the Mach number dependence of the iteration.

The ensuing analysis develops a finite element variational method in conjunction with such a perturbation technique for these compressible flows. Exploitation of the perturbation relations and the resulting matrix properties allow an efficient and very fast algorithm for both lifting and non-lifting flows.

9.2 Variational principles as energy functionals

The well-known variational principles of particle mechanics are expressed as the Euler–Lagrange theory and more broadly as Hamilton's principle. For a conservative discrete dynamical system, Hamilton's principle is to make stationary the functional $\int_{t_0}^{t_1} (T - P)\,\mathrm{d}t$, where T is kinetic energy and P is potential energy. The integrand $L = T - P$ is called the Lagrangian. An associated important functional is the Hamiltonian $H = \int_{t_0}^{t_1} (T + P)\,\mathrm{d}t$, the total energy of the system. This latter functional leads to the influential dynamical procedures associated with Hamilton's equations, and Hamilton-Jacobi theory (and the complete integral), for determining the motion of a system. Analysts familiar with Hamilton's principle in dynamics and the analogous virtual work expressions in elasticity, perceive that the functional concerned is typically an energy integral. It is natural to expect, by standard continuum reasoning, that the analogous situation occurs in fluid mechanics. The finite element method is widely used in a variational context based upon these energy principles of equilibrium elasticity theory and dynamics; but the hydrodynamic counterpart is more complicated and intuitively less tangible.

The generalization of the Hamiltonian considers non-conservative systems (essentially the question is that of existence of a potential). The functional is path-dependent and the Hamiltonian is redefined for the discrete dynamics application to $H = \sum_i p_i \dot{q}_i - L$ where p_i and $\dot{q}_i$ are generalized momenta and velocities. There is an extensive literature on the mathematics of particle dynamics using the Hamiltonian. However, dissipative systems in fluid mechanics have not received an equally thorough development from this standpoint. Viscous dissipative problems are the most direct example. The questions of existence and construction of functionals for these flows have been rigorously examined only recently using Fréchet differentials.[8]

Pursuing the energy functional apparoach, the simplest problem corresponds to potential flow. The partial differential equation for incompressible, inviscid, irrotational flow reduces to Laplace's equation, written $\rho\phi_{,ii} = 0$, with potential ϕ and density ρ. The corresponding functional is obtained on integration by parts of $\int_\Omega \rho\phi_{,ii}\,\delta\phi\,\mathrm{d}\Omega$ and is the Dirichlet integral, $\int_\Omega \rho/2(\phi_{,i})^2\,\mathrm{d}\Omega$ or $\int_\Omega \rho/2q^2\,\mathrm{d}\Omega$ where q is the velocity magnitude. The density ρ is included for convenience of interpretation and the functional is clearly the kinetic energy. Flows where work is done by an external force are easily accommodated by inclusion of the corresponding potential energy in the functional in precisely the same manner as the virtual work expressions of elasticity.[1]

The natural extension is to steady compressible flows and the functional to be stationary comprises the sum of kinetic energy and half the pressure. In the ensuing analysis attention will be focused on steady compressible isentropic flows. Prior to this the general flow criterion is stated.

The Navier–Stokes equation may be written in component form as

$$\frac{\partial u_i}{\partial t} + u_{i,j}u_j = -\frac{1}{\rho}p_{,i} + \nu u_{i,jj} + \tfrac{1}{3}\nu u_{j,ji}$$

where $\mathbf{u}$, p, ρ and ν are the velocity, pressure, density and viscosity, respectively. In general this will not admit an equivalent low-order variational statement of the energy type previously encountered and the solution must be sought in the extended domain of the problem and its adjoint form.

In his elegant treatise on mathematical principles of classical fluid mechanics, Serrin remarks (Reference 2, p. 149): 'It is *likely* that one can derive the equations of motion of a viscous fluid by a variational argument', (emphasis added); he later notes that the energy equation must be inserted as a *side condition* without which 'it does not appear possible to obtain the equations of motion of a viscous fluid from Hamilton's principle.' The symmetry condition of Fréchet derivatives for the differential operator has since shown under precisely what limitations a variational principle exists.[3] In particular, difficulties arise for applications in which inertial and viscous terms arise together. Variational principles can be constructed using Fréchet derivatives for flows involving one or another of these features, but not both; that is, only if $\mathbf{u} \times (\nabla \times \mathbf{u}) = \mathbf{0}$ or $\mathbf{u} \cdot \nabla\mathbf{u} = \mathbf{0}$.

Even for ideal fluids, though the existence of a variational form follows from the general argument, the Fréchet construction is dependent on manipulation to an appropriate form of the governing differential equation and may not be straightforward. This amounts to a requirement of self-adjointness for linear problems and analogously a symmetric Fréchet derivative for non-linear problems. It may happen that the partial differential equation and associated side conditions do not permit such a symmetry. This is in fact the case for the simultaneous occurrence of inertial and viscous effects above and, in turbomachines, for compressible rotational through-flows that are not isentropic,[4] to mention just two such instances. For linear problems, integrating factor techniques can be applied to generate self-adjoint equations, but this approach is not so readily applicable to non-linear differential equations.

Attention is now directed to compressible flows, the main thrust of this article. Even for compressible ideal flows the variational principles have been constructed only recently. We begin with stationary principles for time-dependent compressible flows. Again there is an inherent disadvantage: the potential-energy functional is not convex (not positive definite) and numerical stability need not follow automatically, as it does in the positive-definite case.

Introducing the virtual work W due to external forces, then Hamilton's principle for a mechanical system whose energy is completely known is that $\int_{t_0}^{t_1} \int_\Omega (\delta L + \delta W)\, \mathrm{d}\Omega\, \mathrm{d}t = 0$. Writing $L = T - P$ with kinetic energy $T = \frac{1}{2}\rho q^2$ and potential energy $P = \rho U$ where $U(\rho, s)$ is the internal energy and s is entropy, then

$$\Pi = \int_{t_0}^{t_1} \int_\Omega \{\tfrac{1}{2}\rho q^2 - \rho(U + W)\}\, \mathrm{d}\Omega\, \mathrm{d}t$$

is the functional to be made stationary. This statement is due to Herivel[5] and the basic assumption is that $U = U(\rho, s)$ with specific entropy s of each fluid particle constant ($\mathrm{d}s/\mathrm{d}t = 0$) through the motion. In the functional, variations are to be taken with respect to velocity, density and entropy, subject to the constraints

$$\frac{\partial \rho}{\partial t} + (\rho u_i)_{,i} = 0 \quad \text{continuity}; \qquad \frac{\mathrm{d}s}{\mathrm{d}t} = 0 \quad \text{conservation of energy}$$

These constraint conditions can be applied by restricting the trial functions *a priori* to satisfy them (rejected on practical grounds for general finite element applications) or, alternatively, they can be incorporated in an integral sense through Lagrange multipliers, λ_i. The functional then becomes

$$\Pi = \int_{t_0}^{t_1} \int_{\Omega} \left\{ \tfrac{1}{2}\rho q^2 - \rho(U + W) + \lambda_1(\mathbf{x}, t)\left(\frac{\partial \rho}{\partial t} + (\rho u_i)_{,i}\right) - \lambda_2(\mathbf{x}, t)\rho \frac{\mathrm{d}s}{\mathrm{d}t} \right\} \mathrm{d}\Omega\, \mathrm{d}t$$

Taking variations returns the two constraint conditions in addition to the equation of motion $\rho\, \partial u_i/\partial t = \rho f_i - p_{,i}$ where $\{f_i\}$ are external forces and p is the pressure. Herivel's functional aims at determining the equations of motion of a perfect fluid on a variational principle of spatial (Eulerian) type. In reality it determines a Lagrangian (initial reference state) variational principle. The extremals yield solutions for irrotational flows only. That is, with this functional the variational condition determines an isentropic flow solution as an extremal only if the flow is also irrotational. Herivel's functional was finally modified by Lin[6] to include rotational flows by appending the additional constraint of Lagrangian coordinates. A Lagrangian coordinate frame requires $\mathrm{d}\mathbf{a}/\mathrm{d}t = 0$, where $\mathbf{a}(\mathbf{x}, t)$ are the initial coordinates of the particle with current position $\mathbf{x}$ at time t. Then by continued Lagrange multiplier methods to include the reference frame constraint, the Herivel–Lin functional is simply

$$\Pi = \int_t \int_{\Omega} \left\{ \tfrac{1}{2}\rho q^2 - \rho(U + W) + \lambda_1(\mathbf{x}, t)\left(\frac{\partial \rho}{\partial t} + (\rho u_i)_{,i}\right) - \lambda_2(\mathbf{x}, t)\rho \frac{\mathrm{d}s}{\mathrm{d}t} - \boldsymbol{\lambda}_3(\mathbf{x}, t) \cdot \frac{\mathrm{d}\mathbf{a}}{\mathrm{d}t} \right\} \mathrm{d}\Omega\, \mathrm{d}t$$

Taking variations with $\mathbf{u}$, ρ, s and $\mathbf{a}$ now unconstrained, the stationary functional requirement yields four subsidiary equations. Manipulating these equations and applying the requirements $\mathrm{d}s/\mathrm{d}t = 0$ and $\mathrm{d}\mathbf{a}/\mathrm{d}t = 0$, yields the governing equation and so verifies that the stationary curve of this functional determines the equation of motion of the flow.

Finally, steady, compressible, irrotational flows are considered. On the basis of the underlying energy and Hamiltonian motivation that led to the more general functional of Herivel in the preceding discussion, variational principles may again be devised. The principle of Herivel and Lin is a stationary principle, whereas the principles for these more restrictive flows are maximum and minimum principles.

For constant entropy, the internal energy U and thermodynamic pressure p are related by $\mathrm{d}U = -p\,\mathrm{d}V = p/\rho^2\,\mathrm{d}\rho$. With a conservative force and no viscous effects, the equation of motion reduces to $\frac{1}{2}q^2 + \mathrm{d}/\mathrm{d}\rho(\rho U) = 0$. The equivalent variational principle is called the Bateman–Dirichlet principle[2,7] and is to make stationary the functional

$$\Pi_1(\phi) = \int_\Omega p \, d\Omega + \int_\Gamma \phi f \, d\Gamma$$

subject to the constraint conditions: $p = \rho^2 \, dU/d\rho$ (thermodynamic relation); $\frac{1}{2}q^2 + d/d\rho(\rho U) = 0$ (equation of motion); $\mathbf{u} = \phi_{,i}$ (irrotational flow). The Euler equation and natural boundary conditions follow from $\delta\Pi_1 = 0$ and application of the divergence theorem as $(\rho u_i)_{,i} = 0$ in Ω and $\rho n_i \phi_{,i} = f$ on Γ; the stationary functional is a maximum. The complementary (minimal) principle may be similarly constructed. This is termed the **Bateman–Kelvin principle**[2,7] and makes stationary the functional

$$\Pi_2(\mathbf{u}) = \int_\Omega (p + \rho q^2) \, d\Omega$$

among velocity fields satisfying $p = \rho^2 \, dU/d\rho$, $\frac{1}{2}q^2 + d/d\rho(\rho U) = 0$, $(\rho u_i)_{,i} = 0$ in Ω and $\rho n_i u_i = f$ on Γ.

These variational principles then provide upper and lower bounds on the sum of pressure and twice the kinetic energy, $\Pi(\phi, \mathbf{u}) = \int_\Omega (p + \rho u_i u_i) \, d\Omega$, so $\Pi_1(\phi) \leqslant \Pi(\phi, \mathbf{u}) \leqslant \Pi_2(\mathbf{u})$. They apply to subsonic flows and, if it exists, an extremal is unique and provides the flow-field solution. However, there may be no extremal flow-field corresponding to an irrotational subsonic flow and satisfying the given boundary conditions. The variational functional may accordingly be modified to accommodate this situation. The restriction to subsonic flows is a necessary condition for an extremal. Details of proof are given by Serrin (Reference 2, p. 205) together with other subsonic functionals giving rotational isoenergetic and rotational isentropic compressible flows.

It is apparent from the preceding discussion that variational fluid mechanics provides a complex subject area for even moderately restricted flows. Equally evident is the fact that there is no simple functional or class of functionals that will describe the variety of flow situations that frequently occur and, most commonly, classes of flows must be considered individually.

From a finite element point of view, the compressible flow problem is complicated, not only by the form of the variational principle, but also by the number of dependent variables and constraint side conditions. The subsidiary conditions almost always must be included in the variational functional by Lagrange multiplier techniques, rather than constraints the approximate solution satisfies *a priori*. Within the finite element calculations the multipliers are unknown functions of position and must be expanded on the element basis, increasing the algebraic problem complexity and size. Alternative strategies imbed side conditions as subsidiary equations to be satisfied iteratively.[4] Ideally, a simple variational functional with single dependent variable of interest, say the stream function, is sought. The determining relations for pressure, density, entropy and velocity will generally prohibit explicit expression of these quantities in terms of the stream function and its derivatives, except in very special applications or in conjection with special techniques, such as the asymptotic method presented here.

Construction of variational methods to more difficult transonic flows, supersonic flows and flows with shocks and contact surface discontinuities can be attempted, by using Fréchet differential analysis or energy arguments, but are not easily formulated.

9.3 The compressible irrotational flow problem

Consider compressible, non-viscous, irrotational flow in two dimensions. Let $q = u + iv$ be the complex velocity with $\rho u = \rho_\infty U_\infty \psi_y$ and $\rho v = -\rho_\infty U_\infty \psi_x$. Stream function $\psi(x, y)$ is the primary unknown flow variable, $\rho(x, y)$ is the density, and ρ_∞, U_∞ are remote density and uniform velocity.

The irrotational flow requirement implies $\Delta\psi = \nabla\psi \,.\, (1/\rho)\nabla\rho$.

Applying the Bernoulli relation ($\frac{1}{2}q^2 + \int \mathrm{d}p/\rho$ is constant on a stream line), together with the adiabatic equation of state $p = \kappa\rho^\gamma$, yields

$$\frac{1}{\rho}\nabla\rho = -\frac{M_\infty^2}{U_\infty^2}\nabla(\tfrac{1}{2}q^2)\Bigg/\left\{1 + M_\infty^2\left(\frac{\gamma-1}{2}\right)\left(1 - \frac{q^2}{U_\infty^2}\right)\right\}$$

where M_∞ is the incident Mach number.

Combining these two results, the governing non-linear partial differential equation for the stream function is

$$\Delta\psi\left\{1 + M_\infty^2\left(\frac{\gamma-1}{2}\right)\left(1 - \frac{u^2}{U_\infty^2} - \frac{v^2}{U_\infty^2}\right)\right\} = -M_\infty^2\left[\nabla\psi \,.\, \nabla\left\{\frac{1}{2}\left(\frac{u^2}{U_\infty^2} + \frac{v^2}{U_\infty^2}\right)\right\}\right] \tag{9.1}$$

More precisely, the equation is termed quasilinear as the second order derivatives enter in a linear fashion. The equation is elliptic if the flow is subsonic and hyperbolic if the flow is supersonic. At points where the local speed equals the local speed of sound the equation is parabolic.

This analysis determines the governing non-linear equation in terms of the stream function and density. An analogous procedure using the continuity equation yields the equivalent expression in terms of the potential function $\phi(x, y)$ alone,

$$\phi_{,ii}\left\{1 + M_\infty^2\left(\frac{\gamma-1}{2}\right)(1 - (\phi_{,j})^2)\right\} - M_\infty^2\phi_{,i}\phi_{,j}\phi_{,ij} = 0$$

where tensor subscript notation is used for later algebraic convenience. For simplicity, since density variations are not involved, but without loss of generality, the problem of constructing a variational functional will be examined with this potential form.

Writing $N(\phi) = 0$ for the given partial differential equation, direct divergence manipulation of $\delta\bar{\Pi} = \int_\Omega N(\phi)\,\delta\phi\,d\Omega$ will not yield the desired functional so the sufficiency of operator N is tested by Fréchet differentiation.[12] The Fréchet derivative N'_ϕ of operator N in the direction η is defined by

$$DN = N'_\phi\eta = \lim_{\varepsilon\to 0}\frac{N(\phi+\varepsilon\eta) - N(\phi)}{\varepsilon} = \left[\frac{\partial}{\partial\varepsilon}\{N(\phi+\varepsilon\eta)\}\right]_{\varepsilon=0}$$

where DN is termed the Fréchet differential. Substituting for N,

$$N'_\phi\eta = \left[\left\{1 + M_\infty^2\left(\frac{\gamma-1}{2}\right)(1 - (\phi_{,j})^2)\right\}(\ )_{,ii} - 2M_\infty^2\left\{\left(\frac{\gamma-1}{2}\right)\phi_{,i}\phi_{,jj}\right.\right.$$
$$\left.\left. + \phi_{,j}\phi_{,ij}\right\}(\ )_{,i} - M_\infty^2\phi_{,i}\phi_{,j}(\ )_{,ij}\right]\eta$$

where $[(\ )_{,ij}]\eta \equiv \eta_{,ij}$ and symbolically separates the operator N'_ϕ.

Symmetry of N'_ϕ is a sufficient (but not necessary) condition for a functional to exist for operator N. The functional then can be constructed directly as $\Pi(\phi) = \int_\Omega \phi \int_0^1 N(\lambda\phi)\,d\lambda\,d\Omega$. To determine possible symmetry of N'_ϕ, the adjoint operator $\tilde{N}'_\phi$ is formed by integration of parts of $(\xi, N'_\phi\eta)$, and the symmetry criterion $N'_\phi = \tilde{N}'_\phi$ is tested. Evaluating $\tilde{N}'_\phi$,

$$(\xi, N'_\phi\eta) = \int_\Omega \xi\left[\left\{1 + M_\infty^2\left(\frac{\gamma-1}{2}\right)(1 - (\phi_{,j})^2)\right\}\eta_{,ii} - 2M_\infty^2\left\{\left(\frac{\gamma-1}{2}\right)\phi_{,i}\phi_{,jj}\right.\right.$$
$$\left.\left. + \phi_{,j}\phi_{,ij}\right\}\eta_{,i} - M_\infty^2\phi_{,i}\phi_{,j}\eta_{,ij}\right]d\Omega$$

Collecting coefficients of η derivatives,

$$(\xi, N'_\phi\eta) = \int_\Omega \xi\left[\left\{\left[1 + M_\infty^2\left(\frac{\gamma-1}{2}\right)(1 - (\phi_{,j})^2)\right]\delta_{ij} - M_\infty^2\phi_{,i}\phi_{,j}\right\}\eta_{,ij}\right.$$
$$\left. - 2M_\infty^2\left\{\left(\frac{\gamma-1}{2}\right)\phi_{,i}\phi_{,jj} + \phi_{,j}\phi_{,ij}\right\}\eta_{,i}\right]d\Omega$$

Divergence manipulations yield $(\xi, N'_\phi\eta) = (\eta, \tilde{N}'_\phi\xi)$ + boundary terms, with

$$(\eta, \tilde{N}'_\phi\xi) = \int_\Omega \eta\left[\left\{\left[1 + M_\infty^2\left(\frac{\gamma-1}{2}\right)(1 - (\phi_{,j})^2)\right]\delta_{ij} - M_\infty^2\phi_{,i}\phi_{,j}\right\}\xi_{,ij}\right.$$
$$- 2M_\infty^2\left\{(\gamma-1)\phi_{,j}\phi_{,ji} - \frac{\gamma-2}{2}\phi_{,i}\phi_{,jj}\right\}\xi_{,i}$$
$$\left. - 2M_\infty^2\left\{\left(\frac{\gamma-2}{2}\right)(\phi_{,j}\phi_{,ji} - \phi_{,i}\phi_{,jj})_{,i}\right\}\xi\right]d\Omega$$

and on term by term comparison of N'_ϕ and $\tilde{N}'_\phi$, symmetry holds if and only if we have the relation $(\phi_{,ii})^2 = (\phi_{,ij})^2$ or equivalently $(\phi_{xy})^2 = \phi_{xx}\phi_{yy}$, and does not distinguish any of the usual flow categories.

Since N'_ϕ is not symmetric, then the equation in its present form does not allow the Fréchet construction of a variational functional. There remain several recourses. A trial approach using undetermined coefficients in the 'plateau problem' functional will yield the desired functional. Alternatively, a different form of the operator N can be tried and this is carried out successfully later in the chapter. The variational functional can be formed in the space of the operator N and its adjoint $\tilde{N}$ so that Π is then $(\zeta, \tilde{N}\phi)$ for functions ϕ and ζ satisfying the associated boundary conditions. The detractions of this approach from a finite element viewpoint are evident: the functional requires element expressions in two unknown functions $\phi(x, y)$ and $\zeta(x, y)$; the element basis functions must be differentiable to an order commensurate with $\tilde{N}$, so that the elements are necessarily high-order elements. The remaining alternative is to formulate a finite element weighted-residual method and the detractions of this approach have been stated earlier.

The occurrence of the non-dimensional Mach number as a parameter suggests that, particularly for subsonic flows, a perturbation procedure might be worth while. Asymptotic expansions of this type have proved useful for similar analytical solutions in classical fluid mechanics.[9]

9.4 Perturbation and variational analysis

Let $\varepsilon = M_\infty^2$ denote the perturbation parameter. Introduce expansions $\psi(x, y) = \psi^{(0)}(x, y) + \varepsilon\psi^{(1)}(x, y) + \varepsilon^2\psi^{(2)}(x, y) + O(\varepsilon^3)$ and $\rho(x, y) = \rho_\infty(1 + \varepsilon\rho^{(1)}(x, y) + \varepsilon^2\rho^{(2)}(x, y) + O(\varepsilon^3))$ into Equation 9.1. Collecting terms, the sequence of perturbation equations is

$$
\begin{aligned}
\varepsilon^0: \quad & \psi^{(0)}_{,ii} = 0 \\
\varepsilon^1: \quad & \psi^{(1)}_{,ii} = -\psi^{(0)}_{,i}\psi^{(0)}_{,j}\psi^{(0)}_{,ij} \\
\varepsilon^2: \quad & \psi^{(2)}_{,ii} = -\left(\frac{\gamma - 1}{2}\right)(1 - \psi^{(0)2}_{,k})\psi^{(1)}_{,ii} - \psi^{(0)}_{,j}(2\psi^{(1)}_{,i}\psi^{(0)}_{,ij} + \psi^{(0)}_{,i}\psi^{(1)}_{,ij}) \\
& \qquad - \psi^{(0)}_{,j}(-\rho^{(1)}_{,i}\psi^{(0)2}_{,i} - 2\rho^{(1)}\psi^{(0)}_{,i}\psi^{(0)}_{,ij}), \quad \text{with } \rho_1 = \tfrac{1}{2}(1 - \psi^{(0)2}_{,i})
\end{aligned}
\tag{9.2}
$$

and so forth. The leading term $\psi^{(0)}(x, y)$ is the incompressible flow solution and satisfies the prescribed boundary conditions. The higher-order terms $\psi^{(1)}(x, y), \psi^{(2)}(x, y), \ldots$, satisfy non-homogeneous linear equations of elliptic type and homogeneous boundary conditions. The general equation for $\psi^{(n)}(x, y)$ involves all prior solution functions $\psi^{(0)}(x, y), \psi^{(1)}(x, y), \ldots, \psi^{(n-1)}(x, y)$ in the non-homogeneous expression, and with increasing complexity as n increases.

Since the operators in Equation 9.2 are self-adjoint their equivalent functionals can be constructed directly. In particular, the Laplacian generates the Dirichlet integral as the leading convex quadratic functional in each perturbation problem. The remaining portions are linear functionals corresponding to the non-homogeneous terms and dependent on the previously calculated lower-order, perturbation solutions.

For the zero-order perturbation term $\psi^{(0)}(x, y)$, the equivalent variational functional is

$$\Pi^{(0)} = \int_{\Omega} \tfrac{1}{2}\psi^{(0)}_{,i}\psi^{(0)}_{,i}\,\mathrm{d}\Omega \tag{9.3}$$

Taking variations, $\delta\Pi^{(0)} = 0$ yields the Euler equation $\Delta\psi^{(0)} = 0$, with $\partial\psi^{(0)}/\partial n = 0$ as natural boundary condition. As derivatives in the functional $\Pi^{(0)}$ are at most first order, the three-node triangle with linear basis functions is applicable.

The associated functional for $\psi^{(1)}(x, y)$ is

$$\Pi^{(1)} = \int_{\Omega} (\tfrac{1}{2}\psi^{(1)}_{,i}\psi^{(1)}_{,i} - \psi^{(0)}_{,i}\psi^{(0)}_{,j}\psi^{(0)}_{,ij}\psi^{(1)})\,\mathrm{d}\Omega \tag{9.4}$$

Observe that the occurrence of $\psi^{(0)}_{,ij}$ in the linear functional portion requires that $\psi^{(0)}(x, y)$, the incompressible solution, be *twice* differentiable on an element. If we hope to use the same idealization for incompressible solution and compressibility correction, this suggests that a higher-order element, say quadratic element basis, is necessary for the previous incompressible solution. Analytically, computationally and practically, such a procedure is not desirable. It is avoided in the following crucial step of the analysis by divergence manipulation of the linear functional.

Let Π^* denote the linear functional and manipulate to divergence form as follows:

$$\begin{aligned}
\Pi^* &= \int_{\Omega} (\psi^{(0)}_{,i}\psi^{(0)}_{,j}\psi^{(0)}_{,i}\psi^{(1)})_{,j}\,\mathrm{d}\Omega - \int_{\Omega} \{(\psi^{(0)}_{,i}\psi^{(0)}_{,j}\psi^{(1)})_{,j}\psi^{(0)}_{,i}\}\,\mathrm{d}\Omega \\
&= \int_{\Omega} (\psi^{(0)}_{,i}\psi^{(0)}_{,j}\phi^{(0)}_{,i}\psi^{(1)})_{,j}\,\mathrm{d}\Omega - \int_{\Omega} \psi^{(0)}_{,ij}\psi^{(0)}_{,j}\psi^{(0)}_{,i}\psi^{(1)}\,\mathrm{d}\Omega \\
&\quad - \int_{\Omega} \psi^{(0)}_{,i}\psi^{(0)}_{,i}\psi^{(0)}_{,jj}\psi^{(1)}\,\mathrm{d}\Omega - \int_{\Omega} \psi^{(0)}_{,i}\psi^{(0)}_{,j}\psi^{(1)}_{,j}\psi^{(0)}_{,i}\,\mathrm{d}\Omega
\end{aligned}$$

Transposing the second integral, $\Pi^* = \int_{\Omega} \psi^{(0)}_{,ij}\psi^{(0)}_{,j}\psi^{(0)}_{,i}\psi^{(1)}\,\mathrm{d}\Omega$, and eliminating the third integral since $\psi^{(0)}_{,jj} = 0$ in Ω,

$$2\Pi^* = \int_{\Omega} (\psi^{(0)}_{,i}\psi^{(0)}_{,j}\psi^{(0)}_{,i}\psi^{(1)})_{,j}\,\mathrm{d}\Omega - \int_{\Omega} \psi^{(0)}_{,i}\psi^{(0)}_{,j}\psi^{(1)}_{,j}\psi^{(0)}_{,i}\,\mathrm{d}\Omega$$

Applying Gauss's divergence theorem,

$$\Pi^* = \tfrac{1}{2}\int_\Gamma \psi_{,i}^{(0)}\psi_{,i}^{(0)}\frac{\partial\psi^{(0)}}{\partial n}\psi^{(1)}\,\mathrm{d}\Gamma - \tfrac{1}{2}\int_\Omega \psi_{,i}^{(0)}\psi_{,i}^{(0)}\psi_{,j}^{(0)}\psi_{,j}^{(1)}\,\mathrm{d}\Omega$$

and

$$\Pi^{(1)} = \tfrac{1}{2}\int_\Omega \psi_{,i}^{(1)}\psi_{,i}^{(1)}\,\mathrm{d}\Omega - \int_\Gamma \psi_{,i}^{(0)}\psi_{,i}^{(0)}\frac{\partial\psi^{(0)}}{\partial n}\psi^{(1)}\mathrm{d}\Gamma + \int_\Omega \psi_{,i}^{(0)}\psi_{,i}^{(0)}\psi_{,j}^{(0)}\psi_{,j}^{(1)}\,\mathrm{d}\Omega \tag{9.5}$$

is the reduced form of the functional for the second perturbation term $\psi^{(1)}(x, y)$. Taking variations of $\Pi^{(1)}$ in Equation 9.5 returns the original differential equation and natural boundary condition $\partial\psi^{(1)}/\partial n = 0$ on that part of the boundary where $\psi^{(1)}(x, y)$ is not prescribed. Since the integrand expressions in $\Pi^{(1)}$ now involve derivatives to at most first order, the three-node triangle with linear element basis functions is appropriate for both incompressible and compressible flow calculations. The same finite element idealization can be used for $\psi^{(0)}(x, y)$ and $\psi^{(1)}(x, y)$ calculations. As the leading quadratic functionals are the same for $\Pi^{(0)}$ and $\Pi^{(1)}$, finite element analysis generates the identical, positive definite, symmetric coefficient matrix for the two solution functions.

This derivative reduction procedure can be continued indefinitely to higher-order solution functionals at the expense of increasingly complicated divergence manipulations. For example, the linear functional contributions for $\psi^{(2)}(x, y)$ produce the reduced linear functional

$$\Pi^{**} = \int_\Omega \left\{ \left(\frac{3-\gamma}{4}\right)\psi_{,i}^{(0)2}\psi_{,j}^{(0)}\psi_{,j}^{(2)} + \left(\frac{10+\gamma}{8}\right)\psi_{,i}^{(0)2}\psi_{,k}^{(0)2}\psi_{,j}^{(0)}\psi_{,j}^{(2)} + \left(\tfrac{3}{2}\right)\psi_{,i}^{(0)2}\psi_{,j}^{(1)}\psi_{,j}^{(2)} + \psi_{,i}^{(0)}\psi_{,j}^{(0)}\psi_{,i}^{(1)}\psi_{,j}^{(2)} \right\} \mathrm{d}\Omega$$

9.5 Finite element formulation

The lowest-order finite element model is presented on a triangular element. Then piecewise interpolation is linear over each element, and solution values at the three vertex nodes define the element interpolant. For linear Lagrange interpolation on a triangular element,

$$\psi^{(0)}(x, y) = \mathbf{L}^{\mathrm{T}}\boldsymbol{\psi}^{(0)}, \qquad \psi^{(1)}(x, y) = \mathbf{L}^{\mathrm{T}}\boldsymbol{\psi}^{(1)} \tag{9.6}$$

where $\mathbf{L}(x, y)$ is the vector of Lagrange interpolants.

The variational problem for $\Pi^{(0)}(\boldsymbol{\psi}^{(0)})$ becomes, in the element basis with $\Pi_e^{(0)}$ the integral contribution of element e,

$$\delta\Pi^{(0)} = 0 = \delta \sum_e \Pi_e^{(0)} = \sum_e \delta\Pi_e^{(0)} = \sum_e \left(\frac{\partial \Pi_e^{(0)}}{\partial \psi_k^{(0)}} \delta\psi_k^{(0)} \right), \qquad \text{node } k$$

This expansion determines the complete difference equation $\sum_e \partial\Pi_e^{(0)}/\partial\psi_k^{(0)} = 0$ at node k in the idealization. Accumulating contributions sequentially by element, the final linear system for $\boldsymbol{\psi}^{(0)}$ is, for element area A_e,

$$\sum_e A_e(\mathbf{L}_x\mathbf{L}_x^{\mathrm{T}} + \mathbf{L}_y\mathbf{L}_y^{\mathrm{T}})\boldsymbol{\psi}_e^{(0)} = \sum_e (A\mathbf{M}\boldsymbol{\psi}^{(0)})_e = \mathbf{0} \tag{9.7}$$

In the finite element representation, only boundary elements will contribute to the surface integral in $\Pi^{(1)}$. The prescribed and natural boundary conditions on Γ make these contributions zero. Then redefine $\Pi^{(1)}$ over an arbitrary element e by

$$\Pi_e^{(0)} = \tfrac{1}{2}\int_{\Omega_e} \psi_{,i}^{(1)}\psi_{,i}^{(1)}\, \mathrm{d}\Omega + \tfrac{1}{2}\int_{\Omega_e} \psi_{,i}^{(0)}\psi_{,i}^{(0)}\psi_{,j}^{(0)}\psi_{,j}^{(1)}\, \mathrm{d}\Omega$$

with $\Pi^{(1)} = \sum_e \Pi_e^{(1)}$. A similar approach to $\Pi^{(1)}$ and $\psi^{(1)}(x, y)$, as that employed for $\Pi^{(0)}$ and $\psi^{(0)}(x, y)$, produces for $\boldsymbol{\psi}^{(1)}$ the finite element system

$$\sum_e \left\{ \frac{A}{2}(2\mathbf{M}\boldsymbol{\psi}^{(1)} + \mathbf{M}\boldsymbol{\psi}^{(0)}\boldsymbol{\psi}^{(0)\mathrm{T}}\mathbf{M}\boldsymbol{\psi}^{(0)}) \right\}_e = \mathbf{0} \tag{9.8}$$

The coefficient matrices in Equations 9.7 and 9.8 are symmetric, sparse, positive definite and, provided that the nodes are ordered judiciously, neatly banded. It follows that Cholesky's method is highly appropriate for algebraic solution. With the preceding formulations on a given discretization, the coefficient matrices for $\boldsymbol{\psi}^{(0)}$ and $\boldsymbol{\psi}^{(1)}$ are identical, and Cholesky triangular decomposition need be carried out for the incompressible solution alone. The solution for $\boldsymbol{\psi}^{(1)}$ resolves to inexpensive forward and backward substitution sweeps, using the previously decomposed triangular matrices.

Similarly, it holds that the solution for $\boldsymbol{\psi}^{(n)}$, $n = 2, 3, \ldots$ on the invariant idealization requires only recalculation of the reduced linear functional contribution, followed by fore and aft substitution sweeps. In view of this degeneration of the higher order solutions by Cholesky's method to simple substitution sweeps, the computational algorithm is extremely fast.

An alternative technique is to transpose the non-linear terms in the partial differential equation and iteratively solve the finite element variational problems for $\psi_{,ii}^{(p+1)} = f(\psi^p)$. This is mathematically equivalent but computationally inferior to the perturbation approach and does not utilize the perturbation relations to reduce the derivative order in the linear functionals arising. Moreover, it obscures the series dependence on M_∞^2 which characterizes the convergence of the iteration.

9.6 Results

Case 1. Rayleigh's problem: Fully-infinite compressible flow past a cylinder This problem is a 'chestnut' of classical theory. Its examination is merited as it admits an exact derivation of the perturbation solutions by complex variable methods and thus affords a reliable accuracy comparison.

The first quadrant of the flow field is modelled by a triangulated extensive annulus with outer radius = 8 × inner radius. A mixed boundary-value problem formulation is used with natural boundary condition automatically satisfied on the vertical centreline, $x = 0$ and uniform horizontal flow U_∞ on the outer radius. Contour plots exhibit the incompressible solution $\psi^{(0)}(x, y)$ and compressibility correction $\psi^{(1)}(x, y)$ in Figure 9.1. The contours agree

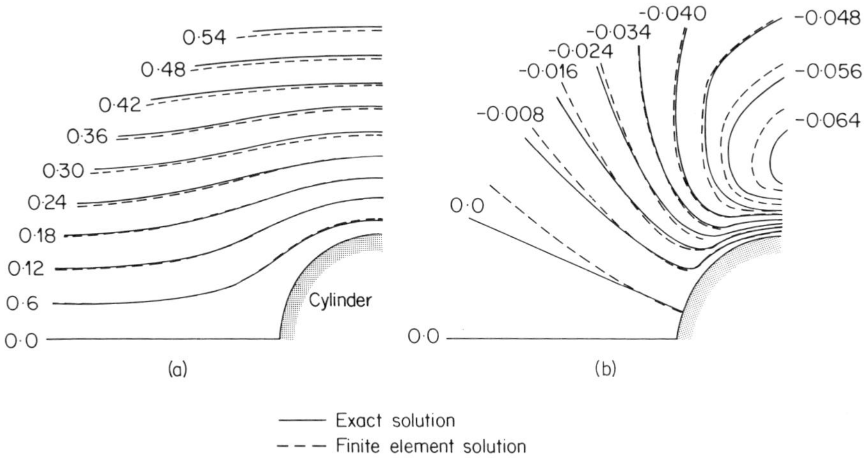

Figure 9.1 (a) Stream function contours of $\psi^{(0)}(x, y)$, the incompressible contribution in the vicinity of the cylinder. (b) First-order compressibility correction $\psi^{(1)}(x, y)$ to the stream function

closely with the exact two-term perturbation solution,

$$\psi(r, \theta) = \left(r - \frac{R^2}{r}\right) \sin \theta + M_\infty^2 \left\{ \left(-\frac{7}{12}\frac{R^2}{r} + \frac{R^4}{2r^3} + \frac{R^6}{12r^5} \right) \sin \theta + \left(\frac{R^2}{4r} - \frac{R^4}{4r^3} \right) \sin 3\theta \right\} + \mathrm{O}(M_\infty^4)$$

where $R = 2$ is the cylinder radius in this application. There is a slight shift in the $\psi^{(1)}(x, y)$ contours due to approximation as a finite domain. This error can be reduced by using a larger domain, or by including in the finite element variational problem an exterior basis function with the asymptotic far-field behaviour, that is, effectively incorporating the far-field solution discrepancy.

Case 2. Circular cylinder in a wind tunnel The problem considers steady uniform flow with remote velocity U_∞ past a cylinder of two units radius placed centrally between parallel walls eight units part. The first flow quadrant is idealized with uniform flow conditions on a remote upstream boundary. The incompressible term produces close agreement with the result of Martin.[11] Compressible stream function and velocity profiles agree reasonably with results of other less efficient numerical techniques.

A significant effect is the influence of wind tunnel interference on the lower critical Mach number, $M_{\infty cr}$, the value of M_∞ at which sonic velocity is attained on the cylinder. For fully infinite flow past a cylinder, $M_{\infty cr}$ is to third-order accuracy, 0·404. The wall influence is to reduce this critical value. Since $q/U_\infty = q^{(0)} + M_\infty^2 q^{(1)} + O(M_\infty^4)$, then for $c_{max} \sim c_\infty$,† the speed of sound in free stream,

$$\frac{q}{c_{max}} = \frac{U}{c_{max}}(q^{(0)} + M_\infty^2 q^{(1)} + O(M_\infty^4))$$

At the critical value $q/c_{max} = 1$ solving the cubic $1 = M_\infty(q^{(0)} + M_\infty^2 q^{(1)})$, with $q^{(0)} = 2{\cdot}32$ and $q^{(1)} = 6{\cdot}3$ from the finite element calculation, gives an estimate to $M_{\infty cr}$ of 0·332 and agrees reasonably with extrapolated experimental results (see Reference 10, p. 368).

9.7 The potential function formulation

It is slightly simpler algebraically to use the potential function rather than the stream function, as the relation between stream function and velocity is complicated by variations in density. The variational integrand derivative reductions hold in the same way as for the stream function analysis. Also, the treatment of prescribed and natural boundary conditions is similar, allowing a single quadrant finite element idealization for the symmetrical flow examples considered.

At the expense of slight algebraic complexity, the stream function formulation was developed, since a stream line depiction of the flow field is physically more satisfactory. More importantly, for generality it is better to work with the stream function as it alone is defined for rotational flows. Although the present irrotational solution scheme is not readily extended to general rotational flows, it is applicable to slightly rotational flows by imbedding a second perturbation analysis, in terms of a new small perturbation parameter for the rotationality. For example, consider a slight shear flow with remote upstream velocity $U(1 + \alpha y/R)$ where $0 < \alpha \ll 1$ is the perturbation parameter and R is the cylinder radius. The vorticity is $\zeta = -\alpha U/R$ and the

† C. Kaplan[12] shows that the upper limit for $c_{max} - c_{least}/c_\infty$ is $O(10^{-1})$ so that $c_{max} \sim c_{least}$ to a first approximation.

governing equation now becomes $N(\psi) = \zeta$.[13] Writing $\psi = \psi(x, y; \varepsilon, \alpha) = \psi_{\{0\}}^{(0)}(x, y) + \varepsilon\psi_{\{0\}}^{(1)}(x, y) + \alpha\psi_{\{1\}}^{(0)}(x, y) + \varepsilon\alpha\psi_{\{1\}}^{(1)}(x, y)$ + higher-order terms, a new perturbation analysis includes such limited rotational flow fields.

From a practical standpoint, calculation of the first few terms for irrotational flow is approached equally well with either stream or potential function. If several perturbation terms are desired, the potential function analysis is algebraically less complicated and is recommended.

9.8 The general non-linear functional

In the quest earlier for a non-linear functional corresponding to $N(\phi) = 0$, it is observed that a direct divergence manipulation of $\int_\Omega N(\phi)\,\delta\phi\,\mathrm{d}\Omega$ is unsuccessful. However, if the continuity form of the governing equation is used directly, then a non-linear functional can be devised. Let $\delta\bar{\Pi} = \int_\Omega (\rho\phi_{,i})_{,i}\,\delta\phi\,\mathrm{d}\Omega$. The divergence theorem gives,

$$\delta\bar{\Pi} = \int_\Gamma \rho\frac{\partial\phi}{\partial n}\,\delta\phi\,\mathrm{d}\Gamma - \int_\Omega \tfrac{1}{2}\rho\,\delta((\phi_{,i})^2)\,\mathrm{d}\Omega$$

Using the relation $\delta\int_0^u f(u)\,\mathrm{d}u = f(u)\,\delta u$, with $u = (\phi_{,i})^2$ and $f(u) = \rho$,

$$\delta\bar{\Pi} = \int_\Gamma \rho\frac{\partial\phi}{\partial n}\,\delta\phi\,\mathrm{d}\Gamma - \tfrac{1}{2}\int_\Omega \delta\int_0^{t=(\phi_{,i})^2} \rho(\sqrt{t})\,\mathrm{d}t\,\mathrm{d}\Omega$$

This implies that the desired functional is

$$\Pi = \tfrac{1}{2}\int_\Omega\int_0^{t=(\phi_{,i})^2} \rho(\sqrt{t})\,\mathrm{d}t\,\mathrm{d}\Omega \tag{9.9}$$

where

$$\rho(\sqrt{t}) = \left[1 + \left(\frac{\gamma-1}{2}\right)M_\infty^2(1-t)\right]^{1/(\gamma-1)}$$

The finite element approximant $\phi_e(x, y) = \mathbf{N}^{\mathrm{T}}\boldsymbol{\phi}_e$ implies $(\boldsymbol{\phi}_{,i})_e^2 = \boldsymbol{\phi}_e^{\mathrm{T}}\mathbf{M}\boldsymbol{\phi}_e$ and the element contribution for linear $\phi_e(x, y)$ is

$$\frac{\partial\Pi_e}{\partial\boldsymbol{\phi}_e} = A_e\left[1 + \left(\frac{\gamma-1}{2}\right)M_\infty^2(1 - \boldsymbol{\phi}^{\mathrm{T}}\mathbf{M}\boldsymbol{\phi})\right]_e^{1/(\gamma-1)}\mathbf{M}_e\boldsymbol{\phi}_e = K_e(\boldsymbol{\phi})\boldsymbol{\phi} \tag{9.10}$$

where A_e is the element area. Combining element contributions nodally, $\sum_e \partial\Pi_e/\partial\boldsymbol{\phi}_e = \mathbf{0}$ is the final non-linear finite element system.

Using the non-linear functional thus determined, the perturbation variational problems can be deduced by substitution of a perturbation expansion for $\phi(x, y)$.

9.9 Extension to lifting profiles

In the foregoing theory and analysis, non-lifting profiles have been considered. The analysis can be extended to problems with non-zero circulation $\mathscr{C}$ by means of a straightforward, easily implemented procedure. This technique is developed first for the incompressible lifting solution. The generalization to successive higher-order perturbation solutions follows immediately.

Let $\phi^{(0)}(x, y)$ be the incompressible solution potential for flow past an airfoil with unspecified circulation $\mathscr{C}^{(0)}$ to be determined from the Kutta condition. A branch cut emanates from the trailing edge. Jump conditions at the branch cut are chosen such that the line integral of the tangential velocity around any simple closed contour containing the airfoil determines the circulation (Figure 9.2).

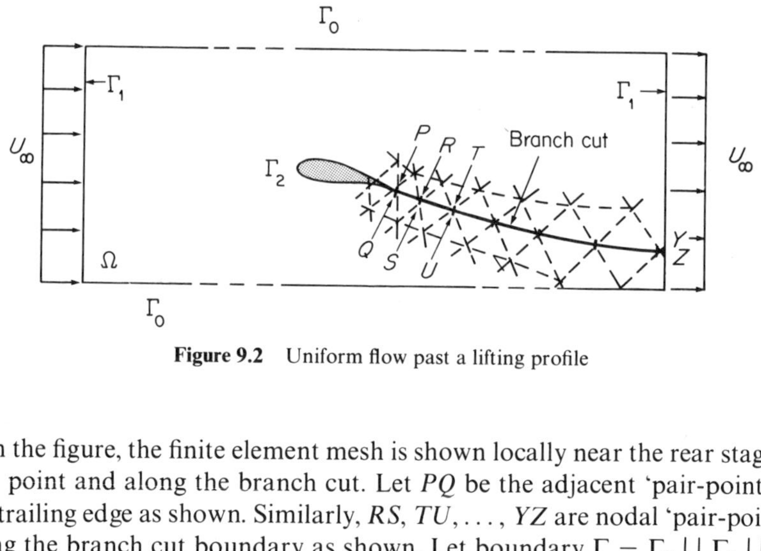

Figure 9.2 Uniform flow past a lifting profile

In the figure, the finite element mesh is shown locally near the rear stagnation point and along the branch cut. Let PQ be the adjacent 'pair-point' at the trailing edge as shown. Similarly, RS, $TU, \ldots, YZ$ are nodal 'pair-points' along the branch cut boundary as shown. Let boundary $\Gamma = \Gamma_0 \cup \Gamma_1 \cup \Gamma_2$ and $\bar{\Gamma}_2 = \Gamma_2$ excluding the point PQ on the cut. The circulation for tangential velocity $q_s^{(0)}$ is

$$\mathscr{C}^{(0)} = \oint_{\Gamma_2} q_s^{(0)}\, \mathrm{d}\Gamma_2 = \phi^{(0)}(P) - \phi^{(0)}(Q) = [\phi^{(0)}]_Q^P = [\phi^{(0)}]_L^U$$

where $[\phi^{(0)}]_L^U$ denotes the potential jump discontinuity between upper (U) and lower (L) edges of the cut.

Without loss of generality, setting $\phi^{(0)} = \mathscr{C}^{(0)}$ on the upper edge and $\phi^{(0)} = 0$ on the lower edge, the problem statement is

$$\Delta\phi^{(0)} = 0 \text{ in } \Omega, \quad \text{with } \frac{\partial \phi^{(0)}}{\partial n} = \begin{cases} 0 & \text{on } \Gamma_0 \text{ and } \bar{\Gamma}_2 \\ U_i n_i & \text{on } \Gamma_1 \text{ (surface normal } n_i) \end{cases}$$

and

$$\phi^{(0)} = \begin{cases} \mathscr{C}^{(0)} & \text{on the upper edge of the branch cut} \\ 0 & \text{on the lower edge of the branch cut} \end{cases}$$

where $\mathscr{C}^{(0)}$ is unknown and is to be determined by the Kutta condition.

The problem can be equivalently posed as a superposition:[15] Let $\phi^{(0)}(x, y) = \mathscr{C}^{(0)}\xi^{(0)}(x, y) + \eta^{(0)}(x, y)$ with $\xi^{(0)}(x, y)$ and $\eta^{(0)}(x, y)$ to be found satisfying

$$\Delta\xi^{(0)} = 0 \text{ in } \Omega \quad \text{with } \frac{\partial\xi^{(0)}}{\partial n} = 0 \text{ on } \Gamma_0, \Gamma_1 \text{ and } \bar{\Gamma}_2,$$

and

$$\xi^{(0)} = \begin{cases} 1 & \text{on the upper edge of the cut} \\ 0 & \text{on the lower edge of the cut} \end{cases};$$

$$\Delta\eta^{(0)} = 0 \text{ in } \Omega, \quad \text{with } \frac{\partial\eta^{(0)}}{\partial n} = \begin{cases} 0 & \text{on } \Gamma_0 \text{ and } \bar{\Gamma}_2 \\ U_i n_i & \text{on } \Gamma_1 \end{cases}$$

and $\eta^{(0)} = 0$ on both edges of the branch cut.

The circulation $\mathscr{C}^{(0)}$ does not appear directly in these subsidiary problems, which can be solved for $\xi^{(0)}(x, y)$ and $\eta^{(0)}(x, y)$. Let $v^{(0)}(x, y)$ and $w^{(0)}(x, y)$ be the corresponding x velocity fields. The Kutta condition requires parallel flow from the trailing edge stagnation point S ($\equiv P$ or Q). Differentiating $\phi^{(0)}(x, y)$ with respect to x,

$$u^{(0)}(S) = \mathscr{C}^{(0)}v^{(0)}(S) + w^{(0)}(S) = 0, \quad \text{so that } \mathscr{C}^{(0)} = -w^{(0)}(S)/v^{(0)}(S)$$

determines $\mathscr{C}^{(0)}$ and the superimposed solution is $\phi^{(0)}(x, y) = \mathscr{C}^{(0)}\xi^{(0)}(x, y) + \eta^{(0)}(x, y)$.

This analysis can be incorporated with the perturbation-variational theory. Write $\phi(x, y) = (\mathscr{C}^{(0)}\xi^{(0)}(x, y) + \eta^{(0)}(x, y)) + M_\infty^2(\mathscr{C}^{(1)}\xi^{(1)}(x, y) + \eta^{(1)}(x, y)) + M_\infty^4(\mathscr{C}^{(2)}\xi^{(2)}(x, y) + \eta^{(2)}(x, y)) + O(M_\infty^6)$. The general term $\phi^{(i)}(x, y) = \mathscr{C}^{(i)}\xi^{(i)}(x, y) + \eta^{(i)}(x, y)$ can be calculated by the prior technique and the variational finite element computations become:

(1) Solve $\Delta\xi^{(i)} = 0$, with $\partial\xi^{(i)}/\partial n = 0$, on Γ except on the branch cut where $\xi^{(i)} = 1$ and $\xi^{(i)} = 0$ for upper and lower edges respectively. This solution need be made only once (for $i = 0$, say, since $\xi^{(0)}(x, y) = \xi^{(1)}(x, y) = \ldots$) and the decomposed coefficient matrix is retained.

(2) Solve $\Delta\eta^{(i)} = f^{(i)}$, with $\eta^{(i)} = 0$ on the branch cut and

$$\frac{\partial\eta^{(i)}}{\partial n} = \begin{cases} \left.\begin{array}{ll} 0 & \text{on } \Gamma_0 \text{ and } \bar{\Gamma}_2 \\ U_j n_j & \text{on } \Gamma_1 \end{array}\right\} i = 0 \\ 0 \quad \text{on } \Gamma_0, \Gamma_1 \text{ and } \bar{\Gamma}_2, i > 0 \end{cases}$$

Numerical solution for successive terms requires forward and backward substitutions with the decomposed matrix of Step (1) above. The x velocity fields $v^{(i)}(S)$, $w^{(i)}(S)$ of $\xi^{(i)}$ and $\eta^{(i)}$ determine $\mathscr{C}^{(i)}$. Thus the only additional significant computation necessary for lifting profiles is the single forward and backward substitution in Step (1) and the method is computationally very attractive.

9.10 Conclusion

In the preceding analysis, the general variational question for compressible fluid flows was first examined and discussed from the viewpoint of finite element applications. Those situations under which a variational principle of energy type could be constructed were developed. The theoretical limitations and degree of difficulty of the variational problem were investigated. When no variational principle could be constructed the adjoint operator inclusion and finite element weighted residual methods provided an avenue for solution. These are, perhaps, computationally comparable with existing finite difference methods but treat boundary shapes more easily. The value of the variational finite element formulation lies in its inherent structure and the speed of the contingent solution algorithms. This led directly to the concept of transforming the non-linear flow equations to a form permitting the desirable variational structure.

The finite element formulation in conjunction with a perturbation technique admitted such a convenient variational expression. An important step was the application of the perturbation relations in divergence manipulation to reduce derivative order in the functionals for compressibility corrections. Resulting computational features enhanced rapid calculation of the first few perturbation terms developed. The constraint to low and moderate Mach number flows arises from the form and range of validity of the perturbation series. An interesting point, however, is that the perturbation solutions are accurate in the presence of supersonic bubbles in the flow.[14] That is, despite the change in the canonical form of the governing equation from elliptic to hyperbolic in these zones, the perturbation solutions to linear elliptic problems are still adequate. For mixed and supersonic flows at high incident Mach number, the weighted residual or nonlinear functional finite element analyses apply, but the economics of computation are far less appealing.

Approximate solution of difficult flow problems by finite element methods requires an integrated approach to the various interwoven processes of theory, analysis and computation. Where possible analytic techniques may be incorporated advantageously. This was evident in the perturbation transformation of the initial problem and in the divergence manipulations that maintained low-order basis functions for all solution terms. Without such an analysis, higher-order basis functions or a similar device becomes inevitable and the simple elegance and efficiency of the approach suffer.

Acknowledgements

The author expresses his appreciation to Professor Carl E. Pearson for his helpful discussion.

References

1. H. C. Martin and G. F. Carey, *Introduction to Finite Element Analysis*: McGraw-Hill, 1973.
2. J. Serrin, 'Mathematical principles of classical fluid mechanics', in Handbuch der Physik, Vol. 8, pp. 125–262, 1959.
3. B. A. Finlayson, 'Existence of variational principles for the Navier-Stokes equation', *Phys. Fluids*, **15**, 6, 963–967 (1972).
4. G. C. Oates and G. F. Carey, 'A variational formulation of the compressible throughflow problem', *Tech. Rept. No. AFAPL-TR- 74–78*, Aero Propulsion Laboratory, Wright-Patterson AFB, Ohio, 1974.
5. J. W. Herivel, *Proc. Camb. Phil. Soc.*, **51**, 344 (1955).
6. C. C. Lin, *Proc. Intl. School Physics*, Varenna, New York, 1963.
7. H. Bateman, *Proc. Nat. Acad. Sci. U.S.A.*, **16**, 816 (1930).
8. E. Tonti, 'Variational formulation of nonlinear differential equations I and II', *Bull. Acad. Roy. Belg.* (*Class Sci.*), **5**, 55, 137–165, 262–278 (1969).
9. Lord Rayleigh, 'On the flow of compressible fluid past an obstacle', *Phil. Mag. Jnl. Sci.*, **6**, 97–102 (July 1916).
10. A. H. Shapiro, *Dynamics and Thermodynamics of Compressible Fluid Flow*, Vol. 1, The Ronald Press Company, New York, 1953.
11. H. C. Martin, 'Finite element analysis of fluid flows', *Proc., 2nd Conf. Matrix Methods Struct. Mech., Wright-Patterson AFB, AFFDL-TR-68-150, Ohio, October 15–17*, 1968.
12. C. Kaplan, 'Two-dimensional subsonic compressible flow past elliptic cylinders', National Advisory Committee for Aeronautics, *Report No. 624*, Langley Memorial Aeronautical Lab., Feb. 1938.
13. R. von Mises, *Mathematical Theory of Compressible Fluid Flow*, Academic Press, 1958.
14. M. J. Lighthill, 'Higher approximations', in *General Theory of High Speed Aerodynamics*, W. R. Sears (Ed.), Section E, pp. 345–489, Princeton University Press, Princeton, 1954.
15. D. H. Norris and G. de Vries, *The Finite Element Method: Fundamentals and Applications*, Academic Press, New York, 1973.

Chapter 10

An Aerodynamicist Looks at the Finite Element Method

S. F. Shen

10.1 Introduction

Recent rapid advances and the ever-spreading use of the finite element method in the solution of field equations modelling all kinds of practical problems have made it one of the most important numerical techniques. Its application to fluid dynamics has been recently reviewed by Norrie and de Vries,[1] who give numerous references. The only surprise is that it has received only token recognition from the aerodynamicists. An obvious explanation is the furious effort and, on its own, the great progress achieved by workers on computational fluid dynamics, the main thrust being the finite difference method and such techniques as the 'particle-in-the-cell' that are partly Lagrangian and start from the very basic physical laws. Serious computation is now only limited by the capacity of the largest computer, but they do manage to calculate many complicated phenomena that occur and are observed in fluid motion. In contrast, the finite element method in its usual form is primarily designed for the solution of boundary value problems of elliptic partial differential equations in a finite domain. The aerodynamicist must go much further.

The finite element method basically seeks to find an approximate solution in terms of piecewise continuous local interpolation formulas, by controlling the error in some average sense through a variational principle or the Galerkin procedure. As such, there may also be a sense of *déjà vu*. The von Karman momentum integral method in boundary layer theory is evidently a finite element method in primitive form. Attempts to use the variational principle, known in the important case of steady compressible flows since Bateman, have been made seriously, beginning with Wang and coworkers.[2,3,4] For a recent review, see, for example, Rasmussen.[5] Indeed, Greenspan and Jain's calculation[6] with the finite difference formula in the variational integral must be considered as precisely a finite element attack employing the simplest shape function on a regular mesh. However, the strict variational principle leads to transcendental equations. The practical question may be whether the additional accuracy, beyond the straight finite difference method, cannot be better achieved by simply refining the mesh size, the mesh pattern being kept regular.

If the variational principle is dispensed with, the finite element method would then be very close to the 'method of integral relations'[7] which originated with Dorodnitzyn, as a generalization of Karman's moment integral, and has been highly developed by primarily but not exclusively Russian research workers. The success of the 'method of integral relations' in problems of the supersonic flow with detached shock waves and non-equilibrium effects,[8] flows in nozzles with non-equilibrium effects,[9,10] transonic flow over airfoils,[11] etc. is quite impressive. It would naturally suggest that the finite element method might have a similar capability in dealing with such apparently rather difficult aerodynamic problems.

The greatest advantages of the finite element method are the relative ease of the numerical computation and the flexibility of arbitrary mesh pattern to cope with geometrical constraints. The use of local approximation generally produces an algebraic problem involving sparse matrices, convenient for machine calculation, but so does the finite difference method. Much larger irregular elements, in contrast, are possible because the derivatives are no longer approximated. The domain of computation can be covered consequently with much smaller number of unknowns. The mesh pattern can be *designed* efficiently by putting emphasis wherever there appears to be a demand. The finesse and knowledge of the user now plays a greatly enhanced role. The interplay with the human element, in our opinion, is a distinctive feature of the finite element method. For a given computer capacity, an effective human input cannot help but enlarge the horizon of problems that can be treated.

It is from this viewpoint that we wish to examine the finite element method. Left alone will be the discussion of the variational principles, whether they exist or are preferable to the simpler Galerkin procedure. Nor shall we touch the topics of the development of different types of finite elements and shape functions, the formal aspects of the algebraization of the partial differential equation and the computational algorithms. Our main interest lies in the question of how to bring in the maximum amount of *a priori* information, theoretical and analytical, so as to minimize the chore that must be done numerically in the finite element method.

A progress report of our work has been presented earlier.[12] In this paper, we focus our attention on the aerodynamic problems of a steady, compressible, inviscid flow. The domain of interest is often infinite; the governing equation is non-linear and not everywhere elliptic. The concept of 'finite element patching' to handle local peculiarities is first introduced, followed by brief remarks on the difficulties which enter into the extension of the finite element method to hyperbolic and mixed type equations. As examples of the application of finite element patching, results from our study of the incompressible potential flow over airfoils and the compressible flow in a given convergent–divergent nozzle are presented.

10.2 Finite element patching

In view of the various difficulties that may arise in different parts of the domain of interest, it is necessary to devise different local remedies. A comprehensive numerical program must incorporate all these features—then it also becomes unnecessarily unwieldy. We are thus led to a reemphasis of the spirit motivating the finite element method, that is, the assemblage of subdomains in each of which a local simple approximation has been made. It is by no means essential in such a discretization to employ polynomials and control their error by a variational principle, except when it is efficient for convenience and accuracy of the numerical work to do so. Often it is found in finite element work super-elements formed by a subassembly of a large number of elements or in terms of a large number of parameters, as the basis of a modular approach. A natural extension would be to divide up the entire domain of interest into large parcels according to the method of attack that will perform best locally. Some may indeed be treated by a finite element method appropriate to the nature of the local mathematical problem. Others may be better handled by a finite difference method, or analytically. It remains to patch up these localized approximate solutions along the common boundaries for the global solution. To avoid confusion of terminology these generalized super-elements may be referred to as 'patches'. The global solution is, figuratively speaking, obtained by sewing up the patches along the seams. In a sense, the patching procedure may be interpreted as the numerical counterpart of such analytical approaches as the 'matched asymptotic expansions' or the 'method of multiple scales' that seek to join together simpler solutions in different regions constructed by exploiting the consequences of different scales. To achieve analytical matching to higher orders the complexity usually encountered quickly becomes insurmountable. We do not anticipate this type of difficulty in numerical patching.

10.2.1 The infinite domain and local singularity

Consider the case of steady, two-dimensional potential flows of a compressible fluid, governed by the equation

$$\mathscr{D}(\phi) = (\phi_x^2 - a^2)\phi_{xx} + 2\phi_x\phi_y\phi_{xy} + (\phi_y^2 - a^2)\phi_{yy} = 0 \tag{10.1}$$

where ϕ is the velocity potential, and a the local sound speed related to the velocity vector $\nabla\phi$ in a well known manner. In the limit of M (= Mach number) $\to 0$, Equation 10.1 is reduced to the Laplace equation

$$\mathscr{D}_0(\phi) = \phi_{xx} + \phi_{yy} = 0 \tag{10.2}$$

and boundary value problems can be directly attacked by the finite element method in its usual form, based upon a minimum energy principle. As is often the case, however, the aerodynamicist must construct the solution in an infinite domain. Figure 10.1 shows the uniform flow over an airfoil. We seek

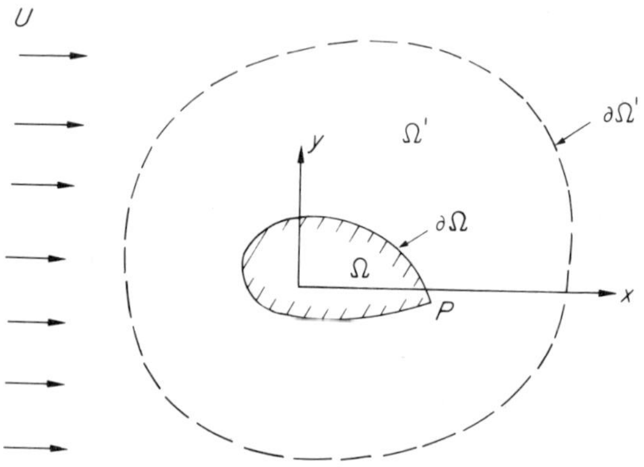

Figure 10.1 Uniform flow over an airfoil

to find ϕ such that

$$\begin{aligned} \mathscr{D}_0(\phi) &= 0 && \text{in } \Omega_\infty\text{–}\Omega \\ \frac{\partial \phi}{\partial n} &= 0 && \text{on } \partial\Omega \\ \phi &\to Vx && \text{as } (x, y) \to \infty \\ \nabla\phi &= 0 && \text{at } P \text{ (the Kutta condition)} \end{aligned} \tag{10.3}$$

where Ω is the region enclosed by the airfoil contour $\partial\Omega$, Ω_∞ the infinite domain, and V the speed of the uniform flow. The standard numerical approach is to replace the infinite domain by a bounded one Ω', with boundary $\partial\Omega'$, sufficiently large to permit the prescription of $\phi = Vx$ on $\partial\Omega'$. For the non-lifting case, the error of the solution from this approximation is clearly $O(\varepsilon^2)$, ε being the inverse of the linear dimension of $\partial\Omega'$ measured in terms of the body size. The lifting case requires a circulation which corresponds to a logarithmic singularity (at infinity) of the complex potential. The same boundary along $\partial\Omega'$ now has an error $O(\varepsilon)$ in the velocity. However, the error of the solution is still $O(\varepsilon^2)$, as shown by considering the special case of a circular cylinder. The general airfoil can be mapped conformally to a circle; and the error is unchanged.

This estimate is not directly applicable to the compressible case where Equation 10.2 must be modified to include non-linear terms. We choose, therefore, to treat the boundary condition on $\partial\Omega'$ more prudently. Back to Equation 10.2 again as the example, a patching concept is proposed, which amounts to the removal of the logarithmic behaviour by analytical means. Let the problem of Equation 10.3 be reformulated in two patches. The inner one is described by

$$
\begin{aligned}
\mathscr{D}_0(\phi) &= 0 && \text{in } \Omega'\text{–}\Omega \\
\frac{\partial\phi}{\partial n} &= 0 && \text{on } \partial\Omega \\
\phi &= \phi' && \text{on } \partial\Omega' \text{ to be determined} \\
\nabla\phi &= 0 && \text{at } P \text{ (the Kutta condition)}
\end{aligned}
\tag{10.4}
$$

and the outer one by

$$
\begin{aligned}
\mathscr{D}_0(\phi) &= 0 && \text{in } \Omega_\infty\text{–}\Omega' \\
\phi &= \phi' && \text{on } \partial\Omega' \\
\phi &\to Ux && \text{as } (x, y) \to \infty
\end{aligned}
\tag{10.5}
$$

The solution of the outer patch can now be written down immediately as a Laurent series,

$$
\begin{aligned}
w &= \phi + i\psi \\
&= Vz + A_0 \ln z + \frac{A_1}{z} + \frac{A_2}{z^2} + \cdots \\
z &= x + iy
\end{aligned}
\tag{10.6}
$$

where w is the complex potential, ψ the stream function, and A_0, A_1, etc. are complex constants. Thus ϕ' is cast in terms of the coefficients A_0, A_1, etc. The inner patch now can be handled by the finite element method in the usual way.

Such an approach is self-evident and by no means novel. If the airfoil has a sharp leading edge, or other slope discontinuities along the contour $\partial\Omega$, again there will be a local singularity whose behaviour is well known. We would then introduce another patch to deal with the singularity separately. The use of a 'special shape function' near a singular point has been made in many applications. The study of Pian, Tong, and Luk[13] of the stress concentration near a crack, for example, is in much the same spirit.

10.2.2 *Compressibility effects*

Let us return to Equation 10.1 without approximations. The variational principles in this case are known.[5] Briefly, for a bounded domain Ω, if the flow is everywhere subsonic, the solution ϕ will maximize, with $\rho(\partial\phi/\partial n)$

prescribed along $\partial\Omega$,

$$\Pi[\phi] = \int_{\Omega} p\,\mathrm{d}A + \oint_{\partial\Omega} \rho\phi\frac{\partial\phi}{\partial n}\,\mathrm{d}s \tag{10.7}$$

where p is the pressure, ρ the density, $\mathrm{d}A$ the area element, $\mathrm{d}s$ the length element along $\partial\Omega$, and $\partial\phi/\partial n$ the velocity component in the outward normal direction. The pressure and density are further related to ϕ through

$$\begin{gathered} p \propto \rho^{\gamma} \\ \frac{\gamma}{\gamma - 1}\frac{p}{\rho} + \tfrac{1}{2}(\phi_x^2 + \phi_y^2) = \text{const.} \end{gathered} \tag{10.8}$$

γ being the ratio of specific heats. If the flow is not everywhere subsonic, the first variation of Equation 10.7, i.e.

$$\delta\Pi[\phi] = 0 \tag{10.9}$$

is still satisfied by the exact solution. However $\Pi[\phi]$ then is only an extremum.

We shall not pursue here the discussion on the finite element aspects of the variational principle based on Equations 10.7 and 10.8. The transcendental nature of the resulting equations at least will be highly tedious for numerical solution. It seems of more basic importance to reflect upon the difference between totally subsonic flows and those which may involve supersonic or transonic regions. For totally subsonic flows, Equation 10.1 is elliptic and, like the degenerate case of incompressible flow governed by Equation 10.2, we are to solve a boundary value problem by prescribing suitable data all along $\partial\Omega$. For supersonic flows, Equation 10.2 is of the hyperbolic type, for which the method of characteristics is a rigorous technique in obtaining numerical solutions. The crucial point is that the boundary values cannot be prescribed everywhere along the contour $\partial\Omega$. Clearly, if one wishes to adapt the finite element method to handle an equation of the hyperbolic type, a routine quite different from the usual one (primarily for equations of the elliptic type) will be necessary.

When the flow is transonic, changing from subsonic to supersonic in the domain of interest, or vice versa, additional complications arise. Equation 20.1 now becomes of the mixed type. In the neighbourhood of the sonic line, even for small perturbations the first approximation has to be solved on the basis of a non-linear equation. Writing

$$\phi = a^*x + \phi' \tag{10.10}$$

where a^* is the local sound speed, ϕ' the perturbation potential and the x axis along the direction of the flow, Equation 10.1 is reduced to

$$-(\gamma + 1)\phi'_x\phi'_{xx} + a^*\phi'_{yy} = 0 \tag{10.11}$$

How to state a well-posed problem for Equation 10.11 is given in the literature. We need to emphasize only that, again, a different numerical routine suggests itself.

A supersonic flow may further go back to subsonic, occasionally in a smooth manner but more often through a shock. The discontinuity across a shock obviously cannot be modelled by the finite element method that presumes continuous variation. It is then necessary to devise a 'shock-fitting procedure' to search out the location of the shock front and accept, again, a possible change of type of the governing equation. In addition, a curved shock generates vorticity which is convected along the streamline. As a result, even Equation 10.1 is no longer valid. Instead the vorticity equation is needed. For two-dimensional flows, it is more conveniently expressed in terms of the stream function:

$$\left(\frac{1}{\rho}\psi_x\right)_x + \left(\frac{1}{\rho}\psi_y\right)_y = -\omega(\psi) \tag{10.12}$$

where $\omega(\psi)$ is the vorticity as a given function of the stream function ψ.

All these complications are familiar to the aerodynamicists and various ways of attacking them numerically have been developed in modern finite difference methods. The question boils down to whether the finite element approach can be tailored so as to be able to resolve all the difficulties that the finite difference method has managed to overcome, while retaining the basic advantage of larger and irregular mesh patterns and, consequently, fewer unknowns left to the computing machine to solve. We have outlined above the need of separate subroutines, which in turn mean that it might be not only desirable, but unavoidable, to construct the solution by the assembling of different patches as the specific problem may require.

10.3 Non-elliptic equations

The finite element method has been widely used in literature in such problems as transient heat conduction, transient response of elastic structures, as well as the unsteady Navier–Stokes equation. In all these applications to non-elliptic partial differential equations, there is always the time variable and generally a distinction is made between the space and time directions. The finite element approximation is ordinarily made in the space variables, resulting in a set of ordinary differential equations for the nodal variable in time. The solution in time may be carried out by either finite differences or numerical integration. This type of procedure will be referred to as the semi-discrete (finite element) method. There can be found also discussions on the space–time finite element technique.[14] A closer look reveals that generally they do not differ from the semi-discrete method but deal with essentially approximate schemes to carry out the convolution integral for linear systems.

It may be of interest to interpret the semi-discrete methods in the following way. Let us symbolically write the governing equations for an unknown function $\psi(x, t)$ as

$$\mathscr{D}(\psi) = 0$$

$\mathscr{D}$ being the operator, not necessarily linear, $\mathbf{x}$ the vector in the space-like coordinates and t the time-like coordinate. We wish to solve $\psi(\mathbf{x}, t)$ for an initial-boundary value problem. The first step is to recast

$$\mathscr{D}(\psi) = \mathscr{D}_t(\psi) + \mathscr{D}_x(\psi) = 0 \tag{10.13}$$

through grouping the terms so that the equation $\mathscr{D}_x(\psi) = F(\mathbf{x}, t)$, for given $F(\mathbf{x}, t)$ and the given boundary conditions, will define a boundary value problem in the $\mathbf{x}$ space, with t playing the role of a parameter. $\mathscr{D}_t(\psi)$ now is the rest of $\mathscr{D}(\psi)$, containing at least all the time derivatives of ψ. The finite element procedure is next applied to solve the boundary value problem *at fixed t* for

$$\mathscr{D}_x(\psi) = -\mathscr{D}_t(\psi) \tag{10.14}$$

treating $\mathscr{D}_t(\psi)$ as the source term $F(\mathbf{x}, t)$. If $\mathscr{D}_x$ is the Laplace operator, for example, at each fixed t and given $\mathscr{D}_t(\psi)$, the finite element procedure may be constructed on the basis of the variational principle for the Poisson equation. The procedure should not be confused with a variational principle for $\mathscr{D}(\psi) = 0$. At any rate, the result becomes a system of ordinary differential equations in t.

Obviously, to proceed in this manner implies at least:

(a) the initial-boundary problem has been *well posed*, including designation of the time-like variable t;
(b) the solution is continuous to the necessary order in the space-like domain, for the finite element approximation in $\mathbf{x}$.

While these conditions are met in usual applications, they remain to be nonetheless fundamental considerations in new problems. The simple wave equation, for instance, permits the propagation of discontinuities along the characteristics and needs caution in any treatment by finite elements even in the space-like direction.

A true space–time finite element method should permit a more or less irregular mesh pattern in space–time without categorical discrimination. Regarding its feasibility, it is important to keep in mind the developments in the finite difference method. Figure 10.2 shows the regular (equal spacing) mesh pattern of the finite difference method. For illustration, consider first the Laplace equation,

$$\psi_{xx} + \psi_{yy} = 0$$

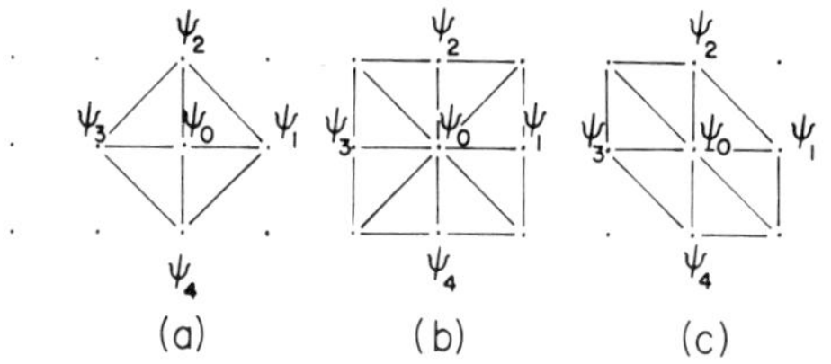

Figure 10.2 Finite elements on regular mesh

The simplest central difference produces the familiar result,

$$\psi_0 = \tfrac{1}{4}(\psi_1 + \psi_2 + \psi_3 + \psi_4) \tag{10.15}$$

Inasmuch as the value ψ_0 is related to only its nearest neighbours, there exists a common spirit between the finite difference and the finite element methods. It is only the details of constructing the local approximation that permit the finite element method to have its advantage of irregular and large meshes. In fact, as is well known, linear triangular elements in the regular mesh, Figure 10.2, yield exactly the same formula, Equation 10.15, for the nodal values. The basic central difference formula also provides the clue that its repeated use at all grid points should enable the simultaneous solution of all the interior nodal values, once the boundary values are assigned.

We now turn to the simple wave equation

$$\psi_{xx} - \psi_{yy} = 0 \tag{10.16}$$

If a Galerkin procedure is adopted, we set

$$\int_\Omega \delta\psi(\psi_{xx} - \psi_{yy})\,\mathrm{d}x\,\mathrm{d}y = 0$$

and with the usual partial integration, it is equivalent to

$$-\tfrac{1}{2}\delta \int_\Omega (\psi_x^2 - \psi_y^2)\,\mathrm{d}x\,\mathrm{d}y + \oint_{\partial\Omega} \delta\psi(\psi_x\,\mathrm{d}y - \psi_y\,\mathrm{d}x) = 0 \tag{10.17}$$

Take again the four triangles of Figure 10.2(a) and assume a linear approximation in each of the triangles. The straightforward application of the finite element practice in assembling the four triangles and varying the resultant area integral with respect to ψ_0 leads to

$$\psi_1 = \psi_2 + \psi_4 - \psi_3 \tag{10.18}$$

Note that Equation 10.18 is no other than the simple explicit one-sided difference formula of the wave equation. In fact, still in regular mesh but with the finite elementations of Figure 10.2(b) or (c) involving many more triangles related to ψ_0 the very same formula Equation 10.18 will result. The other extra nodal values do not enter, just like ψ_0. The nodal values ψ_2, ψ_3 and ψ_4 determine ψ_1, so the boundary values cannot be completely assigned. We have clearly an initial value problem.

Thus, for the wave equation, the simplest local approximations are again identical regardless of whether the finite difference or the space–time finite element technique is used. As in the explicit finite difference method, then, the space–time finite element formulation must seek its solution by marching in the time-like direction. Furthermore, the step-size in the time-like direction must be governed by the Courant–Friedrichs–Lewy criterion.[15] Broadly it says that the time step must be small enough so that the new point, in our example the node of value ψ_1, lies within the zone of influence of the initial data. This guideline should be of prime importance in the designing of irregular space–time finite elements. Furthermore, a knowledge of the zone of influence is equivalent to that of the local characteristic directions. In the case of steady, supersonic flow, the time-like direction is along the streamline. We have to have at least some information on the local velocity direction and magnitude before arranging the space–time mesh pattern. On the other hand, certain implicit one-sided difference formulas are known to be unconditionally stable. It would be of great interest if a finite element approximation can be developed to have the same basic feature, thereby circumventing the difficulty.

Equations of the mixed type, as required in the transonic region, are considerably more difficult. Equation 10.11 shows that, depending on the sign of the perturbation velocity component ϕ'_x, it may be either elliptic or hyperbolic. In the hyperbolic region, the permissible step size in x is seen to be inversely proportional to $(\phi'_x)^{\frac{1}{2}}$, making it impossible to implement a space–time finite element in a simple way. Successful handling of the transonic equations by the finite difference method, e.g. Murman[16] rests on the use of the 'correct' finite difference formula at each point. There must be, in the hyperbolic region, an implicit one-sided difference formula for which no restriction of the time step is necessary, plus a testing of the local ϕ'_x for the selection of the local difference formula. At this moment, it is not apparent how to design a space–time finite element method for transonic flows to contain these features.

However, all is not lost. The method of integral relations, as developed and tested by the Russians, is closely related to the semi-discrete finite element method. Their experience in its applications to transonic flows shows that the ordinary differential equations will in general have a saddle point singularity at the sonic line. In fact, the requirement that the singular point be a saddle point provides an otherwise missing condition to determine the solution. There seems to be little doubt the equations resulting from the semi-discrete finite element method will behave essentially in the same manner for transonic flows.

10.4 The lifting-airfoil in incompressible flow

The case of an isolated airfoil is sketched in Figure 10.1 and the statement of the problem given as Equation 10.3. For more convenient handling of the boundary condition on the airfoil, instead of Equation 10.3, the problem may be recast in terms of the stream function ψ,

$$\begin{aligned} \mathscr{D}_0(\psi) &= \nabla^2\psi = 0 & & \\ \psi &= \psi_0 & & \text{on } \partial\Omega\text{, the airfoil contour} \\ \psi &\to V_y & & \text{as } (x, y) \to \infty \\ \nabla\psi &= 0 & & \text{at } P \text{ (the Kutta condition)} \end{aligned} \tag{10.19}$$

where ψ_0 is a constant to be determined in the solution. For the inner patch described by Equation 10.4, the corresponding statement is

$$\begin{aligned} \mathscr{D}_0(\psi) &= 0 & & \text{in } \Omega'\text{–}\Omega \\ \psi &= \psi_0 & & \text{on } \partial\Omega \\ \psi &= \psi' & & \text{on } \partial\Omega' \text{ to be determined} \\ \nabla\psi &= 0 & & \text{at } P \text{ (the Kutta condition)} \end{aligned} \tag{10.20}$$

The finite element treatment of Equation 10.20 is now straightforward. Using the variation principle Equation 10.7 in its simplified form for incompressible fluid,

$$\begin{aligned} \delta\Pi &= \delta \int_{\Omega'-\Omega} \tfrac{1}{2}(\nabla\psi)^2 \, \mathrm{d}x \, \mathrm{d}y - \oint_{\partial\Omega'} \delta\psi \frac{\partial\psi}{\partial\eta} \, \mathrm{d}s + \delta\psi_0 \oint_{\partial\Omega} \frac{\partial\psi}{\partial n} \, \mathrm{d}s \\ &= 0 \end{aligned} \tag{10.21}$$

Note that $\partial\psi/\partial n$ in the line integrals *should be exactly prescribed.* If $\partial\psi/\partial n$ is evaluated from the finite element approximation, the use of a large Ω' cannot completely eliminate this error.

Turning to the outer patch,

$$\begin{aligned} \mathscr{D}_0(\psi) &= 0 & & \text{in } \Omega_\infty\text{–}\Omega' \\ \psi &= \psi' & & \text{on } \partial\Omega' \\ \psi &\to Vy & & \text{as } (x, y) \to \infty \end{aligned} \tag{10.22}$$

There is no need to proceed numerically for Equation 10.22. The asymptotic solution at large distances to the airfoil is simply, by Equation 10.6

$$\psi \sim Vr\sin\theta + \frac{\Gamma}{2\pi}\ln r - \frac{A_1}{r}\sin\theta + \cdots \tag{10.23}$$

where (r, θ) are the polar coordinates, Γ is the circulation and A_1 the doublet strength, etc. This analytical description of the outer patch, in assembling with the inner patch, provides the data of ψ and $\partial\psi/\partial n$ along $\partial\Omega'$. Depending on the number of terms retained in Equation 10.23 and the size of Ω', the evaluation of $\partial\psi/\partial n$ on $\partial\Omega'$, though not exact because of the truncation, is considerably more accurate than obtainable from the shape functions used to construct the finite elements within Ω'. In addition, along the airfoil contour of Equation 10.21, we may set

$$\oint_{\partial\Omega} \frac{\partial\psi}{\partial n}\,\mathrm{d}s = \Gamma$$

which is exact.

In actual implementation of the finite element method for the inner patch, many details remain that affect the accuracy of the result. For aerodynamic purposes, the pressure distribution over the body is of primary interest and has been used by us as the test of the procedure. The case of a circular cylinder with circulation is first experimented with as a special airfoil. The basic finite element chosen is the simplest linear triangle, yielding a constant velocity, hence constant pressure, in each element. The nodal value of the pressure is obtained, as a first approximation, by arithmetic averaging of the values of the surrounding elements. To conform to the circular boundary, these triangles are laid out in the (r, θ) plane, Figure 10.3, thus actually curvilinear in the physical (x, y) plane. An automatic mesh generating routine is also built in to cover the domain Ω'–Ω. The series Equation 10.23 is truncated at A_1, and $\partial\Omega'$ chosen to be at 5 times the radius.

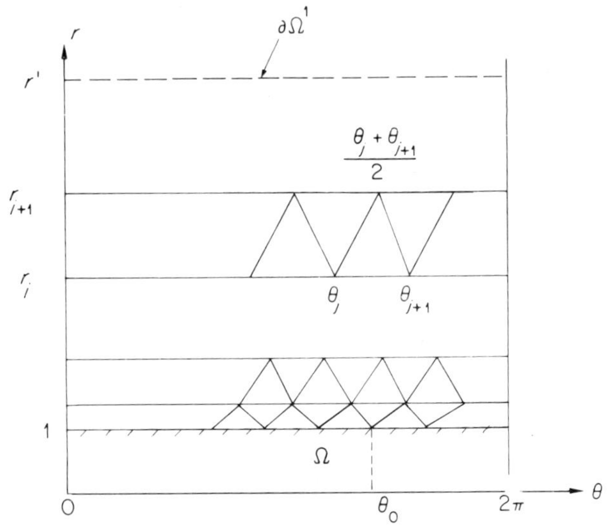

Figure 10.3 Finite elements in (r, θ) plane, airfoil problem

The most important conclusion from our numerical experimentation is that the accuracy is most sensitive to the depth of the first layer of elements next to the body. Typically we find the depth of the first layer elements should be on the order of 1 % of the circle radius to obtain 1 % error of the total lift. Equally important is to restrict the size of the element at the stagnation points. These are but rules of thumb, but it is not hard to understand the reasons behind them. The computed kinetic energy in an element using linear approximation must have a percentage error proportional to the ratio of the mesh size to the local scale. The local scale of the flow near the circle is the circle radius, and an error in the layer immediately next to the circle amounts to effectively a comparable error on the boundary condition. Because of the continuous dependence of the solution on boundary conditions, the same order of percentage error must be reflected in the answer. As for the circumferential direction, the stream function has no gradient at the circle, so much larger spacing than 1 % of the circumference can be tolerated except at the stagnation points which control the circulation through the Kutta condition, in fact as local singularities.

Some details of this study are included in Reference 12, and a complete account will be given elsewhere.

10.4.1 Results for a single airfoil

In the case of an arbitrary airfoil, let us first consider the question of the mesh pattern which may work well for the elongated shape and the large pressure variation near the leading edge. Our answer is to map the airfoil by the simple Joukowski transformation

$$z = \zeta + \frac{l^2}{\zeta}$$
$$z = x + iy, \qquad \zeta = \xi + i\eta \tag{10.24}$$

l being a suitable length, say the half-chord of the airfoil, to render the problem into one of the uniform flow over a near circle in the (ξ, η) plane. This mapping is really only algebraic in nature, to deform the airfoil contour into one easier for handling, similar to the use of polar coordinates for the circle. Now for the near circle, the procedure essentially follows that for the circle. In addition to regularizing the contour shape, we may add, the mapping serves also to spread out the leading and trailing edge regions to permit a relatively even circumferential mesh size.

Calculations are performed for a Joukowski airfoil, Figure 10.4, which is intentionally mapped to a non-circle by an arbitrary choice of the origin in the (ξ, η) plane. The airfoil contour and the chosen mesh pattern are shown in Figure 10.5. With 29 points along the circle and 10 layers of elements to

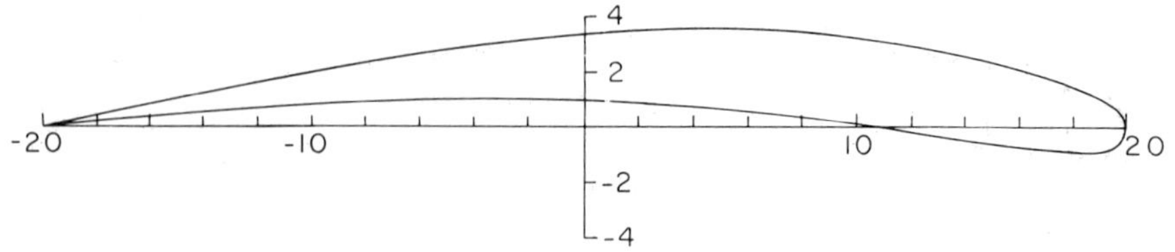

Figure 10.4 A Joukowski airfoil

cover a region Ω' of $r = 5l$ in the (ξ, η) plane, the lift obtained is within 2%. The local pressure distribution, however, is not very satisfactory. By increasing the number of mesh points to 39 on the airfoil, the computed lift is essentially unchanged but there is marked improvement of the pressure distribution, as shown in Figure 10.6.

For the problem on hand, no doubt we could achieve equal accuracy with much less effort by casting the near-circle problem as a perturbation of the perfect circle, for which the exact solution may be utilized. As a test case, we felt it was preferable to have a scheme applicable to rather arbitrary shapes.

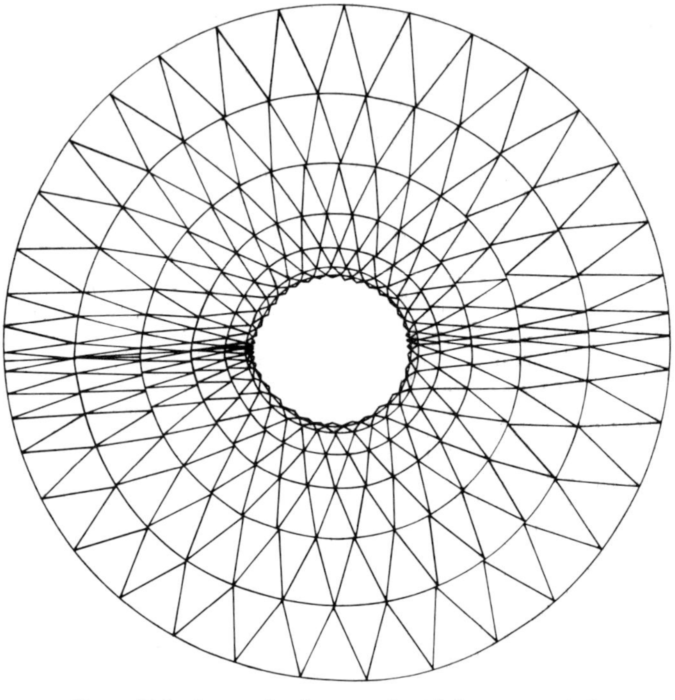

Figure 10.5 Layout for the example airfoil as a near-circle

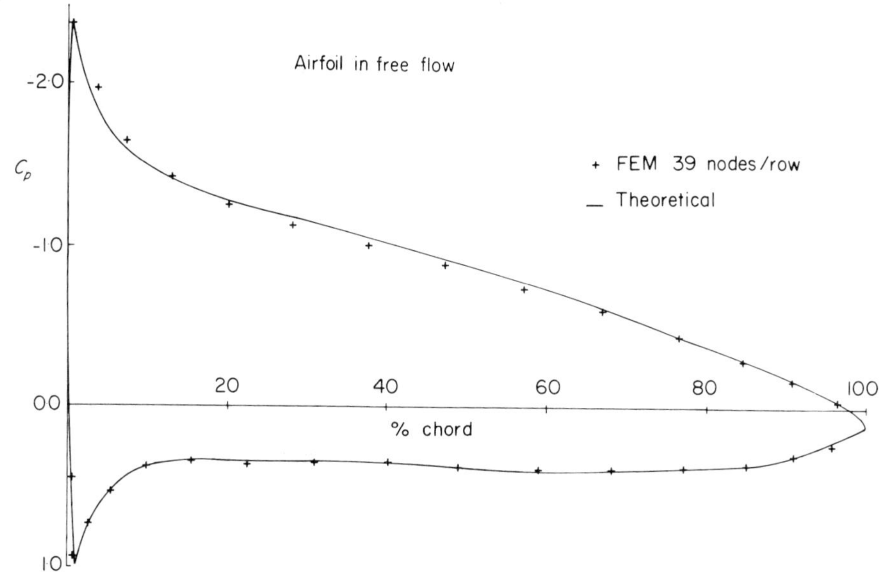

Figure 10.6 Comparson of pressure distribution over an airfoil

10.4.2 Multiple airfoils

Once the basic routine for the inner patch that surrounds the airfoil is available, it can be used as the building block for problems involving multiple airfoils. Take, for example, the unstaggered cascade problem which is an infinite number of the same airfoil stacked up as in Figure 10.7. Because of the periodicity, we choose the domain of computation as a horizontal strip of

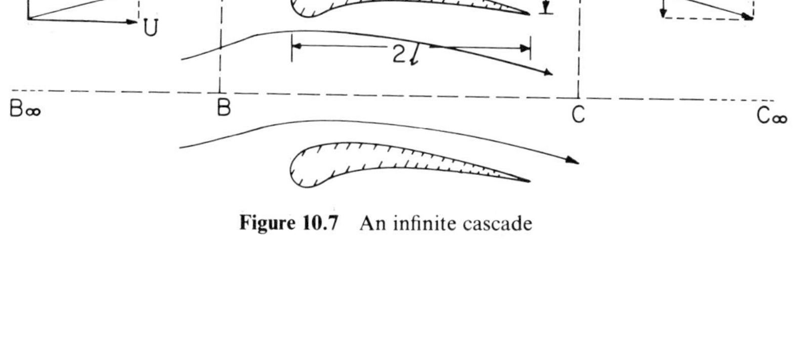

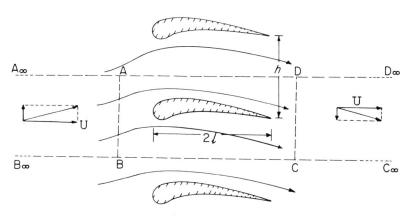

Figure 10.7 An infinite cascade

height $\overline{AD} = \overline{BC} = h$, the pitch. The solution is then obtained by assembling the following patches:

(1) An airfoil patch with boundary $\partial\Omega_1'$ (EFGH) containing the airfoil (contour $\partial\Omega'$), constructed in the same manner as the 'inner patch' for the single airfoil above, Figure 10.8.

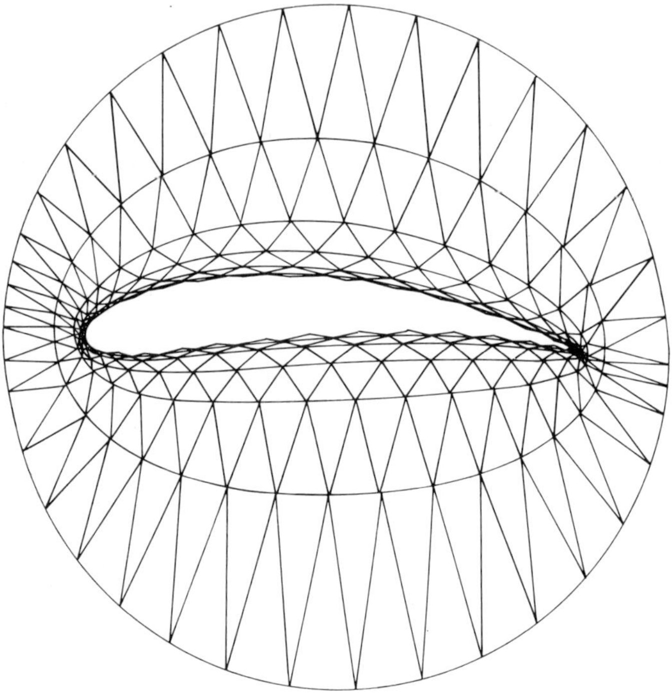

Figure 10.8 Inner patch for the cascade airfoil

(2) A transition patch with boundaries $\partial\Omega_2'$ (ABCD) and $\partial\Omega_1'$ constructed directly in the (x, y) plane with linear triangular elements Figure 10.9.
(3) An upstream outer patch with boundary $A_\infty ADD_\infty$ extending to infinity ahead of the airfoil.
(4) A downstream outer patch with boundary $B_\infty BCC_\infty$ extending to infinity behind the airfoil.

As long as AD and BC are moderately distant from the airfoil, it is again possible to describe rather simply the asymptotic solution in the two outer patches by superposing an infinite number of equally spaced vertices (each of unknown strength Γ) and doublets (each of unknown strength A_1) repre-

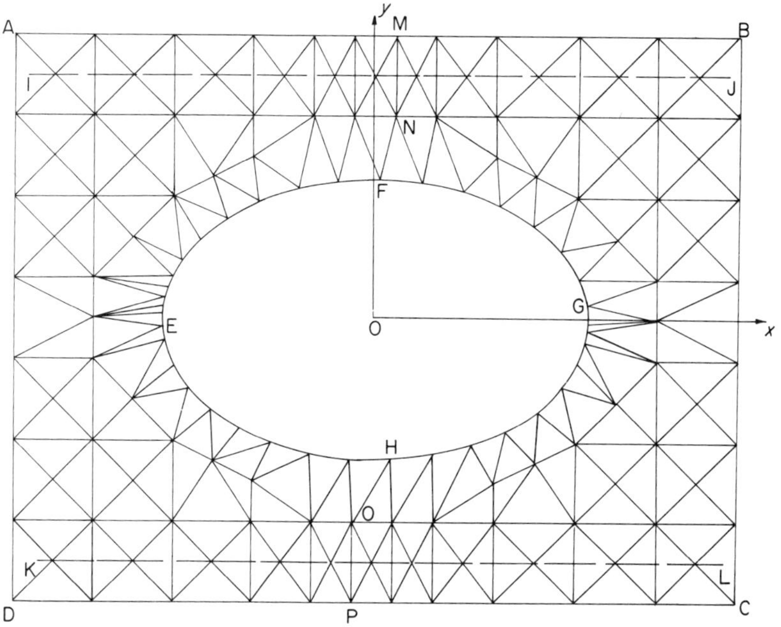

Figure 10.9 Transition patch for the cascade airfoil

senting the airfoils. Thus we get

$$\psi = Vy + \frac{\Gamma}{2\pi}\ln r^* - \frac{A_1 h}{2\pi}\frac{\sin 2\pi y/h}{r^{*2}} + \cdots$$
$$r^* = \cosh\frac{\pi x}{h}\cos\frac{\pi y}{h}\left(\tanh^2\frac{\pi x}{h} + \tan^2\frac{\pi y}{h}\right)^{\frac{1}{2}} \quad (10.25)$$

Equation 10.25 provides the boundary values of ψ and $\partial\psi/\partial n$ along AD and BC, in terms of Γ and A_1, to allow us to carry out the finite element calculations by assembling only the other two patches.

To handle the periodicity of the *velocity* along AB and DC, it is felt that a statement on the nodal values would be less prone to the deficiency of our simple linear element. To this end, we have actually pushed out the boundaries AB and DC so that in Figure 10.9 the distance IK (and JL) is actually the pitch h. The upper strip of elements between N and M is an exact replica of the lower strip between P and O, the corresponding nodal points having their ψ values differ by the constant amount Vh.

Calculations are made for the same Joukowski airfoil of Figure 10.4 at a pitch–chord ratio of 1·5. The pressure distribution on the airfoil is shown as Figure 10.10. The number of unknowns is roughly the same as for the single airfoil.

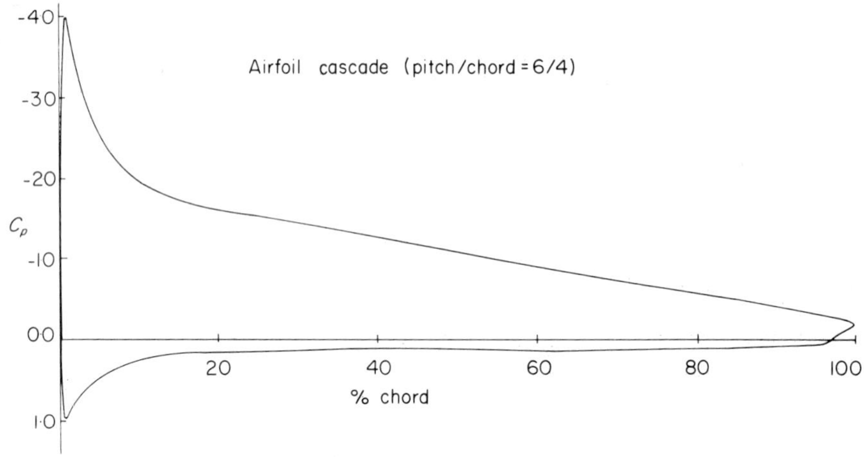

Figure 10.10 Computed pressure distribution over an airfoil in cascade

If we replace the periodicity condition along AB and DC by requiring no penetration, the problem becomes that of the airfoil in a wind tunnel. With a modification on the asymptotic solution, the same patches may be used. We have also programmed for two tandem airfoils. There being no new features, the computed results are omitted here.

10.5 The convegent–divergent nozzle

The convergent–divergent (Laval) nozzle, aside from its practical importance, embodies all the troublesome features of compressible flows, and because of the simple geometry has been extensively treated by analytical as well as numerical means. There is also the one-dimensional treatment to afford a first approximation. We have therefore chosen to study this problem as a test of the finite element method, specifically the two-dimensional hyperbolic nozzle first solved by Emmons[17] using relaxation.

The Emmons nozzle is of such a shape which, for incompressible flow, is described by the complex potential

$$w = \xi + i\eta = \sinh^{-1} z \qquad z = x + iy \tag{10.26}$$

where ξ and η stand for the incompressible velocity potential and stream function, respectively. The nozzle boundary is taken to be the hyperbole

$$(y_w/\sin 0{\cdot}6)^2 - (x_w/\cos 0{\cdot}6)^2 = 1 \tag{10.27}$$

(x_w, y_w) being a point on the nozzle wall. His computation is carried out by mapping the nozzle into a horizontal strip in the (ξ, η) plane, Figure 10.11.

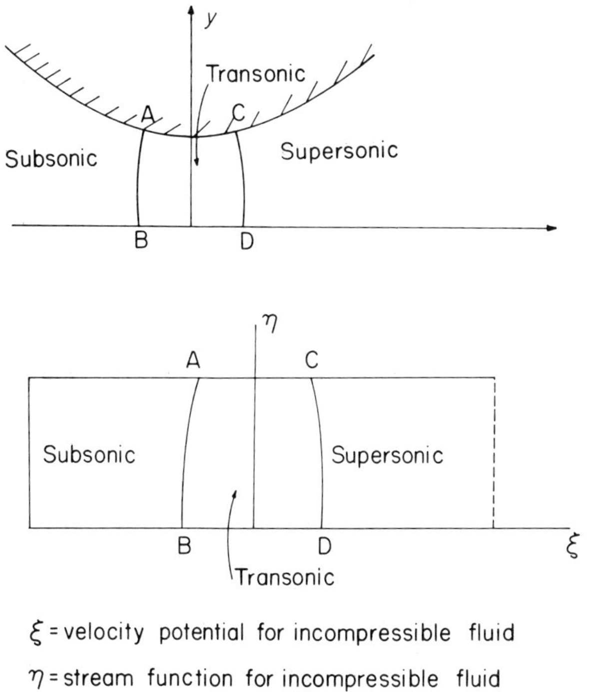

Figure 10.11 The Emmons nozzle mapped to a strip

The nozzle walls become

$$\eta_w = \pm 0{\cdot}6$$

and the centreline becomes $\eta = 0$. For compressible irrotational flows, the equation in terms of the stream function ψ is now

$$\psi_{\xi\xi} + \psi_{\eta\eta} - \psi_\xi\left(\ln\frac{\rho}{\rho_0}\right)_\xi - \psi_\eta\left(\ln\frac{\rho}{\rho_0}\right)_\eta = 0 \tag{10.28}$$

with

$$\begin{aligned} \rho/\rho_0 &= \left[1 - \frac{\gamma-1}{2}\left(\frac{q}{a_0}\right)^2\right]^{1/(\gamma-1)} \\ \rho q &= \rho q_I \psi_0[\psi_\xi^2 + \psi_\eta^2]^{\frac{1}{2}} \end{aligned} \tag{10.29}$$

where q and q_I are the resultant speeds at the point (ξ, η) for the compressible (unknown) and incompressible (known) cases, respectively, ρ is the density, a the sound speed, and subscript 0 denotes the reference value.

10.5.1 The subcritical case

When the flow is everywhere subsonic, Emmons first recast Equation 10.28 as

$$\psi_{\xi\xi} + \psi_{\eta\eta} + C(\psi, \rho) = 0 \tag{10.30}$$

(This is similar to rewriting Equation 10.13 into Equation 10.14, except that here the non-linear terms act as the 'source'.) His procedure is to iterate on the basis of

$$\begin{aligned} \psi^{(n)}_{\xi\xi} + \psi^{(n)}_{\eta\eta} &= -C^{(n-1)} \\ C^{(n-1)} &= C(\psi^{(n-1)}, \rho^{(n-1)}) \end{aligned} \tag{10.31}$$

The superscript n denotes the nth iterate. Thus in each iteration $\psi^{(n)}$ satisfies a Poisson equation, with the source term determined by the last iteration. The initial $\psi^{(0)}$ is taken to be the incompressible solution

$$\psi^{(0)} = \psi_w \eta / \eta_w$$

Argyris and coworkers[18] have used a similar iterative procedure in treating the airfoil in a wind tunnel. With the symmetry with respect to the throat location ($\xi = 0$), for subsonic flow the boundary conditions are

$$\begin{aligned} &\psi = 0 && \text{on } \eta = 0 \\ &\psi_\xi = 0 && \text{on } \xi = 0 \\ &\psi = \psi_w && \text{on } \eta = \eta_w, \text{ given} \\ &\psi \to \psi_w \eta/\eta_w && \text{as } \xi \to \pm\infty \end{aligned} \tag{10.32}$$

In actual computation, Emmons applied the asymptotic condition for $\xi \to \infty$ at the finite distance $\xi' = 1{\cdot}35$.

It is straightforward to mimic the Emmons procedure in the finite element format. At each iteration, the functional to be minimized is simply

$$\Pi[\psi^{(n)}] = \iint \{\tfrac{1}{2}(\nabla\psi^{(n)})^2 - C\psi^{(n)}\}\, \mathrm{d}\xi\, \mathrm{d}\eta$$

since $C(\psi^{(n-1)}, \rho^{(n-1)})$ is assumed as a known function. We have carried out such a calculation using the quadrilateral element composed of four 6-node quadratic triangular sub-elements. The layout of the finite element is shown in Figure 10.12. The results for $\psi_w = 0{\cdot}948$, giving a Mach number of 0·6 at the origin, are directly comparable with one case computed by Emmons and are presented in Figure 10.13. The convergence criterion for our iterations is on the local Mach number M,

$$|M^{(n)} - M^{(n-1)}| < 0{\cdot}001$$

which is more stringent than the same accuracy on ψ. Aside from the slight but systematic deviation from Emmons' results, we find also that the iteration

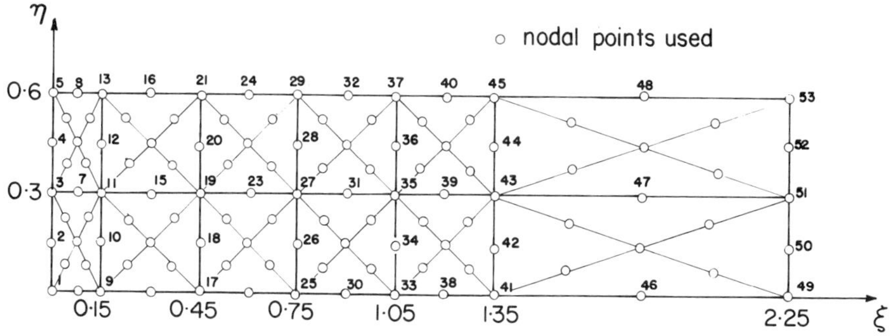

Figure 10.12 Layout for the subcritical Emmons nozzle problem

runs into difficulty as the local maximum Mach number on the wall approaches 1. The problem appears to arise from the local choking condition at $M = 1$, to no surprise. By various special techniques we did manage to get to maximum local Mach number slightly larger than 0·85. However, as the transonic reality must be faced, we consider it more logical to develop a real transonic capability. As a strictly subsonic routine, the procedure has also been somewhat streamlined as shown below.

10.5.2 The supercritical case

For the supercritical case, when the flow goes from subsonic, to sonic in the throat region and finally to supersonic in the divergent part, the stream line pattern ceases to be symmetrical with respect to $\xi = 0$. Emmons[17] has an *ad hoc* procedure for the supersonic part, but a more systematic approach seems preferable.

A look at Figure 10.13, and also at the computed results of Emmons, should convince us that, for the same ψ_w, the stream lines do not deviate significantly from those of the incompressible case. Thus, Equation 10.28 can be simplified somewhat by putting

$$\psi = \psi_w \eta/\eta_w + \psi'$$

and treating ψ' as a perturbation. A little algebra reduces Equation 10.28 into

$$(1 - M^2)\psi'_{\xi\xi} + \psi'_{\eta\eta} + M^2(\ln q_I)_\eta = 0 \qquad (10.33)$$

where M is the local Mach number. Equation 10.33 clearly shows the change of type of the equation as a function of the local Mach number. Compared against Equation 10.31, the explicit $(1 - M^2)$ terms affecting $\psi'_{\xi\xi}$ should account for a large part of the compressibility effects and make the subsonic iteration considerably more effective.

We now recognize the necessity of dividing the region of computation, Figure 10.11, into three patches: a subsonic one ahead of AB, a supersonic

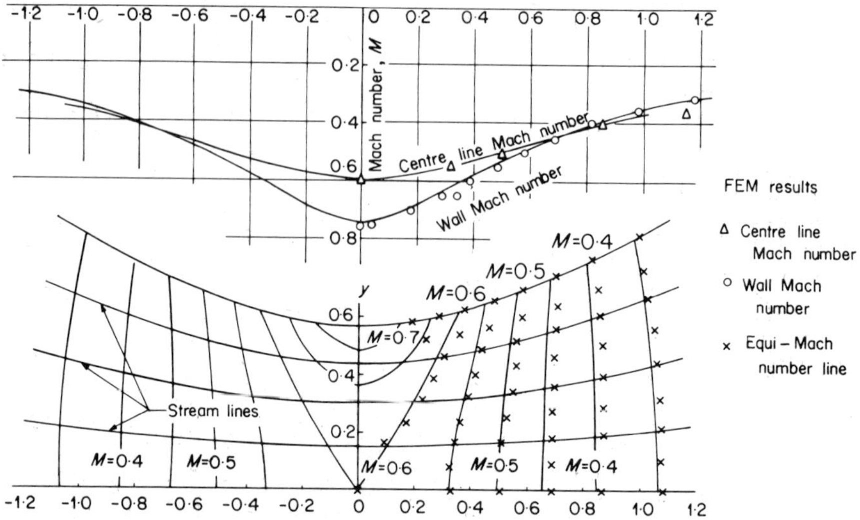

Figure 10.13 Comparison of results for a subcritical nozzle

one after CD and a transonic one bounded by AB and CD and the walls. The location of AB and CD of course is somewhat arbitrary, so long as the sonic line is bracketed between them. The subsonic patch, as a boundary value problem, can be worked out as above except for using Equation 10.33 and replacing the boundary condition on $\xi = 0$ by that on AB. The supersonic part should be constructed with initial data along CD and march downstream. A still different routine must also be devised from the transonic patch.

In this particular problem, it is not necessary to use a general transonic program. A number of analytical solutions for the throat region are in the literature. The Oswatitsch–Rothstein approximation[19] is satisfactory for small throat curvatures, giving

$$
\begin{aligned}
u &\doteq q_s(x)\left\{1 + \frac{1}{2}\left[\frac{y_w''}{y_w} - \frac{y_w'^2}{y_w^2} - \frac{y_w'}{y_w}\frac{q_s'}{q_s}\right][y^2 - y_w^2/3]\right\} \\
v &\doteq q_s(x)\frac{y_w'}{y_w}
\end{aligned}
\tag{10.34}
$$

where $q_s(x)$ is the one-dimensional solution, $y_w(x)$ the channel shape in the throat region and q_s', y_w', etc. are derivatives with respect to x. Hall[20] made a more systematic expansion in terms of the curvature

$$
\begin{aligned}
\frac{u}{a^*} &= 1 + \frac{u_1(\hat{x}, y)}{R} + \frac{u_2(\hat{x}, y)}{R^2} + \cdots \\
\frac{v}{a^*} &= \left(\frac{\gamma + 1}{R}\right)^{\frac{1}{2}}\left[\frac{v_1(\hat{x}, y)}{R} + \frac{v_2(\hat{x}, y)}{R^2} + \cdots\right]
\end{aligned}
\tag{10.35}
$$

where $\hat{x} = x(R/(\gamma + 1))^{\frac{1}{2}}$ and R is the radius of curvature of the channel wall, at the throat, in units of the half throat height. The solutions $u_1, u_2, \ldots, v_1, v_2, \ldots$ can be found in the original paper. Kliegel and Levine[21] showed that the same functions should also appear in an expansion in terms of $1/R + 1$, thus extending the usefulness to at least moderately large curvatures. Their result is in the form

$$\begin{aligned} \frac{u}{a^*} &= 1 + \frac{u_1(\hat{x}, y)}{R + 1} + \frac{u_1(\hat{x}, y) + u_2(\hat{x}, y)}{(R + 1)^2} + \cdots \\ \frac{v}{a^*} &= \left(\frac{\gamma + 1}{R}\right)^{\frac{1}{2}} \left[\frac{v_1(\hat{x}, y)}{R + 1} + \frac{\frac{3}{2}v_1(\hat{x}, y) + v_2(\hat{x}, y)}{(R + 1)^2} + \cdots\right] \end{aligned} \tag{10.36}$$

For a comparison of these approximations against experiment, see Reference 22.

We take now the analytical expression, Equations 10.34, 10.35 or 10.36, for the transonic patch ABCD, in the same spirit as replacing the outer patch by Equation 10.23 in the single airfoil problem before. In particular, the boundary conditions on ψ along AB for the subsonic patch are thus specified. The same iterative program is redone except that Equation 10.33 is used and the local Mach number has to be evaluated also from the last iteration. That the sonic line is solely determined by the local solution is attributable to the assumption of R = const.

What remains to be done is the supersonic patch from CD downstream. Here, again, the initial data along CD are provided by the transonic patch. In the finite element context, the choice to be made is between a space–time procedure, with the ξ axis being evidently essentially time-like, and a semi-discrete one doing the finite element only in the η direction. We have carried out the latter.

For the purpose of demonstration, the finite elementation in the η direction is done in the simplest way corresponding to the linear triangles, i.e. straight line interpolation between nodal points. A Galerkin procedure in the η direction then leads to a set of ordinary differential equations for the nodal values and the marching in ξ direction is by stepwise integration following the Runge–Kutta method. The constant velocity thus evaluated in each element is taken to represent that at the average η value of the element. In this way, the contours for constant Mach number can be sketched in. Our computation is very crude, using only two strips.

Figure 10.14 shows our solution for the Emmons nozzle in a selected case of supercritical flow, by means of the three patches described above. Again there is some discrepancy between our results and those of Emmons. But overall it should be evident that the feasibility of finite element patching to solve the direct nozzle problem appears to be established by the example.

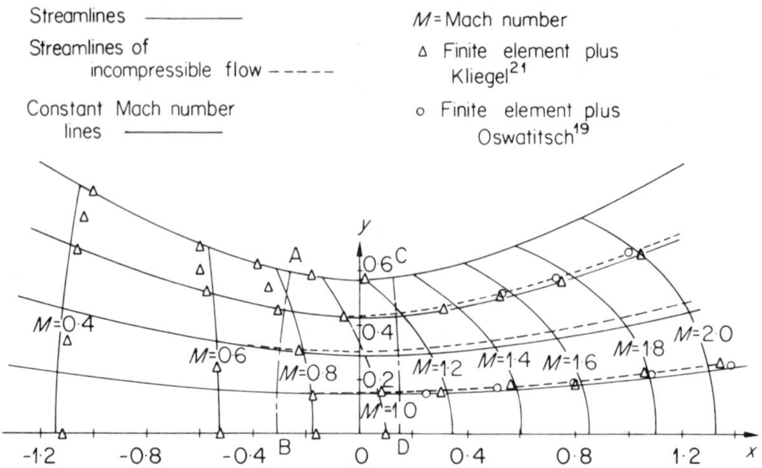

Figure 10.14 Comparison of results for a supercritical nozzle

10.6 Conclusions

Aerodynamic problems often involve infinite domain, local singularities, non-linearity and an equation which may not be elliptic, or changes its type in different parts of the domain of interest. Each of these causes difficulty in the finite element method, some would raise the question of feasibility. We introduce the concept of 'finite element patching', subdividing the domain according to the routine that is necessary, or convenient, in the individual 'patches'. The global solution is then obtained by assembling the patches.

It is always advantageous to bring in a maximum amount of prior knowledge pertaining to a given problem before reducing it for numerical computation. The difficulty of the infinite domain can obviously be removed by utilizing the asymptotic solution. In our terms, it is equivalent to the use of an outer patch that is analytically described. The example of an airfoil in potential flow illustrates how the finite element method in a relatively small inner patch provides accurate results. Once the program for an inner patch with an airfoil is available, the problem of multiple airfoils becomes one of assembling.

For steady supersonic or transonic flows, the semi-discrete finite element method of carrying out the finite elementation only in the space-like directions should be feasible, thanks to the experiences of the closely related 'method of integral relations'. The true space–time finite element method for the wave equation based on linear triangular elements on a regular mesh is shown to be equivalent to the explicit one-sided finite difference method and thus must be subject to the usual Courant–Friedrichs–Lewy criterion. Whether a finite element procedure can be developed to correspond to the unconditionally stable implicit difference formula is an interesting question for further research.

The example of the Emmons nozzle in supercritical flow demonstrates the feasibility of calculating transonic flows by finite element patching. In this case, because of the small throat curvature, we have short-cut the transonic patch by again appealing to analytical approximations.

The problems of a better finite element method for the transonic patch, to cope with large throat curvatures, for example, or for the transonic flow over airfoil, as well as an effective way to handle 'shock-fitting' are left to the future.

Acknowledgements

This work is partially supported by ONR N00014-67-A-0077-0024. The calculation for the airfoils is due to Mr. W. Habashi, and that for the convergent–divergent nozzle is due to Mr. H. C. Chen.

References

1. D. H. Norrie and G. de Vries, *The Finite Element Method—Fundamentals and Applications*, Academic Press, New York, 1973.
2. C. T. Wang, 'Variational method in the theory of compressible fluid', *J. Aero. Sci.*, **15**, 675–685 (1948).
3. C. T. Wang and S. de Los Santos, 'Approximate solution of compressible flows past bodies of revolution by variational method', *J. Applied Mech.*, **18**, 260–266 (1951).
4. C. T. Wang and P. C. Chou, 'Application of variational methods to transonic flows with shock waves', *NACA TN 2539*, 1951.
5. H. Rasmussen, 'A review of the applications of variational methods in compressible flow calculations', *RAE TR 71234*, 1972.
6. D. Greenspan and P. Jain, 'Application of a method for approximating extremals of functionals to compressible subsonic flows', *J. Math. Analy. Appl.*, **18**, 85–111 (1967).
7. O. M. Belotserkovskiy and P. I. Chushkin, 'The numerical solution of problems in gas dynamics', *Basic Developments in Fluid Dynamics*, Vol. 1, M. Holt (ed.), Academic Press, 1965.
8. O. M. Belotserkovskiy, 'Numerical methods in solving steady-state equations of gasdynamics', in *Numerical Methods for Solving Problems of Mechanics of Continuous Media*, O. M. Belotserkovskiy, (Ed.), NASA TTF-667, 1972.
9. M. Holt, 'The design of plane and axisymmetric nozzles by the method of integral relations', *Symposium Transsonicum*, K. Oswatitsch, (Ed.), Springer, 1964, pp. 310–324.
10. N. A. Lukyanov, 'Solution of the direct problem on mixed subsonic and supersonic flow of a gas in a nozzle of finite length', *Doklady*, **17**, 1059–1061 (1973).
11. T. C. Tai, 'Transonic inviscid flows over lifting airfoils with embedded shock wave using method of integral relations', *AIAA Paper No. 73-658*, 1973.

12. S. F. Chen, 'The airfoil problem via the finite element method', *Symp. Application of Computers to Fluid Dynamic Analysis and Design*, Poly. Inst. Brooklyn, January 1973.
13. T. H. H. Pian, P. Tong and C. M. Luk, 'Elastic crack analysis by a finite element hybrid method', *Proc. 3rd. Conf. Matrix Methods in Structural Mechanics*, AFFDL, Dayton, Ohio, 1970.
14. M. Gurtin, 'Variational principles for linear elastodynamics', *Arch. for Rat'l. Mech. and Analysis*, **16**, 34–50 (1969).
15. R. D. Richtmeyer and K. W. Morton, *Difference Methods for Initial Value Problems*, 2nd ed., Interscience, 1967.
16. E. M. Murman and J. D. Cole, 'Calculation of plane steady transonic flows', *AIAA J.*, **9**, 121–141 (1971).
17. H. W. Emmons, 'The theoretical flow of a frictionless, adiabatic perfect gas inside a two-dimensional hyperbolic nozzle', *NACA TN 1003*, 1946.
18. J. H. Argyris, K. E. Buck, J. F. Gloudeman and D. W. Scharpf, 'Some aspects of finite element techniques', *Proc. Symp. Finite Element Techniques*, Univ. Stuttgart, 1969.
19. K. Oswatitsch and W. Rothstein, 'Flow pattern in a converging–diverging nozzle', *NACA TM* 1215, 1949.
20. I. M. Hall, 'Transonic flow in two-dimensional and axially-symmetric nozzles', *Quart. J. Mech. and Appl. Mech.*, **15**, 487–508 (1962).
21. J. R. Kliegel and J. N. Levine, 'Transonic flow in small throat radius of curvature nozzles', *AIAA J.*, **7**, 1375–1378 (1969).
22. R. F. Cuffel, L. H. Back and P. F. Massier, 'The transonic flowfield in a supersonic nozzle with small throat radius of curvature', *AIAA J.*, **7**, 1364–1366 (1969).

Chapter 11

Lubrication Problems—The Selection of Mathematical Models

F. T. Barwell

11.1 Introduction

Generalization about lubrication of bearings (defined here as those situations where force is transmitted between surfaces in relative motion) must be approached with caution because of the very wide range of applied forces, velocities, geometries and lubricants which are encountered in modern engineering. However, from the point of view of the theory of machines these situations can be divided into two classes, lower and higher pairs as illustrated in Figure 11.1. The simplest case of a lower pair, shown in Figure 11.1(a) is familiarly encountered on various machine tool slideways etc.

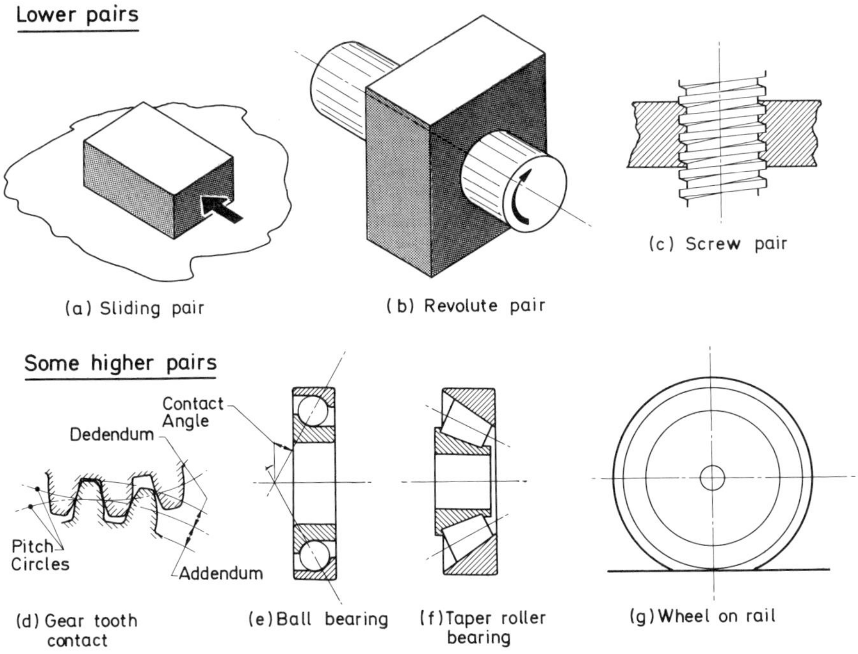

Figure 11.1 Lower and higher pairs

The revolute pair shown in Figure 11.1(b) characterizes the majority of bearings whilst Figure 11.1(c) shows a combination of rotary and linear motion. These three pairs can be referred to as 'sliding pairs', 'turning pairs' and 'screw' pairs respectively. They are the only pairs in which surfaces in continuous contact can slide over each other.

All other pairs must accommodate relative motion between the elements which is partly turning and partly sliding and are known as higher pairs. An important practical condition is that, whereas the lower pairs allow contact to be distributed over a surface, the higher pairs permit only point or line contact. From the tribological point of view this classification can be replaced by 'conformal' and 'counterformal' pairs. Common examples of higher pairs are shown in Figure 11.1. Figure 11.2 illustrates how, in the common case of involute gears, the proportion of sliding to rolling motion changes throughout the cycle of contact.

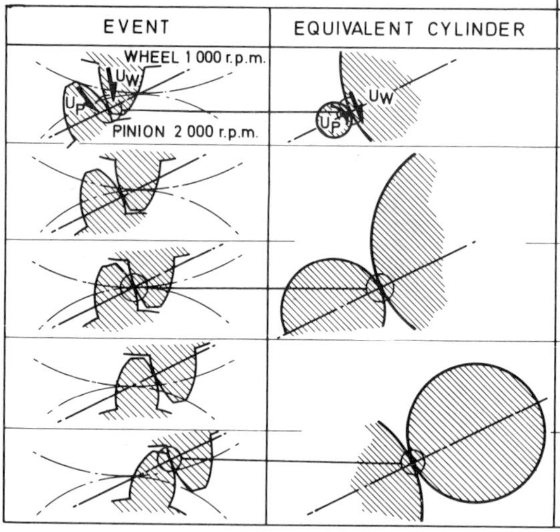

SLIDE/SWEEP RATIO: WHEEL $\frac{U_W-U_P}{U_W}$	SLIDE/SWEEP RATIO: PINION $\frac{U_W-U_P}{U_P}$	SLIDING/ENTRAINMENT RATIO $\frac{U_W-U_P}{U_W+U_P}$
0·65	1·87	0·58
0·49	0·71	0·27
0	0	0
-0·67	-0·40	-0·27
-1·73	-0·63	-0·45

Figure 11.2 Involute gears as higher pairs

The area of conformal pairs is not limited, permitting a low value of intensity of surface stress and enabling soft material to be employed. When one material is softer than the other the combined action of plastic deformation and wear may even enhance the degree of conformity.

Effective lubrication however requires that the surfaces should be nearly but not quite parallel as indicated in Figure 11.3. Sufficient pressure to separate the surfaces is generated by the wedge action. The normal applied force is resisted by fluid pressure thus reducing friction and eliminating wear.

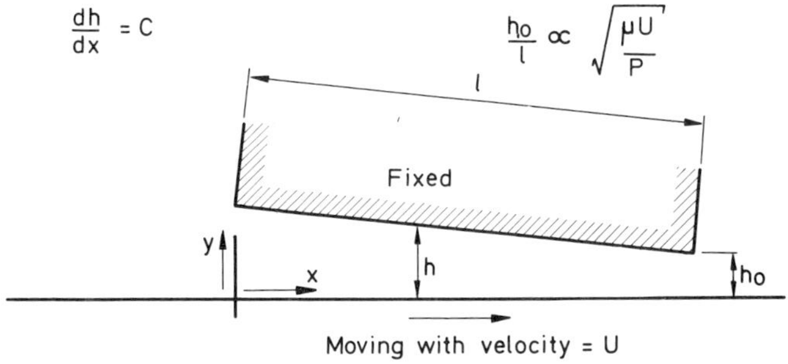

Figure 11.3 Notation for tapered wedge bearing

This is known as 'hydrodynamic lurication' and is characterized by the well known Reynolds equation[1] as follows:

$$\frac{\partial}{\partial x}\left(\frac{h^3}{12\mu}\frac{\partial p}{\partial x}\right) + \frac{\partial}{\partial z}\left(\frac{h^3}{12\mu}\frac{\partial p}{\partial z}\right) = \left(\frac{U_0 + U_1}{2}\right)\frac{\mathrm{d}h}{\mathrm{d}x} + V \qquad (11.1)$$

where x is measured in a direction of relative motion
z is measured in the plane of contact and perpendicular to the direction of motion
h is the distance of separation of the surface (film thickness)
U_0 and U_1 are the translational velocities of the two surfaces in the x direction
V is the relative velocity of approach of the two faces
μ is the viscosity.

A number of assumptions were made in a generation of this equation of which the following are the more important:

(1) that viscosity remains constant throughout the extent of the pressure film;
(2) that the inertia terms governing the motion of the fluid may be neglected;
(3) that the fluid is incompressible;
(4) that the fluid is Newtonian;
(5) that the motion of the fluid in a direction normal to the surface can be neglected in comparison with motion parallel to them;
(6) that there is no pressure gradient through the thickness of the film;
(7) that the surfaces are infinitely rigid.

No explicit account has been taken of variation of temperature with consequential change in viscosity or the distortion of surfaces due to thermal gradients.

An indication of the range of properties of lubricants employed in modern engineering is given in Table 11.1.

Table 11.1 Typical range of lubricant properties

Substance	Absolute viscosity (μ)	Density (ρ)	Kinematic viscosity (ν)	Thermal conductivity (k)	Thermal capacity
	N s/m^2 or Pa s	kg/m^3	m^2/s	W/(m degC)	J/(kg degC)
Range of lubricating oils	0·01 to 0·5	875	$5{\cdot}6 \times 10^{-5}$ to $1{\cdot}1 \times 10^{-6}$	0·123 to 0·133	1850 to 2150
Water at 20 °C	$1{\cdot}01 \times 10^{-3}$	1000	1×10^{-6}	0·605	4170
Air at 20 °C and 760 mm pressure	$1{\cdot}80 \times 10^{-5}$	1·29	$1{\cdot}46 \times 10^{-5}$	0·0254	1005

To take into account all the factors enumerated above would lead to impossibly complicated equations and progress can only be made by selection of the important features of any situation neglecting those likely to have a minimal effect. One possible way of distinguishing those factors which are important from those of less consequence is to consider various non-dimensional parameters such as Reynolds number. The duty parameter $P/\mu U$ where P is per unit width and $U = (U_1 + U_2)/2$, a measure of the load intensity or its reciprocal form, $\mu U/P$ can be employed to separate out certain features of importance as indicated in Figure 11.4. Thus in line 2 where speeds are low and pressures high, it is indicated that film thickness will be low and that the fluid will be constrained to flow in a laminar manner so that inertia terms may be neglected. On the other hand high-speed bearings are likely to be turbulent but pressures will be low so that the surfaces can be taken as being infinitely rigid. Again, because pressure will be low, change of viscosity with pressure can be neglected. Where the value of the parameter is low it may mean that pressures are high and that the distortion of the surfaces and the effect of pressure on viscosity will be of major significance. This will be referred to later in the treatment of 'elasto-hydrodynamic lubrication', Section 11.7. Then again, in a heavily loaded contact, film thicknesses will be thin, the volume of lubricant passing will be small and the amount of heat conducted through the walls will be of great significance. Conversely, in a high-speed bearing working with a thick film, the heat will be generated in the oil and most of it convected away so that only an insignificant proportion is conducted through the walls of the bearing.

ASSUMPTION	NON DIMENSIONAL SPEED FACTOR $\mu U/P$		
	LOW	MEDIUM	HIGH
1) VISCOSITY REMAINS CONSTANT THROUGHOUT PRESSURE FILM	NO	DOUBTFUL	NO
2) INERTIA TERMS CAN BE NEGLECTED	YES	YES	NO
3) FLUID IS INCOMPRESSIBLE	YES	YES	YES
4) COMPONENT OF MOTION NORMAL TO SURFACE CAN BE NEGLECTED	YES	YES	DOUBTFUL
5) NO TRANSVERSE PRESSURE GRADIENT	YES	YES	DOUBTFUL
6) SURFACES INFINITELY RIGID	NO	YES	YES
7) SURFACES DO NOT DISTORT DUE TO THERMAL GRADIENTS	DOUBTFUL	YES	YES
8) HEAT CONDUCTED INTO WALLS	YES	NO	NO
9) HEAT CONVECTED AWAY BY LUBRICANT	NO	YES	YES

Figure 11.4 Applicability of various assumptions used in derivation of Reynolds equation to different operating conditions

The majority of hydrodynamic journal and thrust bearings of the conformal type used in traditional engineering occur in the medium range. Here useful design predictions have been made by using Reynolds equation in a simplified form,[2,3] notably by neglect of the pressure induced components of the velocity profile in the x direction.

However, under extreme conditions of loading, speed or other conditions, one or other of the assumptions enumerated ceases to be applicable and a consequential development of the theory necessitates the use of numerical methods. The mineral oils generally used as lubricants are characterized by values of viscosities which fall off as temperature is raised as shown in Figure 11.5. Every bearing involves the irreversible degradation of mechanical energy into heat to a greater or lesser degree and it must therefore be treated as a thermal system. Viscosity cannot therefore be regarded as constant although average values may be sufficiently accurate for a useful range of design.

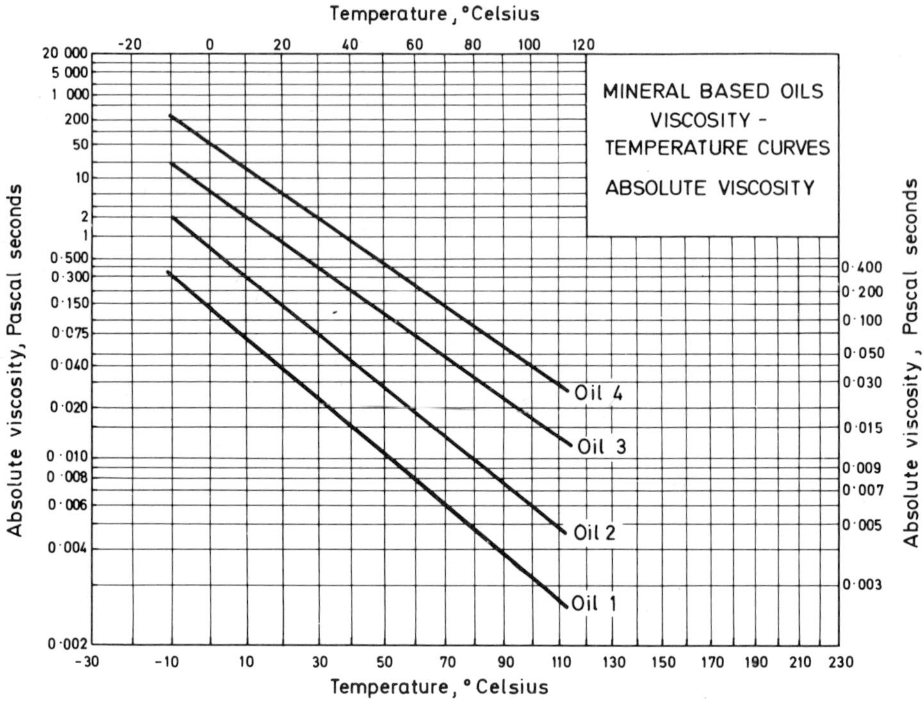

Figure 11.5 Variation of viscosity with temperature

11.2 Plane thrust bearings

If the inclination of the surfaces to each other can be written as $dh/dx = C$ which is constant, then Equation 11.1 becomes

$$\frac{\partial^2 p}{\partial x^2} + \frac{3}{x}\frac{\partial p}{\partial x} + \frac{\partial^2 p}{\partial z^2} + \frac{\sigma \mu U}{C^2 x^3} = 0 \qquad (11.2)$$

which was solved by Michell in 1905 by substituting a trigonometrical series for p and evaluating it using Bessel functions.[4] Total load was then obtained by graphical or numerical integration. Frictional drag was found to be directly proportional to bearing width, the analytical expressions for infinitely wide bearings being directly applicable. Equation 11.2 has also been solved for the sector shaped pad by Mrs. W. L. Wood[5] by the introduction of the Struve function for imaginary argument.

Heat will be developed in an element of fluid undergoing shear and it can be shown that in a tapered wedge bearing there is a considerable variation of temperature throughout the film, the hottest parts being nearest to the moving surface at entry and near the stationary surface at the exit. Figure 11.6, taken from Michell's book, shows the result of his calculations of the rate of heat generation at each point in the lubricant film.

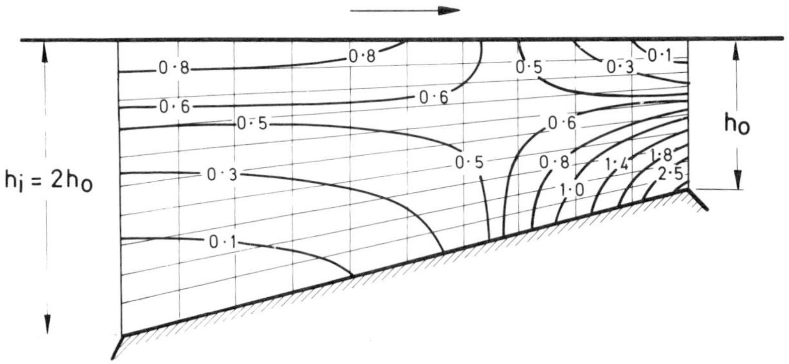

Figure 11.6 Heat generated in a tapered wedge bearing

Whilst it was recognized that the corresponding change in viscosity would markedly affect the velocity distribution it was considered sufficiently accurate for design purposes to assume a constant viscosity based on a mean temperature. However, experiments by Fogg[6] indicated that pressure could be generated between parallel surfaces in relative motion and he postulated a 'thermal wedge' hypothesis to explain this. Whilst it is the writer's view that thermal and elastic distortion can account for the observed effects,[7] nevertheless a number of workers were stimulated to reexamine the hydrodynamic lubrication film using numerical methods. A representative treatment is that published by Cope[8] who simultaneously solved the continuity equation, the Navier–Stokes equations and the Energy equation. Using a self-consistent system of units he derived the values reproduced in Table 11.2, which shows that the dissipation terms were several orders of magnitude greater than the dilatation terms. The inertia terms and conduction in the x direction were also negligible.

Table 11.2 Order of magnitude of derivatives

	$\rho u \dfrac{\partial u}{\partial x}$	$\mu \dfrac{\partial^2 u}{\partial y^2}$	$\mu \left(\dfrac{\partial u}{\partial y}\right)^2$	$p\Delta$	$\dfrac{\partial p}{\partial x}$	$k\dfrac{d\theta}{dx}$
Gas	10^3	10^8	10^{12}	10^9	10^8	10^{-1}
Liquid	10^5	10^9	10^{15}	10^{10}	10^9	10

u = velocity of a particle of fluid in the x direction.
Δ = dilatation.

Table 11.3 presents some realistic quantities relating to a typical oil lubricated high-speed bearing.

Table 11.3 Realistic quantities applicable to a high-speed thrust bearing

R	h	ρ	μ	U	$\dfrac{\partial p}{\partial x}$	P (maximum)	$\partial u/\partial y$ (average)
	m	kg/m^2	Pa s	m/s	N m/m^2	Pa	
0·04	$2{\cdot}54 \times 10^{-5}$	875	7·0	732	10^8	$4{\cdot}2 \times 10^6$	$2{\cdot}9 \times 10^7$

Other workers[10,11,12] have provided solutions to the hydrodynamic equations by conventional means using similar assumptions but Cameron[13] reported a net load carrying capacity between parallel surfaces in relative motion attributable to transverse variation in viscosity. Zienkiewicz[14,15] had combined a step-by-step integration with a relaxation process to take into account the variations in viscosity within the thickness of the film. He considered the equilibrium of an element of oil within the film

$$\frac{\partial p}{\partial x} = \frac{\partial}{\partial y}\left(\mu \frac{\partial u}{\partial y}\right) \tag{11.3}$$

and assumed that, in general, both the density and the viscosity varied exponentially with temperature, thus $\mu = \mu_0 \, \mathrm{e}^{-\beta\theta}$ and $\delta = \rho_0 \, \mathrm{e}^{-\sigma\theta}$.

Transverse velocity and inertia terms were neglected. Two cases were evaluated, one in which the solid surfaces were taken to be at the same temperature as the incoming lubricant and one in which the moving boundary was at a constant temperature above that of the incoming oil. In both cases the results demonstrated the existence of a net load carrying capacity between parallel surfaces in relative motion. However, Huebner,[16] applying finite element methods to the solution of the problem in three dimensions, predicts a negative load carrying capacity for parallel bearing surfaces.

Neal[17] points out that in many actual bearings a layer of hot 'carry over' oil will adhere to the runner surface which, in association with fresh cool oil superimposed thereon will lead to a temperature gradient across the lubricant film at entry with consequent viscosity variation. He presents tabular data which indicates that a serious loss of load carrying capacity may result. This provides some theoretical support for Coles's experimental observations.[17]

Tieu,[18] who also publishes solutions of the two-dimensional case including the effect of variation of viscosity with temperature, compares the finite element method with the finite difference method previously used. He finds the main advantage of the finite element method to be the general flexibility in the field conditions and the ease of precribing boundary conditions. The element size and orientation can be adjusted to suit critical function gradients and complex curved boundaries in a bearing which may be distorted thermally and elastically. A drawback of the method is the lengthy manual preparation of large amounts of input data to specify the elements and nodes. Difficulties are also expected in treatment of higher-order approximations of the field variables.

Brand[19] derives expressions for a sector shaped bearing pad using cylindrical coordinates wherein R is the radius of the runner, h = average film thickness and ω its angular velocity. For inertia terms to be important,

$$\rho\frac{\omega h}{\mu} \approx \frac{1}{h} \tag{11.4}$$

Multiplying both sides by R gives, for the left-hand term $\rho/\mu U_h$ which may be regarded as a Reynolds number for the bearing. Simplifying, the equation/inequality becomes

$$\rho\frac{U}{\mu} \approx \frac{R}{h^2} \tag{11.5}$$

and inserting values from Table 11.3 we have $9{\cdot}3 \times 10^2$ for the left-hand side and $5{\cdot}44 \times 10^7$ for the right-hand side. Thus, in the typical case considered, inertia terms can be neglected. For journal bearings r and c should be substituted for R and h.

The problem of variation of viscosity along the film is usually met by assuming a mean value. Thus Neal[20] assumes a linear rise in temperature along the film calculated according to Equation 11.18 and, inserting corresponding viscosity, completes performance predictions using a finite difference method. The results agree with experimental data.

11.3 Journal bearings

Here the wedge action which determines the generation of pressure derives from the tendency of the shaft to take up a position which is eccentric to the bearing shell as depicted in Figure 11.7.

It is impossible to even list the vast literature on the subject and only a few examples will be noted. The main theoretical difficulties posed by this type of bearing are as follows: (a) definition of the limits of integration of the pressure curve; determination of the downstream limit is particularly difficult because of the initiation of cavitation; (b) variation of viscosity with temperature; (c) complexity of the three-dimensional case; (d) the effect of inertia terms. Generally the working fluid may be taken as incompressible but this may be unsatisfactory where the lubricant is a gas. Sommerfeld[21] by ignoring all the above difficulties solved the two-dimensional case by

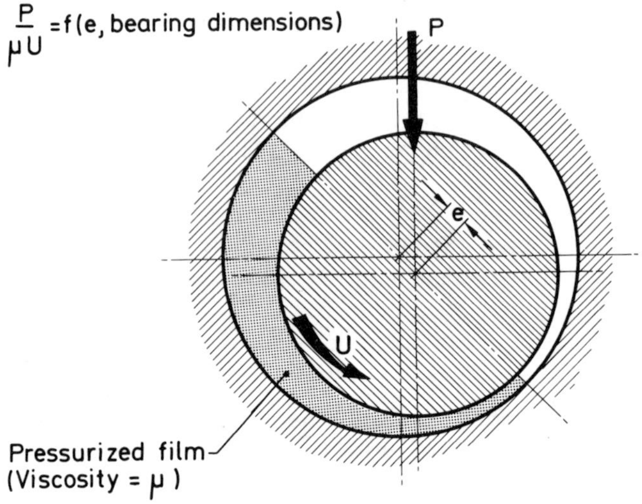

Figure 11.7 Diagram of a simple journal bearing

implying that the fluid was able to support unlimited negative pressures. This provided a misleading description of the shape of the locus of the shaft centre but introduced a generally useful non-dimensional parameter which governed eccentricity ratio and bearing torque as follows:

$$S = \left(\frac{r}{c}\right)^2 \frac{\mu U}{P} \tag{11.6}$$

where r is shaft radius and c denotes radial clearance. Christophenson[9] has provided a solution to Equation 11.1 using Southwell's relaxation method for both wedge and partial journal bearings taking into account the variation of μ in the x and z directions. He concluded that the iso-viscous solution gave good results for the load, the quantity of flow through the bearing and the temperature rise but poor results for the frictional resistance and load position. Cameron and Wood[10] treated the full journal bearing and Sassenfeld and Walther[22] and Jakobsson and Floberg[23] have extended the available data based on iso-viscous treatment.

In contrast to the above workers who used finite difference methods, Reddi[24] claims that, once a computer program with sufficient input flexibility has been generated, it becomes simpler to use the finite element method.

The selection of the correct boundary conditions governing the extent and shape of the pressure film presents much difficulty. This question is discussed by Tipei[25] who invokes the observation of Cole and Hughes.[26]

In all the above treatments the applied load has been assumed to be constant in magnitude and direction as well as the relative velocity of the interacting surfaces. Many practical situations exist where all three quantities vary, notably in the big end bearings of engines. Therefore graphical and numerical treatments have been evolved to enable design predictions to be made. An excellent treatment has been published by Hornsell and McCallion.[27] Milne[28] has applied the finite element method to transient conditions taking cavitation into account.

Where the lubricant is a gas, variation of viscosity with temperature may be neglected but compressibility may become important. Castelli and Pirvics[29] review a number of numerical methods in gas bearing film analysis. Reddi and Chu[30] have postulated two variational formulations for the steady state compressible Reynolds equation which they solve using finite element methods.

11.4 Externally pressurized bearings

The wedge action in both tapered and journal bearings is the direct product of speed and viscosity. Therefore hydrodynamic lubrication does not occur naturally when a mechanism is required to start from rest. However, if pressure can be supplied from an external source, surfaces can still be separated by pressurized fluid film so as to minimize friction and wear. Thus when it is desired that performance should not be dependent on speed or where it is desired to use a fluid, notably a gas, with a very low viscosity, externally pressurized bearings can be used. The main features of such bearings can be calculated very simply but their performance is very much determined by supply conditions. Thus stiffness depends on the relationship between the quantity of fluid provided by the supply system and the back pressure created in the bearing by the applied load. Suitable 'feedback' controls may even provide a negative stiffness to compensate for deflections occurring elsewhere in the machine structure.

Where the bearing surfaces are of simple shape and the applied conditions are not extreme, simple calculations suffice as the basis for design.[31] When, however, the form of the bearing is complicated, where loads are sufficiently intense to cause significant elastic deformation of the interacting surfaces or where the lubricating fluid becomes heated, numerical methods become essential. This is also the case for hybrid bearings, i.e. bearings which are purely hydrostatic at low speeds but wherein hydrodynamic action occurs at higher speeds to augment the lift.[32]

Finite element methods are applied by dividing the bearing area into grids of finite spacing and iterating until both the continuity and the Reynolds equations are satisfied simultaneously.

Where the relationship between cavity pressure and rate of supply is simple, for example when there is a constriction in the supply line to the cavity which takes the form of a capillary tube, a double iteration process may be made to incorporate the finite element analysis as demonstrated by Allan.[33]

11.5 Inertia effects, turbulent conditions

Although the film conditions within a bearing may be laminar the possibility exists that inertia effects may be important at the entry. Milne[34,35] introducing the stream function and considering this as a series, derives a correction due to inertia effects in laminar flow between inclined planes. His final expressions are more general than those previously obtained and indicate that the effect of lubricant inertia is to cause a slight increase in the load capacity of a bearing.

The increase in size of electrical turbo-alternators[36] and the use of unconventional lubricants such as water and liquid metals has led to the occurrence of turbulent flow within bearings. Sir Geoffrey Taylor[37] showed that vortex conditions occur in an annular space bounded by a stationary cylinder on the outside and a rotating cylinder on the inside when $\omega = 15{\cdot}28\,(\eta/p)(1/d^2)(r/c)^{\frac{3}{2}}$. The effect of eccentricity has been explored by Cole[38] using model experiments with the following result:

$$U > 7{\cdot}64(\mu/\rho)(1/d)(r/c)^{\frac{3}{2}}(1 + 0{\cdot}89\varepsilon^2) \qquad (11.7)$$

where d = diameter of shaft = $2r$.

Some experimental work has been conducted in this area although it has been usually restricted to gross measurements such as torque and load capacity.[39] The detailed investigation of the turbulent lubricant film, in which flow parameters such as velocity, temperature and pressure distributions may be examined, has remained the prerogative of the theorists who have adopted similar approaches to those used for laminar flow analysis in that the convective inertia terms in the Navier–Stokes equations have been neglected thus reducing them to the same equations as for the laminar regime except for additional Reynolds stress terms. The equations are then recast in the form of a turbulent Reynolds equation in which the absolute viscosity is replaced by its turbulent counterpart. The success of such a theory of turbulent lubrication now depends on an accurate description of this latter quantity which in the present context is two-dimensional in nature. Two proposals that have come to the forefront in the past decade have used either a mixing length or eddy viscosity concept. Both, however, have been based on conditions of fully developed turbulent flow in pipes (essentially a one-dimensional approach). These analyses have been reasonably successful in predicting characteristics of bearings operating in this regime. However, such an achievement may be attributed to the fact that the turbulent flow data used by them contained adjustable constants that can be chosen to fit the existing experimental data.

If theoretical analyses of turbulent lubrication situations are to be more realistic, a fresh approach is required which will have to include the convective inertia effects, since these are of immense importance at the higher Reynolds numbers, as well as more accurate turbulent exchange properties. In this respect an analysis[40] has been developed in which the complete momentum and continuity equations describing the two-dimensional flow field have been reduced by introducing stream function and vorticity as variables in place of velocity components and pressure. Furthermore, with the particular model of turbulence adopted in the analysis (the Kolmogenov–Prandtl model), the equation for the energy of turbulence was also derived. The three equations have been solved simultaneously using a finite difference procedure while using only the minimum of relevant experimental data to exhibit the two-dimensional nature of the turbulent exchange properties.

Figure 11.8 shows the results of the calculations of Medwell. It will be noted that the onset of turbulence has a profound effect on the velocity profile of the lubricant film.

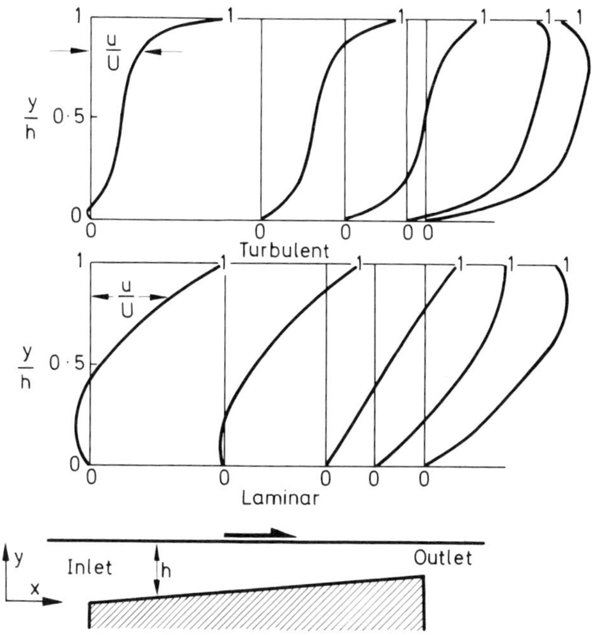

Figure 11.8 Comparison of laminar with turbulent flow

It is urgently necessary for two-dimensional treatment to be extended to three dimensions so that the thermal problems of actual bearings can be simulated in a realistic manner. It is here that the finite element method may score over the finite difference method used up to the present.

11.6 Non-Newtonian behaviour—rheodynamic lubrication

The assumption that a fluid is Newtonian is obviously inapplicable to grease lubrication and may not apply at the high pressures and shear rates applicable to counterformal contacts. The simple case of a Bingham body has been explored by Milne[41,42] and, as shown in Figure 11.9 the 'core flow' in a capillary has its counterpart in dead zones in wedge and journal bearings. This resulted in a slight reduction in load carrying capacity.

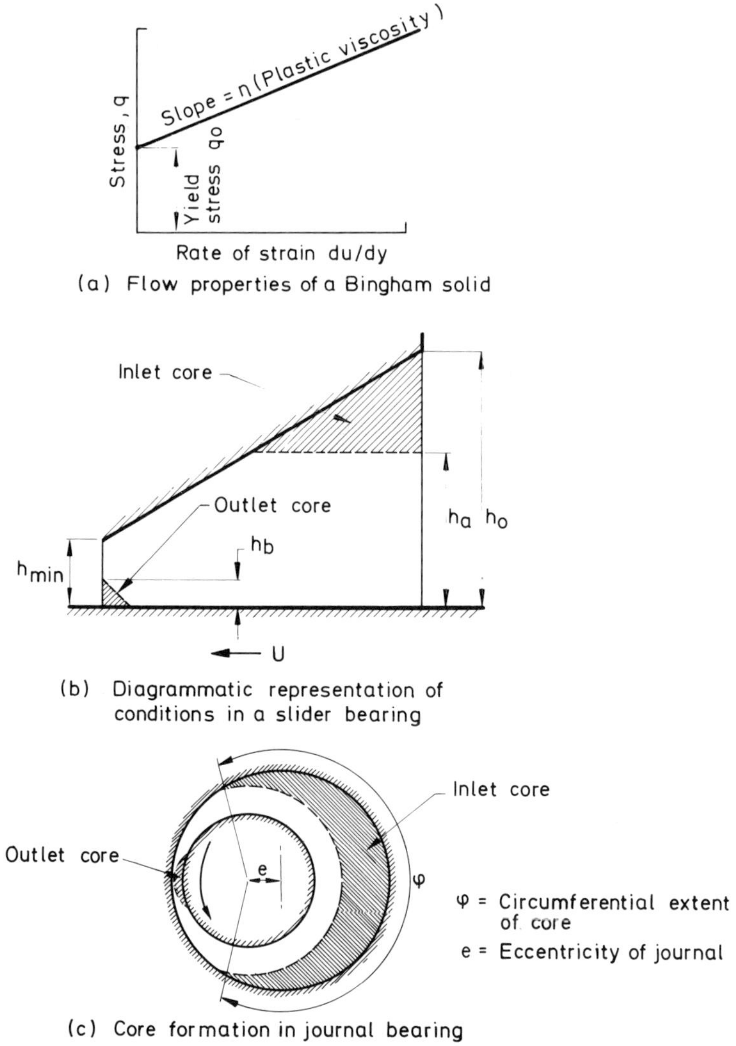

Figure 11.9 Flow of Bingham body in thrust and journal bearing

More complex behaviour is to be expected from polymeric liquids such as mineral oil which are all visco-elastic to a greater or less degree.[43] Some workers have applied the Maxwell model but the writer prefers the Barlow and Lamb model which involves the concepts of reactance and impedance borrowed from electrical engineering.[44]

Shear compliance is composed of two elements, J' and J'' which will be in quadrature.

$$J'(\omega) = \frac{1}{G(\infty)} + \frac{\sqrt{2}}{\sqrt{\omega\mu(0)G(\infty)}}$$
$$J''(\omega) = \frac{1}{\omega\mu(0)} + \frac{\sqrt{2}}{\sqrt{\omega\mu(0)G(\infty)}} \qquad (11.8)$$

where ω represents the angular frequency of oscillatory shear. $G(\infty)$ is the value of shear modulus at infinite frequency of oscillation, derived by extrapolation of the results of tests at various frequencies; $\mu(0)$ is the viscosity at zero shear rate.

Dyson[45] has applied this model to the lubrication of rolling discs and obtained agreement over that range of sliding speed which is low enough to avoid the complication of thermal effects.

11.7 Concentrated contact—elastohydrodynamic lubrication

The foregoing considerations all relate to bearing systems wherein the two interacting surfaces conform to each other so that load is spread over a sufficient area for pressures to be limited. However, instances abound in engineering, notably the flanks of gear teeth or cams and cam followers, where the interacting surfaces are counterformal and the load is concentrated on a limited area tending towards a line or a point. Until comparatively recently it was thought that it would be impossible for a fluid film to exist under the conditions of speed and load encountered in practice, solutions of Reynolds equation indicating ridiculously low film thicknesses.[46] The main advances in the subject during recent years have derived from the realization of two factors not originally included in Reynolds equation, notably the facts that the surfaces conform more closely because of local elastic deformation and that viscosity can be very greatly increased under the action of high local pressures. The term 'elastohydrodynamic lubrication' thus implies the existence of high local pressures up to say 10^9 N m^{-2} (150,000 lb/in^2). Shear rates may be up to 10^7 S^{-1} and the time of interaction may be from 1 to 100 μs. Film thickness may vary from 0·1 to 1 μm (4 to 40 μin), temperatures may reach 400 °C but the length of the contact region is unlikely to exceed 0·1 mm.

Theory must take into account the shape of the surfaces as deformed under pressure, this pressure itself depending upon the deformed shape and the viscosity. The viscosity will in turn depend on the pressure. Therefore the only possibility of solving the equation is by an interactive process and a great deal of work has been reported in recent years[47] leading to a reasonable agreement between theoretical predictions and experimental measurements of film thickness.

Johnson has published a chart[48] which enables the relative contributions of variable viscosity and elasticity to be assessed. He points out that variations in viscosity with pressure become important when

$$(\alpha^4 \mu_0 U E^3/r)^{\frac{1}{4}} > 0{\cdot}1 \tag{11.9}$$

where $1/E = \frac{1}{2}[(1 - \gamma_1^2)/E_1 + (1 - \gamma_2^2)/E_2]$ where γ_1 and γ_2 and E_1 and E_2 are the Poisson ratios and elastic moduli of the two surfaces respectively, and $r = r_1 r_2/r_1 + r_2 . \alpha =$ exponent in expression for viscosity where $\mu = \mu_{0\varepsilon}^{\alpha P}$ and $U = (U_1 + U_2)/2$. When this quantity exceeds 2·2, theory indicates the existence of a sharp pressure peak on the exit side of the nip which has not yet been observed experimentally.

At very light loads the effect of both variable viscosity and elastic deformation may be neglected. Quantitatively when

$$\left(\frac{\alpha^2 P^3}{\mu_0 U_r^2}\right)^{\frac{1}{2}} < 1 \quad \text{and} \quad \left(\frac{\alpha^2 PE}{r}\right)^{\frac{1}{2}} < 1{\cdot}5 \tag{11.10}$$

The finite element method appears to be particularly appropriate to the iteration process necessary for the solution of the elastohydrodynamic equations and Taylor and O'Callaghan have used the 'cubic quadrilateral element' because this can accommodate rapid variations in geometry.[49] Starting from an assumed film geometry and boundary conditions they calculated a pressure distribution from which a surface deformation pattern was obtained. These deformations were then added back to the previous surface profile to yield a new film shape. This was repeated until compatibility of pressure and displacement was achieved. The pressure dependence of viscosity was then introduced into the convergence scheme and the results agreed with the most heavily loaded case previously published.[50]

The results of these calculations are reproduced in Figure 11.10 in comparison with those of Reference 50. Also included is the Hertz pressure distribution which would occur if the same load were applied in the absence of lubricant and the curves resulting from the neglect of enhancement of viscosity with pressure or the elastic deformation of the contacting bodies. Further work is necessary to secure convergence at the highest pressures. The program was extended to the calculation of sub-surface stresses for comparison with photoelastic dynamic studies using synthetic sapphire.

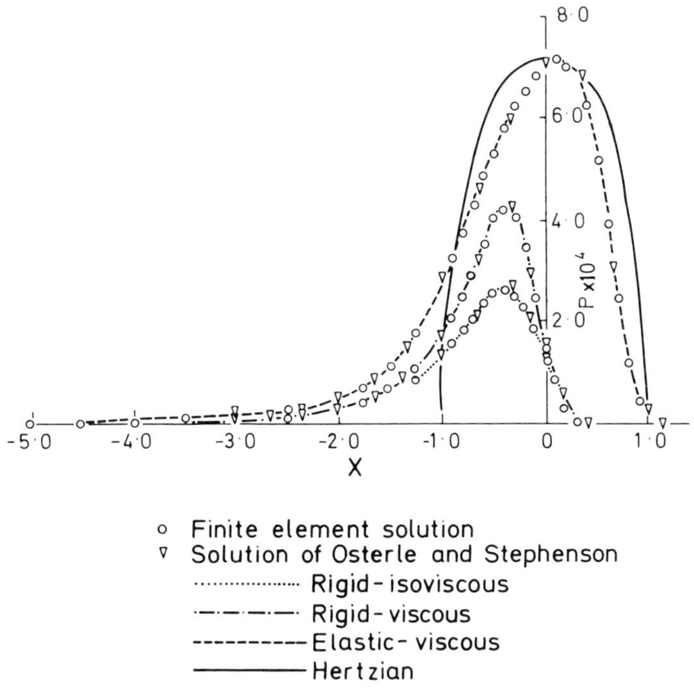

Figure 11.10 Elastohydrodynamic solution

The combination of grease type properties of fluids with hydrodynamic lubrication conditions has been studied by Kauzlarich[51] who uses power law flow model[52] as follows:

$$\tau = \pm[\tau_y + \phi|D|^n] \tag{11.11}$$

where τ = shear stress,
τ_y = limiting shear stress (yield stress),
ϕ = plastic viscosity,
D = shear rate,
n = a constant.

The grease yield stress is found to have a negligible effect on E.H.L. performance.

Further work is still required to enable theoretical and experimental determination of the magnitude of surface tractions in elastohydrodynamic lubrication are to be reconciled. The solution may lie in the recognition of non-Newtonian flow properties of the lubricant.

11.8 Computer-aided design

The complexity of some tribological systems, ball bearings for example,[53] coupled with the wide choice of variables in the equations themselves tend to inhibit the full application of research results to engineering practice. However, by employing the computer[54] mathematical models may be set up which embody the relevant research results. The predictions of such a model may be compared with experimental results and the theory refined until agreement is satisfactory. When this has been achieved, the model may be used to design any number of different bearings provided the relevant parameters lie within the range of the model. In favourable circumstances it is possible to develop optimum designs, for example Blount,[55] for the case of thrust bearing and Ashton[56] for journal bearings.

References

1. D. M. McDowell and J. D. Jackson, (eds.), *Osborne Reynolds and Engineering Science Today*, Manchester University Press, Barnes and Noble, New York, 1970.
2. F. W. Ocvirk and G. B. DuBois, 'Analytical derivation and experimental evaluation of short bearing approximations of full journal bearings', *N.A.C.A. Tech. Rep. 1157* (1953)
3. F. T. Barwell, 'Optimization of bearing design', *Proc/ International Symposium of Plain Bearings, Tatra, Czechoslovakia*, 1972, pp. 82–123.
4. A. G. M. Michell, 'The lubrication of plain surfaces', *Zeit Math. und Physik*, **52**, 123–137 (1905); see also *Lubrication*, Blackie, London, 1950.
5. Mrs. W. L. Wood, 'Note on a new form of the solution of Reynolds equation for Michell rectangular and sector shaped pads', *Phil. Mag.*, 7th series, **XL**, 220–226 (1949).
6. A. Fogg, 'Fluid film lubrication of parallel thrust surfaces', *Proc. I. Mech. E.* **155**, 49–53 (1946).
7. J. A. Cole, 'Experimental investigation of power loss in high-speed plain thrust bearings', *Proc. I. Mech. E. Conf. on Lubrication and Wear*, 1957, pp. 158–163.
8. W. F. Cope, 'The hydrodynamical theory of film lubrication', *Proc. Royal Soc.* (*A*), **197**, 201–217 (1949).
9. D. G. Christophenson, 'A new mathematical method for the solution of film lubrication problems', *Proc. I. Mech. E.*, **146**, 3, 126–135 (1941).
10. A. Cameron and Mrs. W. L. Wood, 'The full journal bearing', *Proc. I. Mech. E.*, **161**, 59–72 (1949).
11. F. Charnes, F. Osterle and E. Saibel, 'On the energy equation for fluid-film lubrication', *Proc. Royal Soc.* (*A*), **214**, 133–136 (1952).
12. G. Vogelpohl, 'Der Ubergang der Reibungswarme von Lagern', *V.D.I. Forschungsheft*, **425** (1944).
13. 'Hydrodynamic lubrication of rotating discs in pure sliding. A new type of oil film formation., *J. Inst. of Petrol.*, **37**, 471 (1951).
14. O. C. Zienkiewicz, 'Notes on the theory of hydrodynamic lubrication of parallel surface thrust bearings', *Proc. Ninth International Congress of Applied Mechanics, Brussels*, 1956.

15. O. C. Zienkiewicz, 'Temperature distribution within lubricating films between parallel bearing surfaces and its effect on the pressure developed', *Proc. I. Mech. E. Conf. Lubrication and Wear*, 1957, pp. 135–137.
16. K. H. Huebner, 'I.F.E.M. in application to lubrication problems', *International Symposium on Finite Element Methods in Flow Problems, University College, Swansea, January 1974*; (Chapter 12 of this book).
17. P. B. Neal, 'Influence of film inlet conditions on the performance of fluid film bearings', *Journal of Mechanical Engineering Science*, **12**, 2, 153 (1970).
18. A. K. Tieu, 'Oil-film temperature distribution in an infinitely wide slider bearing: an application of the finite element method', *Journal of Mechanical Engineering Science*, **15**, 4, 311 (1973).
19. R. J. Brand, 'Inertia forces in lubricating films', *Journal of Applied Mechanics*, **22**, 363–364 (1955).
20. P. B. Neal, 'Analysis of the taper-land bearing pad', *Journal of Mechanical Engineering Science*, **12**, 2, 73 (1970).
21. A. Sommerfeld, 'Zur hydrodynamischen Theorie der Schmiermittelreibung', *Zeit. angew. Math. u. Physik*, **50**, 97–155 (1904).
22. H. Sassenfeld and A. Walther, 'Gleitlagerberechnungen', *V.D.I. Forschungsheft*, **441** (1954).
23. B. Jakobsson and L. Floberg, 'The finite journal bearing considering vaporization', *Trans. of Chambers University of Technology, Sweden, No. 190* (1957).
24. M. M. Reddi, 'Finite element solution of the incompressible lubrication problem', *J. of Lub. Technology, Trans. A.S.M.E.*, Series F., **91**, 3, 524–533 (1969).
25. N. Tipei, in *Theory of Lubrication*, W. A. Gross (ed.), Stanford, 1962.
26. J. A. Coles and C. J. Hughes, 'Oil flow and film extent in complete journal bearings', *The Engineer*, **201**, 255–263 (1956).
27. R. Hornsell and H. McCallion, 'Prediction of some journal bearing characterization under static and dynamic loading', *Proc. I. Mech. E., Lubrication and Wear Convention 1963*, pp. 126–138.
28. A. A. Milne, 'Transient variations of oil film extent', private communication (forthcoming).
29. V. Castelli and J. Pirvics, 'Review of numerical methods in gas bearing film analysis', *J. of Lubrication Tech., Trans. A.S.M.E.*, Series F, **90**, 4, 777–792 (1968).
30. M. M. Reddi and T. Y. Chu, 'Finite element solution of the steady state compressible lubrication problem', *J. of Lubrication Tech., Trans. A.S.M.E.*, Series F, **92**, 495–503 (1970).
31. F. T. Barwell, 'Hydrostatic lubrication in steelworks', in *Tribology in Iron and Steel Works*, The Iron and Steel Institute, London, 1970, p. 118.
32. V. Shapiro, 'Steady state and dynamic analyses of gas-lubricated hybrid journal bearings', *Trans. A.S.M.E., Journal of Lubrication Technology*, p. 171 (Jan. 1969).
33. T. Allan, 'The application of finite element analysis to hydrodynamic and externally pressurized pocket bearings', *Wear*, **19**, 169–206 (1972).
34. A. A. Milne, 'A contribution to the theory of hydrodynamic lubrication', *Wear*, **1**, 32 (1957).
35. A. A. Milne, 'On the effect of lubricant inertia in the theory of hydrodynamic lubrication', *Trans. A.S.M.E. Journal of Basic Engineering*, Series D(2), **81**, 239 (1959).
36. P. Ramsden, 'Review of published data and their application to the design of large bearings for steam turbines', *Proc. I. Mech. E.*, **182**, 3A, 75–81 (1967).
37. Sir Geoffrey I. Taylor, 'Stability of a viscous liquid contained between two rotating cylinders', *Trans. Royal Soc.*, **102A**, 541–542 (1923).

38. J. A. Cole, 'Experiments on the flow in rotating annular clearances', *Proc. I. Mech. E. Conf. on Lubrication and Wear*, 1957, pp. 16–19.
39. J. Frene, D. Nicholas and M. Godet, 'Characteristics of plain turbulent bearings', *Proc. International Symposium on Plain Bearings, Tatra, Czechoslovakia*, 1972, pp. 173–186.
40. J. O. Medwell, private communication.
41. A. A. Milne 'A theory of grease lubrication of a slider bearing', *Proc. 2nd Int. Congr. Rheology*, 1953, p. 427.
42. A. A. Milne, 'A theory of rheudynamic lubrication', *Kolloidzchr*, **139**, 96 (1959).
43. J. F. Hutton, 'Theory of rheology', *Proc. Interdisciplinary Approach to Liquid Lubricant Technology, N.A.S.A., Lewis Research Centre, Cleveland, Ohio*, 1972.
44. A. J. Barlow, A. Erginsav and J. Lamb, 'Visco-elastic relaxation of super-cooled liquids', *Proc. Royal Soc.*, **298A**, 481 (1967).
45. A. Dyson, 'Frictional traction and lubricant rheology in elastohydrodynamic lubrication', *Phil. Trans.*, **A266**, 1 (1970).
46. H. M. Martin (anonymously), 'The lubrication of gear-teeth', *Engineering*, **102**, 119–121 (1916).
47. D. Dowson and G. R. Higginson, *Elasto-hydrodynamic Lubrication*, Pergamon Press, Oxford, 1966.
48. K. L. Johnson, 'Regimes of elastohydrodynamic lubrication', *J. of Mech. Eng. Science*, **12**, 1, 9–16 (1970).
49. C. Taylor and J. F. O'Callaghan, 'A numerical solution of the elastohydrodynamic lubrication problem using finite elements', *Journal of Mech. Eng. Science*, **14**, 229–237 (1972).
50. R. R. Stephenson and J. F. Osterle, 'A direct solution of the elastohydrodynamic lubrication problem', *Trans. A.S.L.E.*, **4**, 2, 365–374 (1962).
51. J. J. Kauzlarich, 'Elastohydrodynamic lubrication with Herschel–Buckley model greases', *Proc. A.S.L.E.*, **15**, 269–277 (1972).
52. W. H. Herschel and R. Buckley, 'Measurement of consistency as applied to rubber benzene solutions', *Proc. A.S.T.M.*, **26**, 621–633 (1926).
53. T. A. Harris, *Rolling Bearing Analysis*, Wiley, New York, 1966.
54. H. C. Rippel, 'Designing fluid film bearings by computer', *Mech. Eng.*, **92**, 6, 30–41 (1970).
55. G. N. Blount, 'A study of the optimum design of tilting pad thrust bearings', Ph.D. thesis, University of Leeds, 1973.
56. J. N. Ashton, 'Optimum computerized design of hydrodynamic journal bearings', Ph.D. thesis, University of Leeds, 1973.

Chapter 12

Finite Element Analysis of Fluid Film Lubrication—A Survey

K. H. Huebner

12.1 Introduction

Man has been attempting to reduce friction for a period of time reaching far back into antiquity. On the walls of Egyptian tombs, we can find paintings showing men pouring lubricant under the runners of heavy sledges. In spite of its long history, lubrication did not emerge from the trial-and-error stage until about 90 years ago when Beauchamp Tower[1] discovered that a pressure can be generated in the lubricant film between the two surfaces in relative motion. Osborne Reynolds[2] was fascinated by this effect and decided to approach the phenomenon analytically. In 1886 he presented his classical paper on the theory of hydrodynamic lubrication and the governing partial differential equation for the pressure which now bears his name.

Since the early 1900s, tremendous progress has been made and many complex problems in fluid-film lubrication have been studied. This activity in lubrication theory stems from the fact that every machine with moving parts has one or more bearings and the successful operation of a machine crucially depends on the successful performance of each bearing.

Predicting bearing performance by solving the relevant Reynolds equation usually requires the use of some numerical procedure because practical bearing problems seldom have closed form 'textbook' solutions. In the last 15 years, developments in high-speed digital computers have made possible such numerical solutions and have spawned several approximate numerical analysis methods.

The most commonly used method is the finite difference method. A number of finite difference schemes for solving Reynolds equations are reviewed in Reference 3. In the analysis of practical lubrication problems, however, we often encounter complex geometrical configurations, driving functions and boundary conditions. We may also have to cope with abrupt changes in field properties, such as film thickness. In these cases, the finite difference method becomes inherently difficult to use because irregular meshes should be employed and special auxiliary conditions are required.

The finite element method is ideally suited to overcome these difficulties. It not only possesses all the versatility of the finite difference method, but it also offers a practical means of handling irregularities in geometry, boundary conditions and field properties. Unlike the finite difference model which discretizes the solution region with a rectangular array of grid points, the finite element model discretizes the solution region by diving it into simply-shaped subregions. The model seeks an approximate solution in piecewise functional form and then selects the functions according to a variational principle.

Applications of finite element techniques to lubrication problems began to appear in the mid-1960s[4] after analysts recognized the generality of the method based on a variational principle. Once the appropriate Reynolds equation for a problem is cast into a classical variational form, the procedure for deriving the finite element matrix equations is straightforward. In fact, computer programs for hydrodynamic lubrication problems can follow the closely analogous programs for steady state heat conduction, mass diffusion and other phenomena governed by quasi-harmonic equations.

As the literature summary chart of Figure 12.1 indicates, nineteen publications on finite element lubrication analysis have appeared since 1967. This

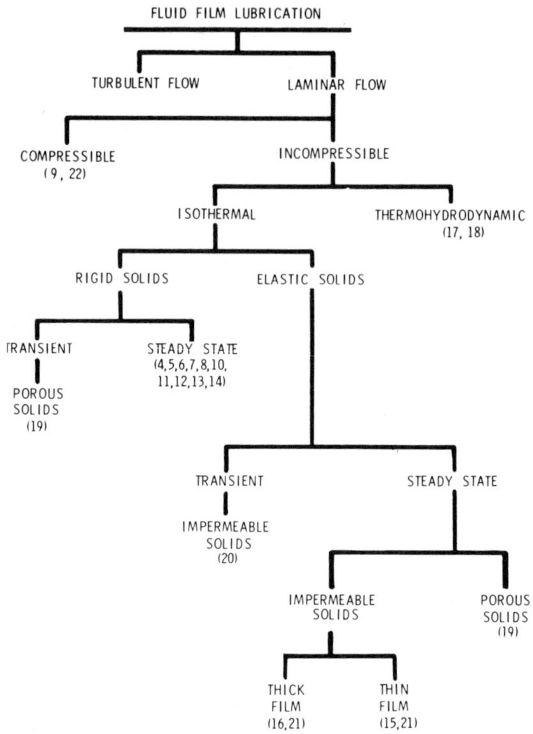

Figure 12.1 Summary of finite element solutions of fluid film lubrication problems (literature reference numbers are in parentheses)

paper aims to present the unifying principles of these analyses. The sample solutions are intended to demonstrate the broad range of problems amenable to finite element analysis.

12.2 Equations governing incompressible hydrodynamic lubrication

12.2.1 Physical mechanisms

The essence of hydrodynamic lubrication theory is that a pressure, and hence a load capacity, can be developed in the fluid film separating two closely spaced surfaces moving relative to one another. There are four physical mechanisms that can generate pressure in the fluid film. These are viscous effects, inertia effects, body force effects and compressibility effects. In some bearings only one of these mechanisms is dominant while, in others, all four may be equally important. For example, in most oil bearings the viscous effects caused by shearing and squeezing the lubricant are the major contributors to pressure generation.

The lubricant flow in high-speed bearings operating with low viscosity lubricants can become 'superlaminar' or 'turbulent'; however, because the majority of bearings do not operate in these regimes, the discussion here will be restricted to the laminar regime.

12.2.2 Reynolds equation

The differential equation governing the generation of pressure in a bearing is derived from the conservation equations of mass and momentum. The approximations and assumptions commonly made in the derivation are

(1) The lubricant is a Newtonian fluid.
(2) The fluid film is so thin that the derivatives of velocity across the film thickness are far more important than any of the other velocity derivatives. (This is the so-called 'thin film approximation'.)
(3) Compared to other effects, inertia effects are negligible.
(4) Curvature of the bearing components introduces only second order, negligible effects in journal bearings.

The general geometry and coordinate system for a pair of film lubricated surfaces are shown in Figure 12.2. In this coordinate system and under the foregoing assumptions, the equation expressing mass conservation is

$$\frac{\partial u}{\partial x} + \frac{\partial v}{\partial y} + \frac{\partial w}{\partial z} = 0 \tag{12.1}$$

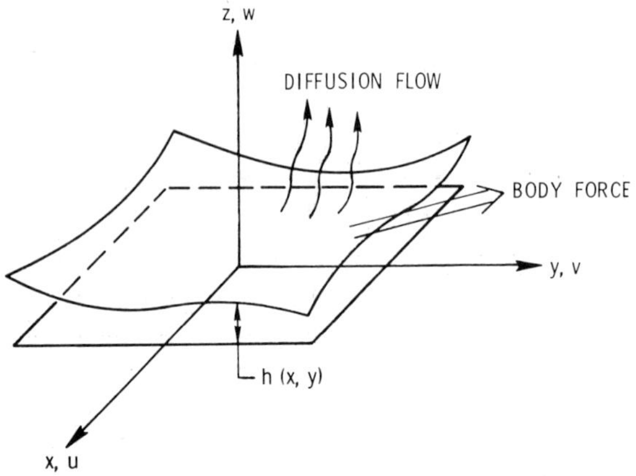

Figure 12.2 Geometry and coordinate system for a lubricated contact

while the equations of motion reduce to

$$\frac{\partial p}{\partial x} = \frac{\partial}{\partial z}\left[\mu \frac{\partial u}{\partial z}\right] + B_x \tag{12.2}$$

$$\frac{\partial p}{\partial y} = \frac{\partial}{\partial z}\left[\mu \frac{\partial v}{\partial z}\right] + B_y \tag{12.3}$$

$$\frac{\partial p}{\partial z} = 0 \tag{12.4}$$

Here p is the pressure in the lubricant film and μ is the lubricant viscosity which is explicitly a function of temperature and implicitly a function of the spatial coordinate x, y, z. B_x and B_y are components of the body force. Equations 12.1–12.4 are subject to the following boundary conditions

$$z = 0, \quad u = U_1, \quad v = V_1, \quad w = W_1$$

$$z = h, \quad u = U_2, \quad v = V_2, \quad w = W_2 \tag{12.5}$$

$p = P(x, y)$ (a specified function on a non-vanishing segment of the boundary)

By integrating Equations 12.2 and 12.3 twice on z (first from 0 to h, then from 0 to z) and applying the boundary conditions expressed in Equation 12.5, we obtain the velocity components

$$u = \frac{\partial p}{\partial x}\left[\int_0^z \frac{z\,\mathrm{d}z}{\mu} - \frac{F_1}{F_0}\int_0^z \frac{\mathrm{d}z}{\mu}\right] + U_1 + \frac{U_2 - U_1}{F_0}\int_0^z \frac{\mathrm{d}z}{\mu} + \bar{B}_x \tag{12.6}$$

$$v = \frac{\partial p}{\partial y}\left[\int_0^z \frac{z\,\mathrm{d}z}{\mu} - \frac{F_1}{F_0}\int_0^z \frac{\mathrm{d}z}{\mu}\right] + V_1 + \frac{V_2 - V_1}{F_0}\int_0^z \frac{\mathrm{d}z}{\mu} + \bar{B}_y \tag{12.7}$$

where

$$F_0 = \int_0^h \frac{dz}{\mu}, \qquad F_1 = \int_0^h \frac{z\, dz}{\mu}$$

$$\bar{B}_{x,y} = \frac{\int_0^z dz/\mu}{F_0} \int_0^h \frac{1}{\mu} \int_0^z B_{x,y}\, dz\, dz - \int_0^z \frac{1}{\mu} \int_0^z B_{x,y}\, dz\, dz$$

From the continuity equation, the velocity component across the film thickness becomes

$$w = -\int_0^z \frac{\partial u}{\partial x} dz - \int_0^z \frac{\partial v}{\partial y} dz + W_1 \tag{12.8}$$

The average fluid film velocity in the plane of the film is defined as

$$\bar{\mathbf{u}} = \left[\frac{1}{h}\int_0^h u\, dz\right]\hat{\mathbf{i}} + \left[\frac{1}{h}\int_0^h v\, dz\right]\hat{\mathbf{j}} \tag{12.9}$$

or

$$\bar{\mathbf{u}} = \bar{u}\hat{\mathbf{i}} + \bar{v}\hat{\mathbf{j}}$$

$\hat{\mathbf{i}}$ and $\hat{\mathbf{j}}$ being unit vectors in x and y directions. Hence, the volume flow q per unit boundary of the film in the xy plane is

$$\mathbf{q} = h\bar{\mathbf{u}} = h\bar{u}\hat{\mathbf{i}} + h\bar{v}\hat{\mathbf{j}} \tag{12.10}$$

When Equations 12.6, 12.7 and 12.9 are combined, we find that

$$\mathbf{q} = G\nabla P + h\mathbf{U}_1 + \frac{\Delta\mathbf{u}}{F_0}\int_0^h \int_0^z \frac{dz}{\mu} dz + \tilde{\mathbf{B}} \tag{12.11}$$

where

$$G = \int_0^h \int_0^z \frac{z\, dz}{\mu} dz - \frac{F_1}{F_0}\int_0^h \int_0^z \frac{dz}{\mu} dz$$

$$\mathbf{U}_1 = U_1\hat{\mathbf{i}} + V_1\hat{\mathbf{j}} \quad \text{etc.}$$

$$\Delta\mathbf{U} = (U_2 - U_1)\hat{\mathbf{i}} + (V_2 - V_1)\hat{\mathbf{j}} = \mathbf{U}_1 - \mathbf{U}_2$$

$$\tilde{\mathbf{B}} = \hat{\mathbf{i}}\int_0^h \bar{B}_x\, dz + \hat{\mathbf{j}}\int_0^h \bar{B}_y\, dz$$

The continuity equation now provides the necessary relation between pressure gradient and volume flow. Integrating Equation 12.1 over the film thickness h gives

$$\nabla\cdot\mathbf{q} + \frac{\partial h}{\partial t} + v_d = 0 \tag{12.12}$$

where v_d is the net outward velocity of a diffusion flow through porous boundary surfaces at $z = 0$ and h. The generalized Reynolds equation, which accounts for three-dimensional viscosity variations in the film, results when Equation 12.11 is substituted in Equation 12.12:

$$-\nabla \cdot G\nabla p = \nabla \cdot (hV_1) + \nabla \cdot (\Delta U F_2) + \nabla \cdot \tilde{\mathbf{B}} + \frac{\partial h}{\partial t} + v_d \qquad (12.13)$$

where

$$F_2 = \frac{1}{F_0}\int_0^h \int_0^z \frac{dz}{\mu}\, dz$$

The Reynolds equation, (12.13), is an elliptic partial differential equation for the pressure in terms of the lubricant properties, density and viscosity, as well as the distributions of film thickness, body forces and diffusion flows. Determining the pressure distribution in the fluid film is not a well posed problem until boundary conditions are specified. If Ω is the region over which a solution is desired, and Γ is the boundary of Ω, then the pressure and flow boundary conditions take the general form

$$p = P(x, y) \quad \text{on } \Gamma_p \qquad (12.14)$$

and

$$Q = \mathbf{q} \cdot \hat{\mathbf{n}} = h\bar{\mathbf{u}} \cdot \hat{\mathbf{n}} \quad \text{on } \Gamma_q$$

where Γ_p and Γ_q are boundary segments of Γ and $\hat{\mathbf{n}}$ is the outward unit normal to the boundary. Segment Γ_p is the *non-vanishing* portion of Γ on which pressure is specified, and Γ_q is the remainder of Γ over which volume flow per unit boundary length is specified. Segment Γ_p must be non-vanishing because unless some pressure boundary condition is specified, a unique solution is impossible. When sufficient pressure and flow boundary conditions are specified, Reynolds equation has a unique solution.

12.2.3 Energy equation

If the variation of the fluid properties along and across the film thickness is taken into account, Reynolds equation must be solved simultaneously with the thermal energy equation and the 'equation of state' for the fluid. The thermal energy equation describing the temperature distribution $T(x, y, z)$ in the lubricant takes the form

$$\rho c\mathbf{V} \cdot \nabla T = k\nabla \cdot \nabla T + \Phi \qquad (12.15)$$

where ρ, c and k are the density, specific heat and thermal conductivity, respectively, of the lubricant and

$$\mathbf{V} = u\hat{\mathbf{i}} + v\hat{\mathbf{j}} + w\hat{\mathbf{k}} \qquad (12.16)$$

When the thin film approximation is applied along with an order of magnitude analysis, the dissipation function Φ reduces to

$$\Phi = \mu(T)\left[\left(\frac{\partial u}{\partial z}\right)^2 + \left(\frac{\partial v}{\partial z}\right)^2\right] \tag{12.17}$$

Sufficient boundary conditions for Equation 12.15 are

(1) $T = T_i$ specified on all surfaces of the control volume through which there is lubricant influx.

(2) $\dfrac{\partial T}{\partial n} = 0$ on all surfaces through which there is lubricant efflux (n is the outward normal on such surfaces).

12.2.4 Viscosity–temperature characteristic

The formulation of the governing equations for the lubrication problem must include the functional dependence of viscosity on temperature; that is, $\mu = \mu(T)$ must be specified. Often an exponential relation is used such as

$$\mu(T) = \mu_0 \, e^{-\beta(T - T_0)} \tag{12.18}$$

where μ_0 is the viscosity at the reference temperature T_0.

12.2.5 Equations for the solids

When heat transfer to the bearing solids is considered, the thermal energy equations for the solids must also be solved simultaneously with the thermal energy equation for the fluid. The lubrication problem becomes further complicated when the bearing solids deform due to thermal and/or mechanical loading. In these cases the film thickness distribution is not known *a priori* but rather must be found from a simultaneous solution of the thermal and mechanical energy equations in the solids.

In the following sections, finite element formulations for particular cases of the general lubrication problem are presented.

12.3 Liquid lubrication in the isothermal state

Most analyses of hydrodynamic lubrication problems assume the lubricant viscosity remains constant. This is equivalent to the assumption that the bearing operates in the isothermal state at some characteristic operating temperature. The isothermal assumption holds for gas lubricated bearings. For liquid lubricated bearings it is simply a convenient contrivance that uncouples the Reynolds and energy equations and simplifies the calculations. Isothermal solutions or constant viscosity solutions to Reynolds equation can be useful for predicting bearing behaviour and developing design data when it is possible to find an 'effective' viscosity based on some representative bearing temperature. Usually this representative temperature is found by estimating the work done on the lubricant, determining from this the temperature rise of the lubricant as it passes through wedge and then adding one half the temperature rise to the given inlet temperature. Such approximate procedures, though convenient, rarely lead to accurate predictions of both bearing load capacity and friction torque.

In this section, a special finite element formulation of Reynolds equation is given for the case where the lubricant viscosity remains constant across the film thickness. A number of investigators[4–8,10–14] have applied the finite element techniques to this problem, but the treatment here follows References 12 and 13.

12.3.1 Reynolds equation

When the lubricant viscosity is independent of z, the coordinate across the film thickness, the general Reynolds equation, (12.13), reduces to

$$\nabla \cdot \left(\frac{h^3}{12\mu}\nabla p\right) = \nabla \cdot \left(h\bar{\mathbf{U}} + \frac{h^3}{12\mu}\mathbf{B}\right) + \frac{\partial h}{\partial t} + v_d \tag{12.19}$$

where

$$2\bar{\mathbf{U}} = (U_1 + U_2)\hat{\mathbf{i}} + (V_1 + V_2)\hat{\mathbf{j}}$$

The boundary conditions in this case become

$$p = P(x, y) \quad \text{on } \Gamma_p$$

$$Q = \mathbf{q} \cdot \hat{\mathbf{n}} = h\bar{\mathbf{u}} \cdot \hat{\mathbf{n}} = h\left(\bar{\mathbf{U}} + \frac{h^2}{12\mu}\mathbf{B} - \frac{h^2}{12\mu}\nabla p\right) \cdot \hat{\mathbf{n}} \quad \text{on } \Gamma_q \tag{12.20}$$

12.3.2 Equivalent variational principle

The linear self-adjoint elliptic boundary value problem of Equations 12.19 and 12.20 has a corresponding variational principle. According to this principle, the solution of Equations 12.19 and 12.20 is the function p that satisfies the pressure boundary condition and minimizes the functional

$$\Pi(p) = \int_A \left[\left(\frac{h^3}{24\mu}\nabla p - h\bar{\mathbf{U}} - \frac{h^3}{12\mu}\mathbf{B}\right) \cdot \nabla p + \left(\frac{\partial h}{\partial t} + v_d\right)p\right] \mathrm{d}A + \int_{\Gamma_q} Q_p \, \mathrm{d}\Gamma \tag{12.21}$$

12.3.3 Finite element formulation

The isothermal incompressible lubrication problem may be stated as follows. Given the parameters or forcing functions μ, h, $\partial h/\partial t$, $\bar{v}$, $\mathbf{B}$, V_d, and the bearing geometry, find the pressure distribution $p(x, y)$ in the lubricant film. The finite element method provides an expedient means of obtaining an approximate solution to this problem regardless of how irregular the bearing geometry may be.

Step one of the solution procedure is to divide the solution domain (the region of the bearing over which the pressure distribution is sought) into polygonal subdomains or elements which are interconnected at nodes located on the element boundaries. Then the pressure and forcing functions are expressed in terms of assumed interpolation functions within each element. Nodal values and interpolation functions† completely define the behaviour of the pressure and the forcing functions within the elements.

More explicitly, the finite element discretization process involves subdividing the solution domain in the xy plane into M polygonal elements of r nodes each. Then within each element the pressure and the distributions of the various forcing functions are related to their r nodal values by r interpolating functions $N_i(x, y)$ as follows:

$$p = \sum_{i=1}^{r} N_i(x, y)p_i = \mathbf{Np} \tag{12.22}$$

$$U = \sum N_i(x, y)U_i = \mathbf{NU} \quad \text{etc.} \tag{12.23}$$

In Equation 12.22 the nodal parameters p_i may be nodal pressures and pressure derivatives, but for convenience we denote all the nodal degrees-of-freedom for the pressure simply as p_i.

When Equations 12.22 and 12.23 are inserted in the functional of Equation 12.21, there results the discretized form of the functional for one element. If the functional for the system of elements is to be a minimum, the functional for one element contributing to the system must also be a minimum. To minimize the element functional with respect to the nodal pressures of the element, we set

$$\frac{\partial \Pi}{\partial p_i} = 0 \quad \text{for } i = 1, 2, \ldots, r \tag{12.24}$$

Equation 12.24 results in a set of element equations of the form

$$\begin{aligned} \overset{r\times r}{[\mathbf{K}_p]}\overset{r\times 1}{\{\mathbf{P}\}} &= \overset{r\times 1}{\{\mathbf{q}\}} - \overset{r\times r}{[\mathbf{K}_{\bar{U}}x]}\overset{r\times 1}{\{\bar{\mathbf{U}}^x\}} - \overset{r\times r}{[\mathbf{K}_{\bar{U}}y]}\overset{r\times 1}{\{\bar{\mathbf{U}}^y\}} \\ &\quad - \overset{r\times r}{[\mathbf{K}_B x]}\overset{r\times 1}{\{\mathbf{B}^x\}} - \overset{r\times r}{[\mathbf{K}_B]}\overset{r\times 1}{\{\mathbf{B}^y\}} \\ &\quad - \overset{r\times r}{[\mathbf{K}_{\dot{h}}]}\overset{r\times 1}{\{\dot{\mathbf{h}}\}} - \overset{r\times r}{[\mathbf{K}_{v_d}]}\overset{r\times 1}{\{\mathbf{v}_d\}} = \overset{r\times 1}{\{\mathbf{Q}_R\}} \end{aligned} \tag{12.25}$$

† The interpolation functions should be chosen such that C^0 continuity is preserved; i.e. the pressure field representation should be at least continuous within the elements and across the element interfaces. Also, if polynomials are used, they should be complete to ensure that the field variable representation does not depend on the oreientation of the coordinate system.

where the coefficients in these fluidity matrices are given by†

$$
\begin{aligned}
K_{P_{ij}} &= -\int_A \frac{h^3}{12\mu}\left(\frac{\partial N_i}{\partial x}\frac{\partial N_j}{\partial x} + \frac{\partial N_i}{\partial y}\frac{\partial N_j}{\partial y}\right) \mathrm{d}A \\
K_{\bar{U}_{ij}}x &= \int_A h\frac{\partial N_i}{\partial x} N_j \,\mathrm{d}A \\
K_{\bar{U}_{ij}}y &= \int_A h\frac{\partial N_i}{\partial y} N_j \,\mathrm{d}A \\
K_{B_{ij}}x &= \int_A \frac{h^3}{12\mu}\frac{\partial N_i}{\partial x} N_j \,\mathrm{d}A \\
K_{B_{ij}}y &= \int_A \frac{h^3}{12\mu}\frac{\partial N_i}{\partial y} N_j \,\mathrm{d}A \\
K_{\dot{h}_{ij}} &= -\int_A N_i N_j \,\mathrm{d}A \\
K_{v_{d_{ij}}} &= K_{\dot{h}_{ij}} \\
q_i &= \int_{\Gamma_q} Q N_i \, d\Gamma \ddagger
\end{aligned}
\qquad (12.26)
$$

These element equations have a distinct physical meaning. The right-hand side of Equation 12.25 may be interpreted as a linear combination of flows caused by shear, body force, squeeze and diffusion effects. These nodal flows are then balanced by the nodal flows due to pressure and the externally applied flows. Equations 12.25 and 12.26 provide the complete finite element description of the general isothermal incompressible lubrication problem. The fluidity matrices and element equations can be explicitly evaluated after the element geometry and interpolation functions are specified. Particular fluidity matrices for triangular elements with linear interpolation functions are given in Reference 13. Of course, other more complex elements are possible as shown in Chapter 2.

To evaluate the integrals of Equations 12.26, it is convenient to interpolate the field properties h and μ between their nodal values. Depending on the shape of the element and the complexity of the integrand, numerical integration may be more feasible than closed-form integration.

The finite element analysis of a system is complete when all the element equations are assembled to form the system equations and the system equations are solved for the unknown nodal pressures and flows. Assembly

† Here it is understood that the integrals pertain to only one element.

‡ $\{q\}$ is a boundary matrix stemming only from elements with portions of their boundary coinciding with the boundary of the complete domain or elements where external feeder flows are specified.

of the system equations is routine and relies on matching element pressures and forcing functions to form system pressures and forcing functions, while element flows are summed to form system flows. The system equations are identical in form to the element equations, only they are expanded in dimension to include all the nodes. Hence, the system equations become

$$\overset{n \times n}{[\mathbf{K}_p]_s} \overset{n \times 1}{\{\mathbf{P}\}_s} = \overset{n \times 1}{\{\mathbf{Q}_R\}_s} \tag{12.27}$$

where n is the total number of nodes for the system. Solution of Equation 12.27 is possible after the system fluidity matrix $[\mathbf{K}_p]_s$ is modified to account for the pressure boundary conditions. Flow boundary conditions have already been included in $\{\mathbf{Q}_R\}_s$, the resultant or net vector of nodal flows. Often, to save computer storage space, the procedure used to solve Equation 12.27 takes advantage of the fact that $[\mathbf{K}_p]_s$ is banded and symmetric.

Once the nodal pressures and flows have been found for the system, the bearing load capacity W and the friction force $\mathbf{F}$ can be found from the relations

$$W = \int_\Omega p(x, y)\, \mathrm{d}x\, \mathrm{d}y \tag{12.28}$$

$$\mathbf{F} = \int_\Omega \mu \frac{\partial}{\partial z}(u\hat{\mathbf{i}} + v\hat{\mathbf{j}})|_{z=0,h}\, \mathrm{d}x\, \mathrm{d}y \tag{12.29}$$

Examples showing some applications of the foregoing finite element formulation are given in Section 12.6.

12.3.4 Elastic bearing surfaces

When the bearing solids are compliant and deform under the action of hydrodynamic pressure, the film thickness distribution $h(x, y)$ is not known *a priori*. Hence, the solution of isothermal elastohydrodynamic bearing problems involves the simultaneous solution of Reynolds equation in the fluid film and the linear elasticity equations in the bearing solids. Since realistic bearing configurations often exhibit irregular geometry, the finite element method can be used to advantage in solving the elasticity equations for the solids.

Since the finite element analysis of elasticity problems is detailed in a number of standard texts, we shall discuss only the essence of the methodology as it applies to elastohydrodynamic lubrication problems. The state-of-stress in an elastic solid acted upon by given surface and body forces can be characterized by the principle of minimum potential energy. The principle states that among all admissible displacement fields, the one that satisfies the equilibrium conditions makes the potential energy a minimum. Thus, the expression for the system potential energy provides a suitable variational statement for the application of the finite element method.

Let $\Pi_p(\mathbf{\Delta})$ be the potential energy of the deformed solid where $\mathbf{\Delta}$ is the displacement vector defined as $\mathbf{\Delta}(x, y, z) = \Delta_x(x, y, z)\hat{\mathbf{i}} + \Delta_y(x, y, z)\hat{\mathbf{j}} + \Delta_z(x, y, z)\hat{\mathbf{k}}$ with $\mathbf{\Delta} \equiv 0$ in the undeformed state. Then, at equilibrium we have

$$\delta\Pi_p(\mathbf{\Delta}) = 0 \tag{12.30}$$

Following the usual discretization procedure, we divide the volume of the solid into elements and then approximate the displacement field within each element in terms of known interpolation functions and unknown nodal values of displacement. The minimizing condition of Equation 12.30 then leads to matrix equations for the discretized structure of the form

$$\mathbf{K\Delta} = \mathbf{F}_R \tag{12.31}$$

where $\mathbf{K}$ is the stiffness matrix and $\mathbf{F}_R$ is the vector of resultant nodal forces.

Symbolically, the isoviscous elastohydrodynamic lubrication problem reduces to the simultaneous solution of the discretized Reynolds equation

$$[\mathbf{K}_p(\mathbf{\Delta})]\{\mathbf{P}\} = \{\mathbf{Q}_R(\mathbf{\Delta})\} \tag{12.32}$$

which contains unknown nodal pressures, and the discretized elasticity equations

$$\mathbf{K\Delta} = \mathbf{F}_R(\mathbf{P}) \tag{12.33}$$

which contain unknown nodal displacements.

Taylor and O'Callaghan[15] as well as Oh and Huebner[16] have investigated the solution of these equations. The solution procedure begins by assuming a film thickness profile for a rigid bearing. Then Equation 12.32 is solved for the nodal pressures which are used to calculate the elastic response of the structure via $\mathbf{\Delta} = \mathbf{K}^{-1}\mathbf{F}_R(P)$. These deformations are then used to obtain a new film thickness profile, and the process is repeated. When two successive displacement fields $\mathbf{\Delta}^K$ and $\mathbf{\Delta}^{K+1}$ are uniformly close, convergence is achieved. This direct iterative scheme converges rapidly for lightly loaded bearings (small eccentricity ratio for journal bearings), but for heavily loaded bearings the scheme diverges unless remedial steps are taken.

The remedial steps involve averaging and limiting nodal displacements for the first few iterations while approaching the highly loaded cases from the lightly loaded side. Instead of attempting a solution for a highly loaded case directly, solutions are obtained starting from a lightly loaded case where direct iteration converges. Then for a slightly heavier load (smaller minimum film thickness) the first guess for the displacement field is taken as the average

$$\{\mathbf{\Delta}^{K+1}\} = \frac{1}{n}\left[[\mathbf{K}]\left\{\mathbf{F}_R(\mathbf{\Delta}^K) + \sum_{i=1}^{n-1}\{\mathbf{\Delta}^i\}\right\}\right] \tag{12.34}$$

where $n \leqslant 5$. In addition to this weighting process, the maximum nodal displacement can be limited to avoid excessively large displacement changes if they happen to occur. After a few cycles at a given loading, the constraints are removed and the remaining cycles converge to any desired degree of accuracy. Details of this procedure may be found in Reference 16, and an example is given in Section 12.6.

Recently, Day[20] proposed and utilized an alternative procedure for solving the elastohydrodynamic lubrication problem. Instead of iterating between pressures and deflections, he combines the matrix equations for pressure and displacement and considers an initial value problem in time. Starting with an assumed pressure distribution and a corresponding or compatible displacement distribution, the solution marches forward in time until the steady state $\partial p/\partial t = 0$ is approached.

It was found that this procedure works well for highly compliant bearing surfaces, but difficulties are encountered for relatively 'stiff' bearing materials because the necessary time steps become small. On the other hand, the previous procedure of modified iteration between pressure and displacement works well for stiff bearings, but it encounters difficulties for 'soft' materials. The reader is encouraged to see References 16 and 19 for further details.

12.4 Liquid lubrication in the thermohydrodynamic state

If the bearing solids are assumed to be rigid but the isothermal assumption is dropped, the bearing analyst must take into account viscous heating in the lubricant film and the resulting changes in lubricant viscosity. His analysis must include not only the conservation equations of mass and momentum (Reynolds equation), but also the conservation equation of thermal energy. The Reynolds and energy equations must be solved simultaneously and an iterative procedure is necessary.

12.4.1 Element equations for pressure

The general Reynolds equation given by Equation 12.13 may be cast into variational form to provide a convenient base for deriving the finite element equations. It can be shown that the pressure distribution that satisfies Equations 12.13 and 12.14 also minimizes the functional

$$\Pi_t(p) = \int_\Omega \left[-\tfrac{1}{2} G \nabla p \, . \, \nabla p + (h\mathbf{U}_1) \, . \, \nabla p + \left(\frac{\Delta \mathbf{U}}{F_0} \int_0^h \int_0^z \frac{\mathrm{d}z}{\mu} \, \mathrm{d}z \right) . \, \nabla P + \tilde{\mathbf{B}} \, . \, \nabla p + p \frac{\partial h}{\partial t} + p v_d \right] \mathrm{d}\Omega + \int_{\Gamma_q} p \mathbf{q} \, . \, \hat{\mathbf{n}} \, \mathrm{d}\Gamma \qquad (12.35)$$

After the solution domain is subdivided into elements and the pressure as well as the forcing functions are interpolated in a manner analogous to Equations 12.22 and 12.23, the derivation of the finite element matrix equations follows the standard procedure leading to an equation of a standard form.

Evaluation of these fluidity matrices requires a known viscosity distribution $\mu(T)$. This is available either from an initially assumed or previously calculated temperature distribution.

12.4.2 Element equations for temperature

The temperature distribution in the fluid film results from a solution of Equation 12.15. Assuming that a set of nodal pressures have been calculated, we can calculate the velocity field in the lubricant film from the expressions for u, v and w given in Equations 12.6, 12.7 and 12.8. This permits evaluation of the coefficients and the dissipation function in the energy equation. The non-linearity is the energy equation due to the non-linear viscosity–temperature relation is circumvented by evaluating $\mu(T)$ at the previous temperature distribution.

The energy equation may be expressed in operator form as

$$\mathscr{D}(T) = \rho c\left(u\frac{\partial T}{\partial x} + v\frac{\partial T}{\partial y} + w\frac{\partial T}{\partial z}\right) - k\left(\frac{\partial^2 T}{\partial x^2} + \frac{\partial^2 T}{\partial y^2} + \frac{\partial^2 T}{\partial z^2}\right) - \Phi = 0 \quad (12.36)$$

Though Tieu[1] has treated two-dimensional problems using the Glansdorff–Prigogine local potential, a particularly convenient way to formulate a finite element model of Equation 12.36 is to use the method of weighted residuals with Galerkin's criterion. To apply Galerkin's method, we assume the unknown temperature distribution within a typical three-dimensional element can be approximated by

$$T^e = \sum_{i=1}^{s} N_i(x, y, z)T_i = \mathbf{NT} \quad (12.37)$$

where the N_i are interpolation functions defined over individual elements, s is the number of nodes per element and the T_i are the unknown nodal temperatures. For the whole solution domain, the complete piecewise representation has the form

$$\tilde{T} = \sum_{e=1}^{M'} T^e \quad (12.38)$$

where M' is the total number of elements. Using the appropriate function at the appropriate node, we may write

$$\tilde{T} = \sum_{i=1}^{n'} N_i T_i \quad (12.39)$$

where n' is the total number of nodes in the three-dimensional solution domain.

The trial solution expressed by Equation 12.39 is such that $\tilde{T}$ exactly satisfies all the boundary conditions. When $\tilde{T}$ is substituted into Equation 12.36, an error residual results because the trial solution $\tilde{T}$ does not in general satisfy the differential equation exactly, i.e.

$$R = L(\tilde{T}) \neq 0 \tag{12.40}$$

Galerkin's weighting principle permits a set of discretized equations to be obtained.

$$\int_{\Omega_3} N_i R \, \mathrm{d}x \, \mathrm{d}y \, \mathrm{d}z = 0 \qquad \text{for } i = 1, 2, \ldots, n' \tag{12.41}$$

where the N_i are the same interpolation functions used to represent $\tilde{T}$ and Ω_3 is now the three-dimensional solution domain (the control volume encompassing the lubricant film). Equation 12.42 gives the set of linear simultaneous equations to be solved for the nodal temperature T_i. In matrix form these equations are

$$[\mathbf{K}_T]\{\tilde{\mathbf{T}}\} = \{\mathbf{F}_T\} \tag{12.42}$$

where

$$\begin{aligned} K_{T_{ij}} = & \int_{\Omega_3} \left[\rho c N_i \left(u \frac{\partial N_j}{\partial x} + v \frac{\partial N_j}{\partial y} + w \frac{\partial N_j}{\partial z} \right) \right. \\ & \left. + k \left(\frac{\partial N_i}{\partial x} \frac{\partial N_j}{\partial x} + \frac{\partial N_i}{\partial y} \frac{\partial N_j}{\partial y} + \frac{\partial N_i}{\partial z} \frac{\partial N_j}{\partial z} \right) \right] \mathrm{d}\Omega_3 \\ & - k \int_{\Sigma} N_i \left(\frac{\partial N_j}{\partial x} \hat{\mathbf{i}} + \frac{\partial N_j}{\partial y} \hat{\mathbf{j}} + \frac{\partial N_j}{\partial z} \hat{\mathbf{k}} \right)_{\substack{\text{on} \\ \text{surface}}} \cdot \hat{\mathbf{n}} \, \mathrm{d}\Sigma \end{aligned} \tag{12.43}$$

$$F_{T_i} = \int_{\Omega_3} N_i \Phi \, \mathrm{d}\Omega \tag{12.44}$$

In the derivation of these element equations, integration by parts was used on the second order derivatives. This has the effect of reducing the order of the highest order derivatives appearing in the thermal 'stiffness' matrix and it introduces the boundary conditions for those elements on the bounding surface of the control volume. The surface integral in Equation 12.43 appears only for boundary elements.

Since the integrals in Equation 12.43 contain only first order derivatives, the interpolation functions N_i need only preserve continuity of value and not slope at element interfaces. The same interpolation functions used to represent $\tilde{T}$ can also be used to express u, v, w and Φ in terms of their nodal values. Equations 12.42, as they have been derived, hold for the entire solution domain. But if the N_i are chosen to preserve interelement continuity, the equations may be evaluated for individual elements and then summed in the usual manner to obtain the system equations.

To find compatible pressures and temperatures in the fluid film, Equations 12.35 and 12.42 must be solved simultaneously via iteration as follows:

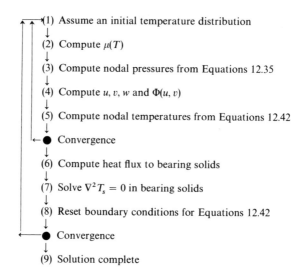

(1) Assume an initial temperature distribution
↓
(2) Compute $\mu(T)$
↓
(3) Compute nodal pressures from Equations 12.35
↓
(4) Compute u, v, w and $\Phi(u, v)$
↓
(5) Compute nodal temperatures from Equations 12.42
↓
● Convergence
↓
(6) Compute heat flux to bearing solids
↓
(7) Solve $\nabla^2 T_s = 0$ in bearing solids
↓
(8) Reset boundary conditions for Equations 12.42
↓
● Convergence
↓
(9) Solution complete

In some cases, it is permissible to use adiabatic boundary conditions at $z = 0, h$ and omit steps 6, 7 and 8.

12.5 Gas lubrication

The derivation of Reynolds equation for compressible lubrication problems follows the same procedure outlined in Section 12.2 except that variable lubricant density must be carried in the equations. Variable viscosity need not be considered because when compared to liquid bearings gas bearings have far lower shear losses. Also, the viscosity of a gas is relatively unaffected by temperature change in range of most bearing operation.

For gas lubrication the relevant Reynolds equation becomes

$$\nabla \cdot \left(\frac{\rho h^3}{12\mu} \nabla p \right) = \nabla \cdot \left(\rho h \bar{\mathbf{U}} + \frac{\rho h^3}{12\mu} \mathbf{B} \right) + \frac{\partial}{\partial t}(\rho h) + \rho v_d \qquad (12.45)$$

with boundary conditions given by Equation 12.20. Though thermal effects are generally not important in gas lubricated bearings, Reynolds equation becomes non-linear because ρ depends on p through the lubricant's equation of state. For an ideal gas we have $p = \rho R T$ and, since the flow is isothermal,

Equation 12.45 takes the form

$$\nabla \cdot \left(\frac{h^3 p}{12\mu} \nabla p \right) = \nabla \cdot \left(ph\bar{\mathbf{U}} + \frac{ph^3}{12\mu} \mathbf{B} \right) + \frac{\partial}{\partial t}(ph) + pv_d \qquad (12.46)$$

Equation 12.46 does not have a classical variational statement, but Galerkin's procedure will yield a suitable finite element form. Reddi and Chu,[9] have used such a semi-variational formulation for an equation similar to Equation 12.46 (body forces, squeeze effects and diffusion flow were absent). When Equation 12.46 is non-dimensionalized, an important parameter λ emerges called the compressibility number and defined as $\lambda = 6\mu UL/h_0^2 p_a$ where L is some characteristic length, h_0 is a reference dimension across the film thickness and p_a is the ambient pressure. Using a perturbation technique, Reddi and Chu set up an incremental solution procedure which involves solving a set of linear finite element equations for each incremental step in λ. A starting solution is found by solving the corresponding incompressible lubrication problem. A sample solution obtained by this procedure is given in Section 12.6.

Oh and Rohde[21] devised a different technique based directly on Equation 12.46. They used Newton's method on the governing non-linear equation (12.46) while employing Galerkin's method to solve the resulting linear equations for the correction term. This procedure with bicubic hermite interpolation functions defined over rectangular elements led to rapid convergence and high accuracy with relatively few elements. Problems with compressibility number as high as $\lambda = 100$ were handled with ease. Experience to date has indicated that the method suggested by Oh and Rohde is the most effective way to treat the general compressible lubrication problem.

12.6 Sample solutions

The finite element method has undoubtedly been used in industry to solve many practical real-life lubrication problems. For proprietary reasons, some of these solutions do not appear in the literature, but enough solutions have appeared to give a good indication of the type of complex lubrication problems being solved. The following sample solutions were selected to demonstrate the versatility, power and potential use of finite element methods in lubrication. Some of the details of these solutions are presented; however, for a complete description of the problem in each case, the reader should refer to the source material.

12.6.1 Incompressible isothermal solutions

12.6.1.1 Rigid bearing surfaces Sometimes it is possible to obtain an accurate prediction of a bearing's load capacity and friction by neglecting the side leakage of lubricant and performing a one-dimensional analysis. This is possible whenever the dimensions of the bearing are such that the ratio of its length (the dimension in the direction of sliding) to its width (the dimension perpendicular to the direction of sliding) is greater than one half. The composite slider bearing[12] depicted in Figure 12.3 satisfies this condition. Linear one-dimensional elements (Figure 12.3(b)) were used to model the solution domain and element boundaries were positioned at places where the slope of film thickness distribution $h(x)$ changes abruptly. Figure 12.3(a) shows the effect of squeeze action and diffusion flow on the pressure distribution. A squeeze velocity and a diffusion flow velocity of only 0·1 per cent of the sliding velocity drastically increases the generated pressure.

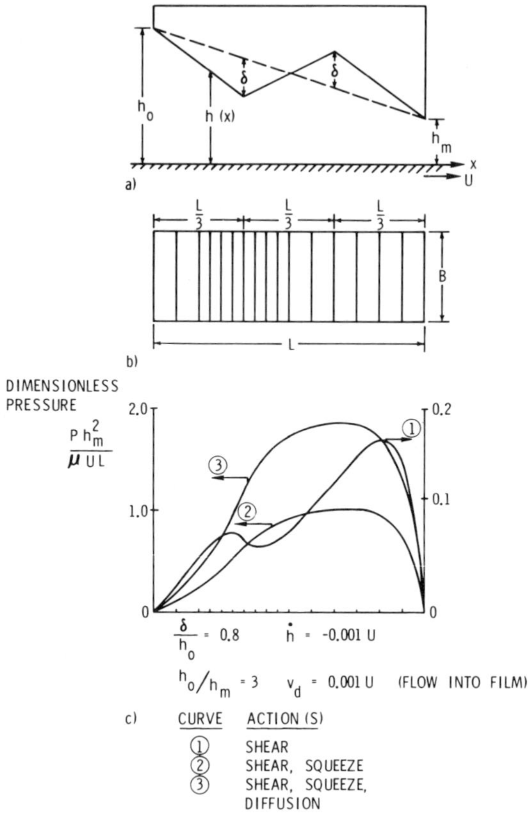

Figure 12.3 Finite element idealization and solution of a composite slider bearing

Determining the load capacity of journal bearings with complex oil feed grooves (Figure 12.4) is a problem of considerable practical interest. Normally these bearings are designed so that the load vector is out of the region of the feed groove, but in some modes of operation it is possible for the load vector to move into the grooved region. In these cases the load capacity of the bearing is diminished, but the questions is: By how much?

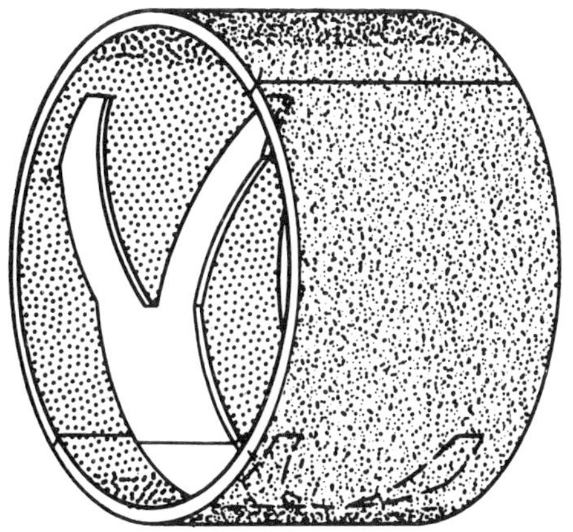

Figure 12.4 Journal bearing with complex oil feed groove

With an analysis tool such as the finite element method, this question can be answered without difficulty. The bearing region can be unwrapped and discretized with a regular triangular mesh, as shown in Figure 12.5. Because of symmetry only half the bearing need be treated. In the groove region, the film thickness is taken to be three orders of magnitude larger than the film thickness anywhere outside the groove. Either feed pressure or flow can be specified at nodes along the boundary of the groove. Figure 12.6 shows the results of an analysis using linear interpolation over triangles. Bearing load degradation is seen to be significant even though the grooved area of the bearing represents only 19 per cent of the total area.

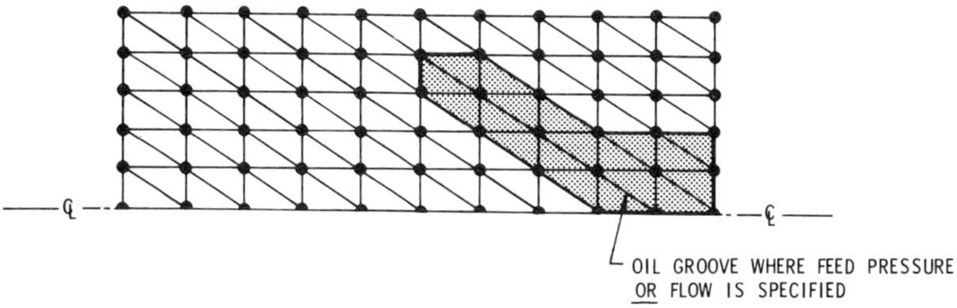

Figure 12.5 Finite element model of unwrapped grooved bearing. The groove helps promote even oil distribution

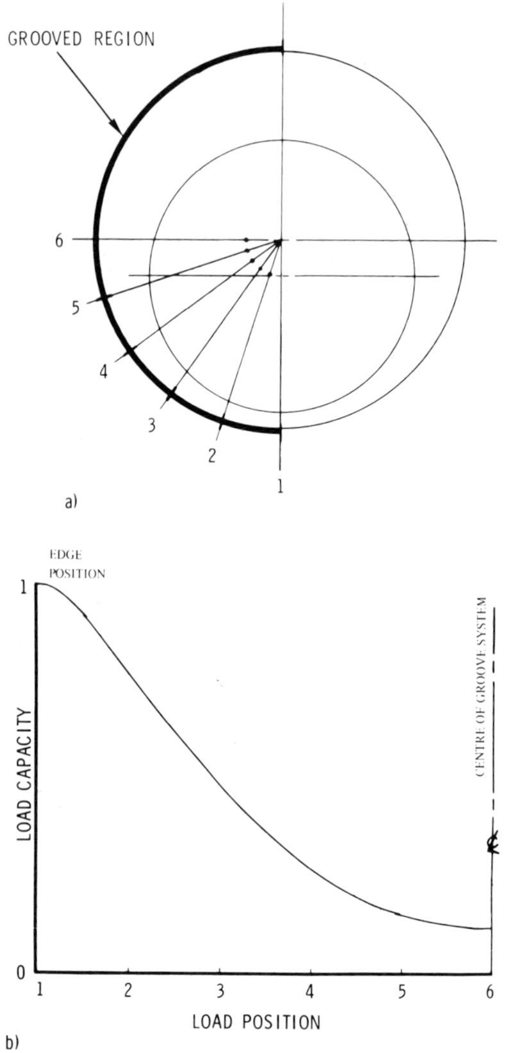

Figure 12.6 Variation of bearing load capacity with groove position

Wada and coworkers[10,11] studied the application of finite element techniques to incompressible lubrication problems and reported case studies using linear as well as higher-order elements for one- and two-dimensional problems. They show that by using bicubic hermite interpolation for the pressure over rectangular elements good accuracy can be obtained with only a few elements. To achieve the same accuracy with linear triangular elements requires, by comparison, many elements.

12.6.1.2 Compliant bearing surfaces To study the effects of elastic distortion on journal bearing performance, Oh and Huebner[16] used the finite element method to analyse the system shown in Figure 12.7. Linear tetrahedral elements assembled to form hexahedral elements were used to model the bearing housing (Figure 12.8.), while linear triangular elements were used for the fluid film. Without deformation of the bearing, the pressure

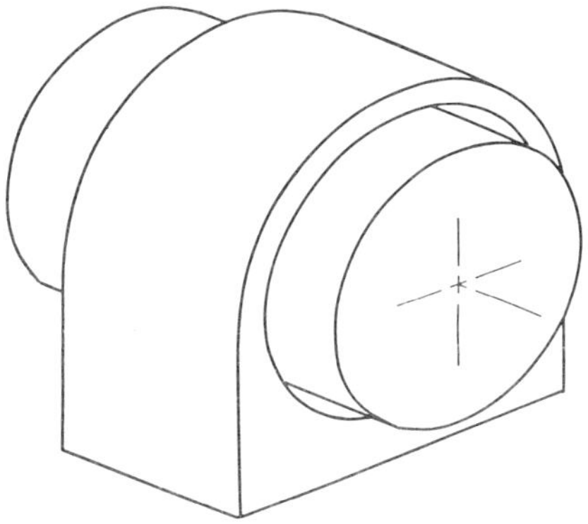

Figure 12.7 Journal bearing with elastic housing

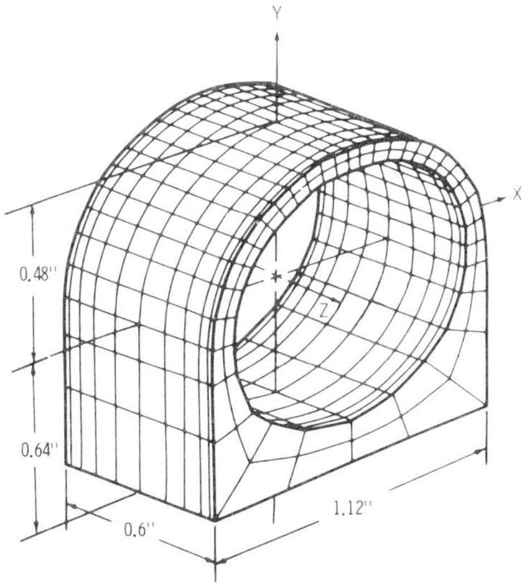

Figure 12.8 Finite element model of the bearing housing

distribution takes the form shown in Figure 12.9(a), but when the bearing housing is allowed to deform (rigidly fixed only at its base) the pressure distribution changes significantly. Figures 12.9(b) and (c) show the resulting

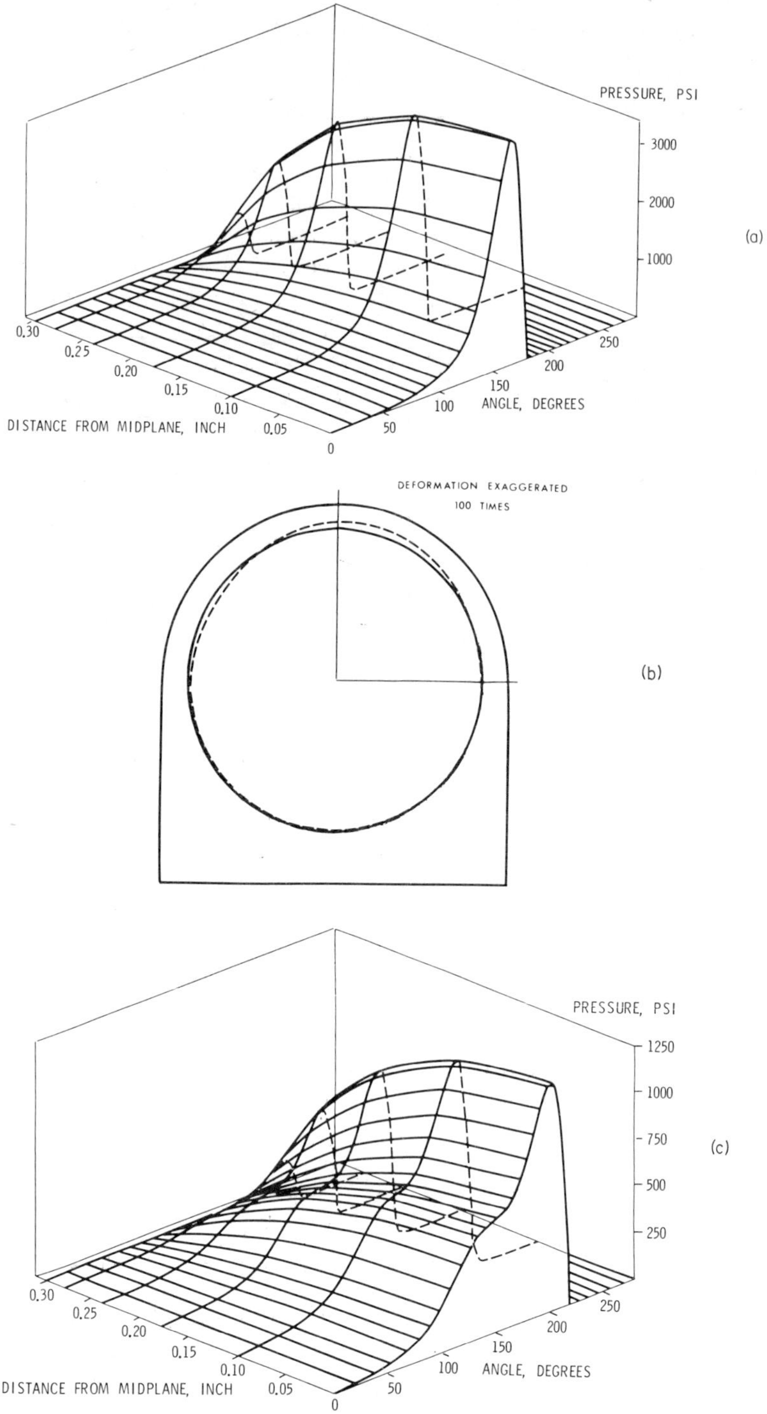

Figure 12.9 Housing deformation and its effect on the pressure distribution

compatible deformations and pressures. The analysis revealed that the elastic housing distributes the load over a wider area and, for the same minimum film thickness and peak pressure, a bearing with an elastic housing has a higher load capacity than the same bearing with a rigid housing. Figure 12.10 shows one of the case studies presented by Day[20] and suggests the form of the deformation to be expected in slider bearings with one compliant surface.

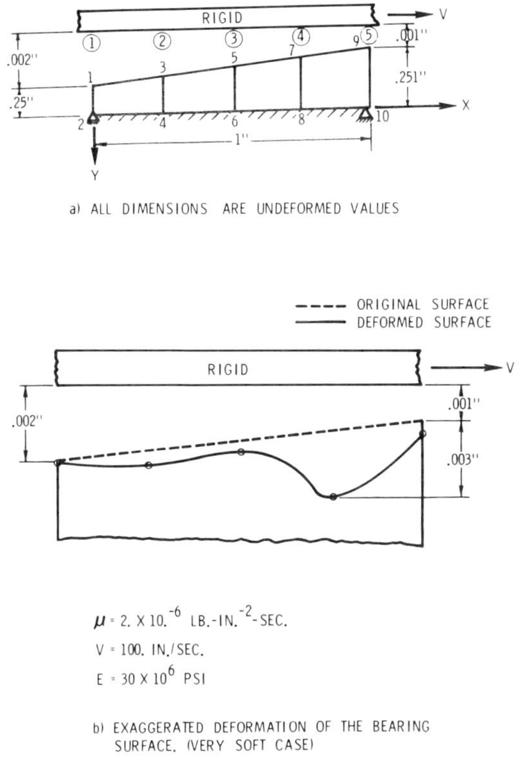

Figure 12.10 Deformation of a compliant slider bearing[19]

12.6.2 Incompressible thermohydrodynamic solutions

Examples of finite element techniques applied to thermohydrodynamic problems have been presented by Tieu[17] for the two-dimensional case and by Huebner[18] for the three-dimensional case. The sector thrust bearing studied by Huebner is shown in Figure 12.11. Only the control volume defined by the dotted lines (with the bearing solids immediately above and below) needs to be considered in the analysis because the complete bearing is actually

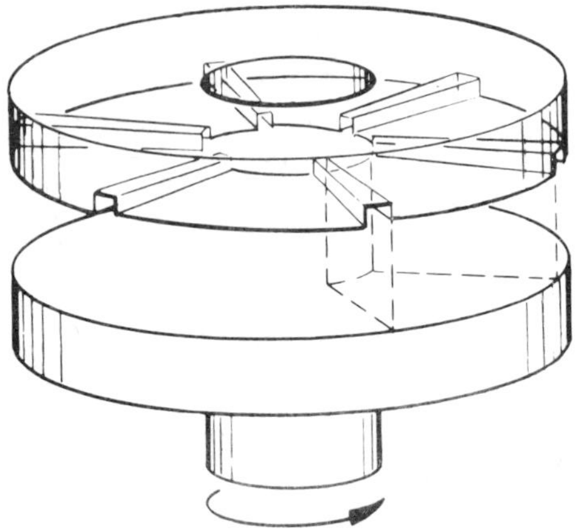

Figure 12.11 Radially grooved sector thrust bearing

an assemblage of these subsystems. Figure 12.12 gives some results obtained using simple linear elements. Comparison of the pressure distributions for isothermal and thermohydrodynamic modes of operation indicates that bearing performance can be significantly affected by viscous heating in the lubricant film.

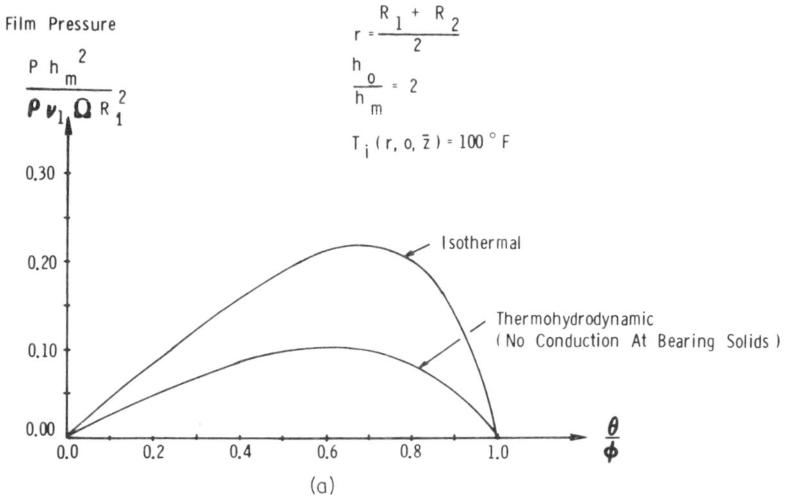

Figure 12.12 Thermohydrodynamic solutions: (a) Pad centreline pressure distributions

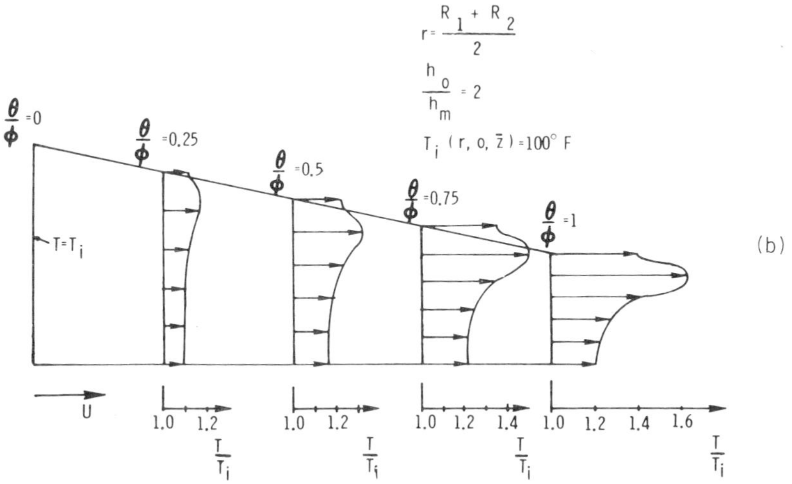

Figure 12.12 (b) Temperature profiles across the film thickness

12.6.3 Compressible solutions

Few examples exist for the finite element analysis of compressible lubrication problems. Reddi and Chu[9] studied several configurations, one of which is the shrouded step bearing shown in Figure 12.13. The element mesh (Figure 12.13(b)) consisted of linear quadrilateral elements refined in areas where

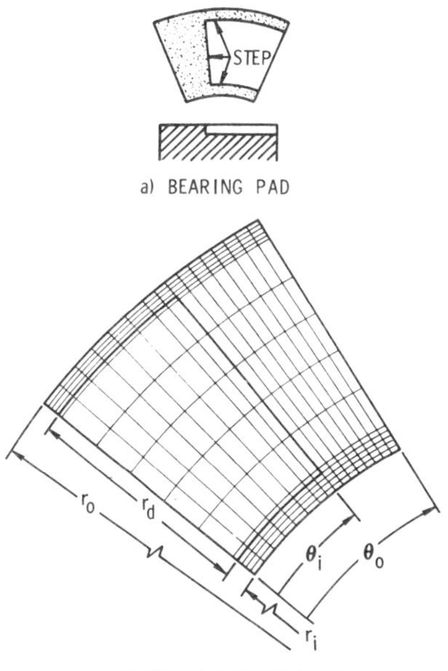

Figure 12.13 Finite element analysis of a shrouded step bearing[9]

pressure variation is expected to be greatest. Figure 12.14 portrays the computed pressure distributions at various circumferential stations. This example illustrates again that abrupt changes in film thickness at the step cause no special difficulties. Continuity of flow at the step is a natural consequence of the matrix assembly technique.

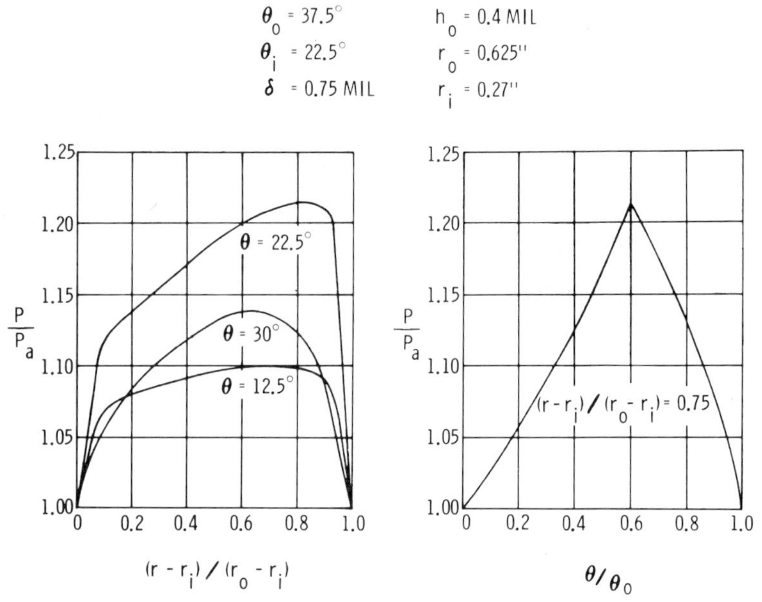

Figure 12.14 Pressure distributions for the shrouded step bearing (after Reddi and Chu[9])

12.6.4 Lubrication with combined effects

Recently, Eidelberg[19] performed a finite element analysis of lubrication in a joint between two bones. This type of lubrication involves porous as well as elastic bearing surfaces (Figure 12.15). Different types of elements were used for the poroelastic cartilage material to represent the elastic distortion and the diffusion flow phenomena. Figures 12.16 and 12.17 show the compatible pressure and film thickness distributions that result under typical operating conditions. Though these results were obtained neglecting the effects of side leakage, they establish the expected trends and demonstrate the viability of the solution technique. Extension to more realistic three-dimensional problems is straightforward, but would involve considerably more computational effort.

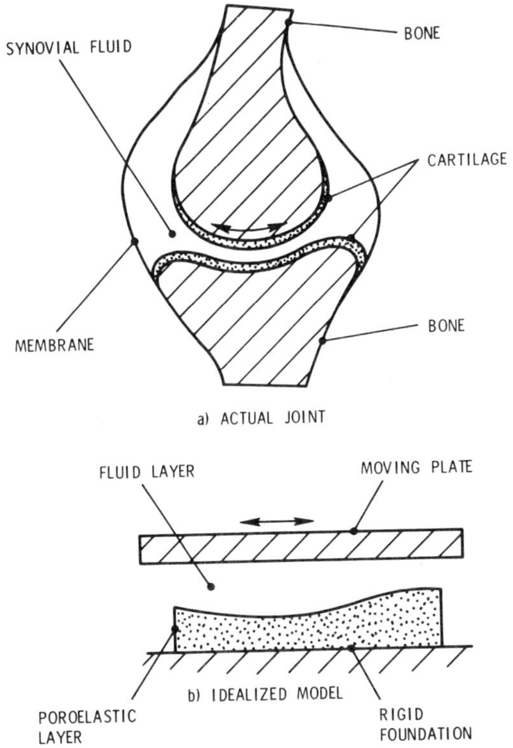

Figure 12.15 The geometry of a natural joint[19]

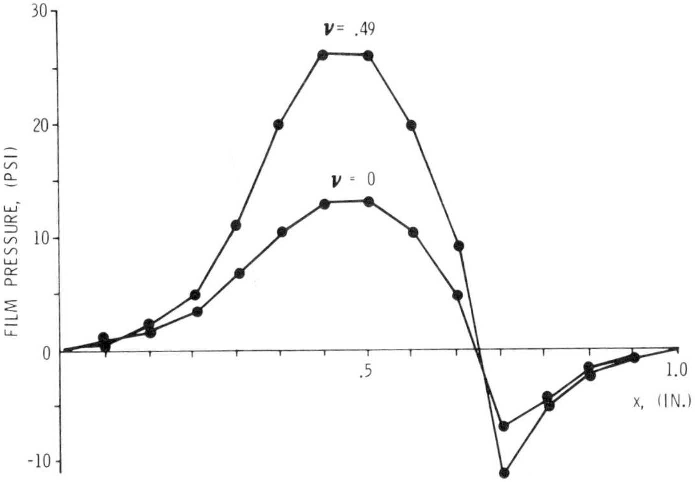

Figure 12.16 Pressure distribution in a natural joint (after Eidelberg[19])

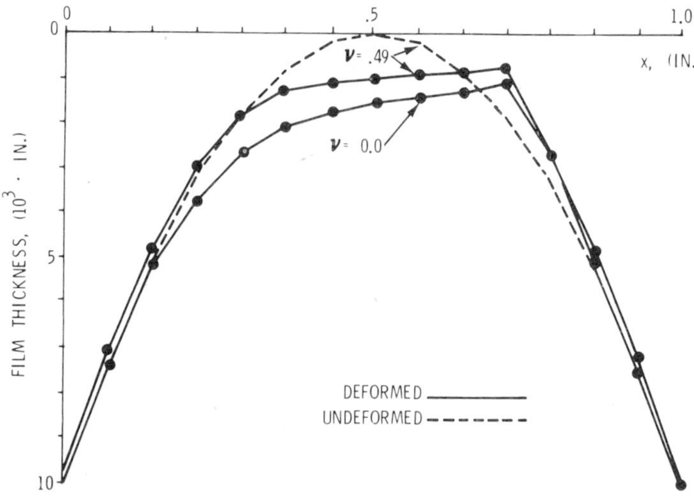

Figure 12.17 Film thickness distribution in a natural joint (after Eidelberg[19])

12.7 Conclusions

This brief survey suggests that the finite element method is now recognized as a potent numerical analysis tool for the solution of complex hydrodynamic lubrication problems. Popularity of the method stems from the ease with which irregular geometries and discontinuous film properties can be handled. The method allows development of one general computer program that can solve a variety of problems simply by accepting different input data.

Several caveats should be mentioned. Application of finite element techniques requires access to relatively large high-speed computers capable of processing and solving large systems of simultaneous equations.

A cumbersome aspect of obtaining a finite element solution is the preparation of the input data cards. Most of the input data consists of a description of the element mesh topology. Typical finite element mesh systems usually contain over 100 elements. Consequently, preparation of the data cards in the majority of cases is time-consuming and tedious because errors must be carefully avoided. Clearly, what every serious user of the method needs is an automatic means of generating the system topology.

Several schemes for accomplishing this have been developed,[23] and it behoves the user to employ one of these or devise his own. When three-dimensional elements are used, such as tetrahedral and rectangular prisms, automatic mesh generation is almost mandatory because the spatial visualization of a body divided into such elements is appallingly difficult. If, for simple problems, the user prepares mesh data by hand, he can greatly diminish the chance of human errors by asking the computer to return a graphical plot of the mesh system. Then a quick visual check of the resulting plot is all that is needed to spot mistakes.

Acknowledgement

The author wishes to thank Professor J. F. Booker of Cornell University for providing material from the these of his students, Messrs. Eidelberg and Day.

References

1. B. Tower, '1st report on friction experiments', *Proc. Inst. Mech. Eng.*, p. 632, Nov. 1883.
2. O. Reynolds, 'On the theory of lubrication and its application to Mr. Beauchamp Tower's experiments including an experimental determination of the viscosity of olive oil', *Phil. Trans.*, **177**(i), 157 (1886).
3. T. Lloyd and H. McCallion, 'Recent developments in fluid film lubrication theory', *Proc. Instn. Mech. Engrs.*, 182, 3A, 36 (1967–68).
4. D. V. Tansea and I. C. Rao, *Student Project Report on Lubrication*, Royal Naval College, Dartmouth, 1966.
5. S. Wada and H. Hayashi, *Conf. of Japan Soc. Lub. Engrs.* (in Japanese), Hiroshima, Oct. 1968.
6. M. M. Reddi, 'Finite element solution of the incompressible lubrication problem', *Journal of Lubrication Technology, Trans. ASME*, Series F., **91**, 3, 524 (1969).
7. T. Fujino, 'Analyses of hydrodynamic and plate structure problems by the finite element method', *Paper No. J5-4, Japan-U.S. Seminar on Matrix Methods of Structural Analysis and Design*, Tokyo, Japan, August 25–30, 1969, p. 725.
8. J. H. Argyris and D. W. Scharpf, 'The incompressible lubrication problem', 12th Lanchester Memorial Lecture, Appendix IV, *The Aeronautical Journal of the Royal Aeronautical Society*, **73**, 1044 (1969).
9. M. M. Reddi and T. Y. Chu, 'Finite element solution of the steady-state compressible lubrication problem', *Journal of Lubrication Technology, Trans. ASME*, Series F, **22**, 3, 495 (1970).
10. S. Wada, H. Hayashi and M. Migita, 'Application of finite element method to hydrodynamic lubrication problems' (Part 1, 'Infinite-width bearings'), *Bulletin of the Japan Society of Mechanical Engineers*, **14**, 77, 1222 (1971).
11. S. Wada and H. Hayashi, 'Application of finite element method to hydrodynamic lubrication problems' (Part 2, 'Finite-width bearings'), *Bulletin of the Japan Society of Mechanical Engineers*, **14**, 77, 1234 (1971).
12. K. H. Huebner, 'Finite element analysis of continuum problems and its application to the incompressible lubrication problem', *General Motors Research Publication GMR-1074*, Feb. 1971.
13. J. F. Booker and K. H. Huebner, 'Application of finite element methods to lubrication: an engineering approach', *Journal of Lubrication Technology, Trans. ASME*, Series F, **94**, 4, 313 (1972).
14. T. Allan, 'The application of finite element analysis to hydrodynamic and externally pressurized pocket bearings', *Wear*, **19**, 169 (1972).
15. C. Taylor and J. F. O'Callaghan, 'A numerical solution of the elastohydrodynamic lubrication problem using finite elements', *J. Mech. Engr. Sci.*, **14**, 4, 229 (1972).

16. K. P. Oh and K. H. Huebner, 'Solution of the elastohydrodynamic finite journal bearing problem', *Journal of Lubrication Technology, Trans. ASME*, Series F, **95**, 3, 342 (1973).
17. A. K. Tieu, 'Oil-film temperature distribution in an infinitely wide slider bearing: an application of the finite-element method', *J. Mech. Eng. Sci.*, **15**, 4, 311 (1973).
18. K. H. Huebner, 'Application of finite element methods to thermohydrodynamic lubrication', *Int. J. Num. Meth. Engr.*, **8**, 1, 139–168 (1974).
19. B. E. Eidelberg, *Finite Element Analysis of Lubrication in Natural Joints*, Ph.D. Thesis, Cornell University, Jan. 1974.
20. C. P. Day, *Transient Elastohydrodynamic Lubrication by Finite Element Methods*, M.S. Thesis, Cornell University, Jan. 1974.
21. S. M. Rohde and K. P. Oh, 'A unified treatment of thick and thin film elastohydrodynamic lubrication problems by higher-order element methods', *Int. J. Num. Meth. Engr.*, 1974.
22. K. P. Oh and S. M. Rohde, 'Analysis of compliant air bearing by higher-order finite element methods' (to be submitted).
23. W. R. Buell and B. A. Bush, 'Mesh generation—a survey', *ASME Paper No. 72-WA/DE-2*, November, 1972.

Chapter 13

Hydromagnetic Stability Studies Using the Finite Element Method

T. J. M. Boyd, G. A. Gardner and L. R. T. Gardner

13.1 Introduction

The research effort in controlled thermonuclear fusion over the past twenty years has had as its goal the containment and heating of a plasma for a period sufficiently long to allow fusion of deuterium nuclei to take place. In considering plasma containment schemes equilibria of various kinds have been examined. Equilibrium is a necessary but insufficient condition since, for fusion to be realized, one must ensure that a perturbation of the equilibrium configuration does not grow indefinitely. From the outset one of the vital questions has concerned the *stability* of the plasma configuration against perturbations. Without stability, the goal of controlled fusion cannot be attained so that it is natural that a great deal of theoretical effort has been expended on stability investigations.

The earliest analyses of plasma stability[1,2,3] centred on the linear pinch configuration in which a plasma column of circular cross-section with a current flowing axially is compressed by the $\mathbf{J} \times \mathbf{B}$ force acting radially inwards. Linear pinches are examples of 'open-ended' systems since the plasma can escape through the ends. An obvious way around this problem is to turn the linear system into a torus. This at least gets rid of the loss from the ends but in fact introduces new problems. An early example of a toroidal system was the ZETA device at Harwell in the 1950s. More recently there has been a resurgence of interest in toroidal containment due largely to promising results obtained in the Soviet Union with a class of devices known as TOKAMAKS. These are axially symmetric toroidal systems in which hot plasma is contained by the magnetic field of a current flowing in it and which have in addition, a strong longitudinal magnetic field parallel to the current. This field is many times stronger than the azimuthal magnetic field produced by the current and it serves to suppress the principal hydromagnetic instabilities.

The standard approaches to plasma stability centre on the use of normal mode analyses[2,3] or a technique based on an energy principle.[4,5] In the former the linearized equations of motion for a perturbation about an equilibrium state are solved and the system is unstable if any solution exists which increases indefinitely in time. The energy principle has as its starting point a variational formulation of the equations of motion. The normal mode stability condition can be replaced by the following variational principle (or energy principle as it is usually called): the necessary and sufficient condition that the system be stable is that

$$\delta W = \int \boldsymbol{\xi} \, . \, F \boldsymbol{\xi} \, \mathrm{d}\tau$$

(in which $\boldsymbol{\xi}$ is a displacement vector and F is a certain self-adjoint operator which will be introduced in Section 13.2) be not negative for every $\boldsymbol{\xi}$ which satisfies the boundary conditions. Physically δW is the change in potential energy of the system resulting from the displacement $\boldsymbol{\xi}$ so that what the energy principle states is that only those equilibrium states are stable for which every possible perturbation is accompanied by an increase in the potential energy. The hydromagnetic stability of the linear pinch configuration has been studied using both the method of normal mode analysis and the energy principle.

The important results from tokamak research[6] have heightened interest in plasma equilibria and stability in toroidal systems. On account of the complex geometry one generally has to resort to numerical methods to examine hydromagnetic stability. The energy principle of Bernstein, Frieman, Kruskal and Kulsrud[4] has been used widely and suggests that a natural way of analysing stability problems numerically is to employ a method based on a variational principle. Such is the finite element method which has developed into a powerful tool in continuum mechanics for the examination of equilibrium configurations and of oscillations and stability.[7–10] The value of this approach in treating hydromagnetic problems has been pointed out independently by Takeda, Shimomura, Ohta and Yoshikawa[11] in their study of a linear pinch with a strong longitudinal magnetic field and ourselves.

Like the Japanese workers we have applied the finite element method to the study of a linear pinch surrounded by a vacuum. This is a system well understood theoretically and for which much of the early work has lately been generalized by Shafranov[12] whose principal conclusions are as follows. In a current-carrying plasma in a strong axial magnetic field B_z ($\gg B_\theta$), hydromagnetic perturbations giving rise to instability may be separated into flute modes localized in the vicinity of the closed lines of force and non-local helical perturbations of the plasma boundary. In the case in which the pinch boundary coincides with the conducting wall surrounding the plasma, helical deformation should not develop. Shafranov showed that in the free boundary case the non-local helical perturbations may be stabilized if the current density falls off sufficiently quickly with radius. We have determined numerically the growth rate of the flute modes in a plasma with fixed boundaries and of the helical modes which occur in a pinch with a free boundary and which are important on account of their significantly higher growth rates.

In Section 13.2 the model of the pinch is described and the functional governing small perturbations of the equilibrium state given, followed in Section 13.3 by a description of the method of solution using the case of a pinch with fixed boundary as an example. In Section 13.4 the free boundary pinch is studied in detail and the solutions, in appropriate limits, are compared with the analytic expressions found by Shafranov.[12] In Section 13.5 a linear pinch with elliptic cross-section has been studied as a preliminary to using the finite element method to analyse toroidal 'belt-pinch' configurations.

13.2 Model

The linear pinch is modelled as a right circular cylindrical column of plasma, radius a, surrounded by an annular vacuum which is bounded on its outer surface, radius b, by a perfectly conducting wall. The plasma is assumed to be incompressible, to have infinite conductivity and to contain a magnetic field $\mathbf{B}$ which has no radial component. A current with current density $\mathbf{J} = \nabla \times \mathbf{B}$ flows within the plasma. The vacuum magnetic field is $\mathbf{B}_v$.

When the plasma undergoes small perturbations $\boldsymbol{\xi}$ the hydromagnetic energy of the system may be written[4,5]

$$W(\boldsymbol{\xi}) = \tfrac{1}{2}\int_p [\mathbf{Q}^2 + \mathbf{J}\,.\,\boldsymbol{\xi} \times \mathbf{Q}]\,\mathrm{d}\tau + \tfrac{1}{2}\int_v \mathbf{Q}_v^2\,\mathrm{d}\tau \tag{13.1}$$

where the quantities $\mathbf{Q}$ and $\mathbf{Q}_v$ are the perturbed magnetic fields within the plasma and vacuum respectively. The first integral ranges over the plasma volume and the second over the vacuum.

For the plasma

$$\mathbf{Q} = \nabla \times (\boldsymbol{\xi} \times \mathbf{B}) \tag{13.2}$$

and for the vacuum $\mathbf{Q}_v$ may be expressed in terms of a vector potential $\mathbf{A}$ by

$$\mathbf{Q}_v = \nabla \times \mathbf{A} \tag{13.3}$$

The potential $\mathbf{A}$ must, at the plasma vacuum interface $r = a$, satisfy the condition

$$\mathbf{n} \times \mathbf{A} = -(\mathbf{n}\,.\,\boldsymbol{\xi})\mathbf{B} \tag{13.4}$$

where $\mathbf{n}$ is the outward normal to the plasma surface and $\mathbf{n} \times \mathbf{A}$ becomes zero at $r = b$.

Introducing cylindrical polar coordinates (r, θ, z) in which the z axis lies along the axis of the plasma column, the displacements ξ, expressed in terms of normal modes are

$$\xi = [\xi_r(r), \xi_\theta(r), \xi_z(r)]\exp(im\theta + ikz + i\gamma t), \tag{13.5}$$

and the equation of continuity becomes

$$\frac{1}{r}\frac{\mathrm{d}}{\mathrm{d}r}(r\xi_r) + \frac{im\xi_\theta}{r} + ik\xi_z = 0 \tag{13.6}$$

For the perturbed motion, the Lagrangian, may be written

$$L = \gamma^2 T - W_p - W_v \tag{13.7}$$

where

$$T = \frac{\pi\rho}{2}\int_0^a (\xi_r^2 + \xi_\theta^2 + \xi_z^2) r\,\mathrm{d}r \tag{13.8}$$

with ρ the plasma density. The plasma hydromagnetic energy W_p is given by[5]

$$W_p = \frac{\pi}{2}\int_0^a r\,\mathrm{d}r\left\{\Lambda + \frac{k^2r^2 + m^2}{r^2}(\zeta - \zeta_0)^2\right\} \tag{13.9}$$

in which

$$\begin{aligned}\Lambda = {} & \frac{1}{k^2r^2 + m^2}\left[(krB_z + mB_\theta)\frac{\mathrm{d}\xi_r}{\mathrm{d}r} + (krB_z - mB_\theta)\frac{\xi_r}{r}\right]^2 \\ & + \left[(krB_z + mB_\theta)^2 - 2B_\theta\frac{\mathrm{d}}{\mathrm{d}r}(rB_\theta)\right]\frac{\xi_r^2}{r^2}\end{aligned} \tag{13.10}$$

$$\zeta_0 = \frac{r}{k^2r^2 + m^2}\left[(krB_\theta - mB_z)\frac{\mathrm{d}\xi_r}{\mathrm{d}r} - (krB_\theta + mB_z)\frac{\xi_r}{r}\right] \tag{13.11}$$

and

$$\zeta = i(\xi_\theta B_z - \xi_z B_\theta) \tag{13.12}$$

The energy contained in the vacuum is

$$W_v = \tfrac{1}{2}\int_v Q_v^2 r\,\mathrm{d}r\,\mathrm{d}\theta \tag{13.13}$$

To permit a direct comparison with earlier work[11,12] we have expressed the results of growth rate calculations in terms of a variable nq defined by

$$nq = \frac{krB_z}{B_\theta} \tag{13.14}$$

This is the key parameter for discussions of pinch stability.[6] The following magnetic fields have been used; in the plasma

$$\mathbf{B} = \left(0, \frac{B_a r}{a}, B_z\right)$$

and in the vacuum

$$\mathbf{B}_v = \left(0, \frac{B_a a}{r}, B_z\right)$$

where B_a and B_z are constant, so that nq is constant over the plasma. The associated current $\mathbf{J}$ is also constant and flows in the axial direction.

At this stage it is convenient to mention an economization technique which will be used in Sections 13.3 and 13.4 to reduce the number of independent variables.[8] The number of independent variables in a problem is reduced by minimizing the potential energy with respect to a subset of the variables, which are thereby eliminated as independent variables from the problem. In the present case W_p can be easily minimized with respect to ζ simply by setting[5]

$$\zeta = \zeta_0 \tag{13.15}$$

The displacements ξ_θ, ξ_z are determined by ζ_r through Equations 13.6 and 13.15 and so may be eliminated from the Lagrangian which then depends solely upon ξ_r.

13.3 Pinch with fixed boundary

In this problem the plasma is assumed bounded by a perfectly conducting wall of radius a; there is no vacuum region. The boundary condition at the wall is

$$\xi_r = 0, \qquad r = a \tag{13.16}$$

To apply the finite element method the plasma radius is divided into N line elements by the (nodal) points $r_0 = 0, r_1, r_2, \ldots, r_N = a$ with the values of ξ_r taken as nodal parameters $\xi_0, \xi_1, \xi_2, \ldots, \xi_N$. Within an element the displacement ξ_r will be obtained by linearly interpolating between the nodal values, so that once the nodal parameters have been found the complete distribution of ξ_r is known.

The hydromagnetic energy for a typical element i can, after integrating along the element, be written in the form

$$W_p^i = (\xi_i, \xi_{i-1})K^i\begin{pmatrix}\xi_i \\ \xi_{i-1}\end{pmatrix} \tag{13.17}$$

where K^i is a 2×2 matrix and corresponds to the element stiffness matrix for solid structures.[8] Similarly we can write the kinetic energy of element i in the form

$$T^i = (\xi_i, \xi_{i-1})M^i\begin{pmatrix}\xi_i \\ \xi_{i-1}\end{pmatrix} \tag{13.18}$$

where M^i is a 2×2 matrix corresponding to the element mass matrix.[8]

The total kinetic and potential energies are now obtained by summing the contributions from each finite element. Assembling the matrices, the Lagrangian has the form

$$L = \gamma^2 \boldsymbol{\delta}^T M \boldsymbol{\delta} - \boldsymbol{\delta}^T K \boldsymbol{\delta} \tag{13.19}$$

where

$$\delta^T = (\xi_0, \xi_1, \xi_2, \ldots, \xi_N) \tag{13.20}$$

is an $(N + 1)$ vector and M and K are $(N + 1)$ square matrices. Minimizing the Lagrangian with respect to each of the nodal parameters produces the matrix equation

$$(\gamma^2 M - K)\boldsymbol{\delta} = 0 \tag{13.21}$$

The stability problem has thus been converted into an eigenvalue problem. The positive eigenvalues, γ^2, give the frequency of the mode and the associated eigenvector $\boldsymbol{\delta}$ the amplitude of the nodal perturbations ξ_r. A negative eigenvalue implies that the associated mode is unstable, with growth rate $\sqrt{|\gamma^2|}$.

We have calculated the growth rate and nodal displacements for the mode defined by $m = -2$, $k = 0{\cdot}2$ using the economized Lagrangian and compared the results with analytic expressions obtained by Shafranov[12] by a normal mode analysis

$$\xi_r \propto \frac{1}{r} J_m\left(z_m \frac{r}{a}\right) \tag{13.22}$$

and

$$\gamma^2 = \frac{-B_a^2}{4\pi\rho a^2}\left[2(m + nq)\frac{ka}{z_m} + (m + nq)^2\right] \tag{13.23}$$

where z_m is the first zero of the mth order Bessel function $J_m(z)$.

From Figure 13.1 it is seen that the distribution of nodal displacements agrees exactly with expression (13.22). Figure 13.2 shows that the calculated growth rates are somewhat lower than those obtained from the analytic expression (13.23), although the regions of instability

$$-m - \frac{2ka}{z_m} < nq < -m < nq < -m + \frac{2ka}{z_m}$$

coincide. The problem was also solved using the Lagrangian without prior economization. Two nodal variables (ξ_r, ζ) are now required, so that effectively the numerical problem is 4 times as large. Within each element the distribution of ξ_r, and also of ζ, is obtained by linear interpolation. The

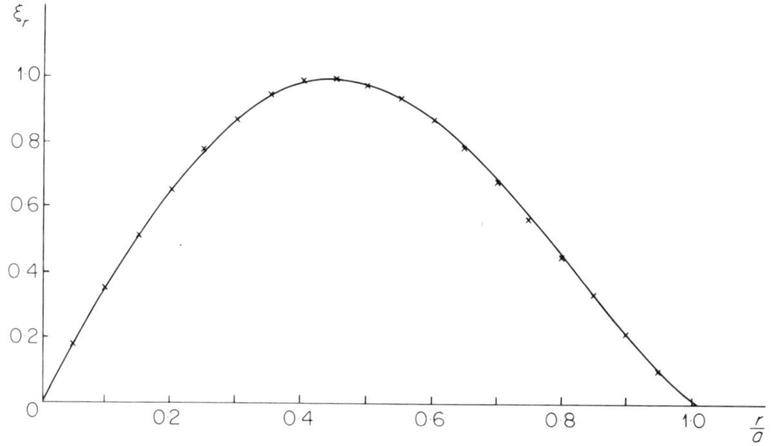

Figure 13.1 Radial variation of the perturbation ξ_r for $m = -2$ in a linear pinch with fixed boundary. The curve is a plot of $\xi_r \propto r^{-1}J_2(z_2 r/a)$ (cf. Equation 13.22) and the points x are the results of a finite element calculation with 20 elements along a radius and $a = 1$, $k = 0{\cdot}2$, $m = -2$

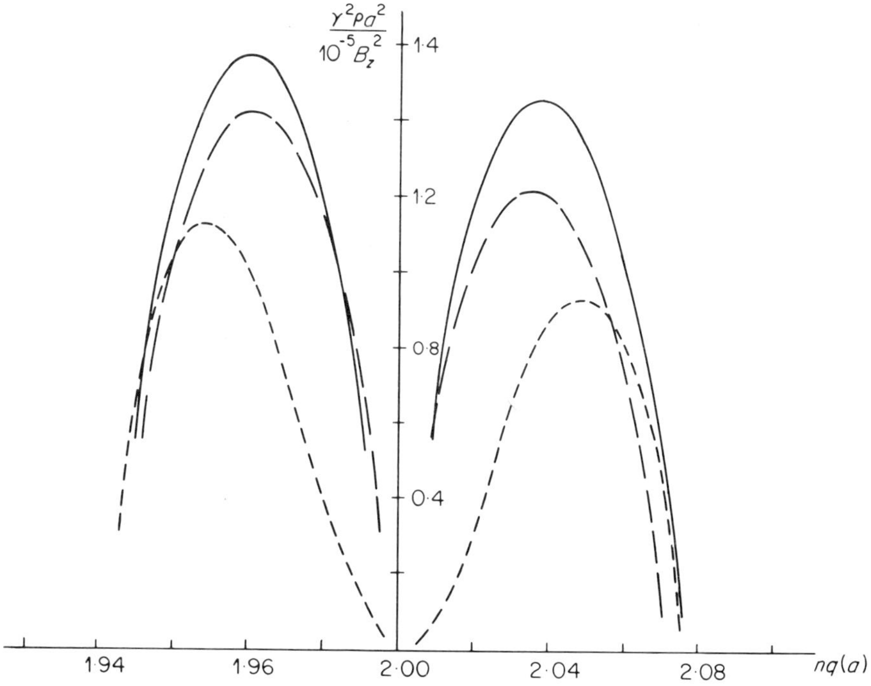

Figure 13.2 Square of the normalized growth rate for the $m = -2$ mode $k = 0{\cdot}2$ in a linear pinch with fixed boundary plotted against $nq(a)$. The continuous curve is the analytic result due to Shafranov. The broken curve is the finite element solution with two nodal variables (ξ_r, ζ) and the dotted curve that with one nodal variable ξ_r. In the finite element computation 8 radial elements were used

Lagrangian is given formally by Equation 13.9 but the vector of nodal parameters

$$\boldsymbol{\delta}^T = (\xi_0, \zeta_0, \xi_1, \zeta_1 \ldots \xi_N, \zeta_N)$$

is a $(2N + 2)$ vector and the matrices M and K are $(2N + 2)$ square. The results of this growth rate calculation are also plotted in Figure 13.2.

These results agree substantially with the finite element solution[11] but do not quite reproduce the analytic values.[12] Takeda and his coworkers suggest that this is in part explained by the analytic result being valid only for $ka \ll 1$, the remaining discrepancy arising from an (unspecified) error in the numerical formulation. However the approximation $ka \ll 1$ should be a good one and it may be that the discrepancy has its origins in that the problems studied by normal mode analysis and the hydromagnetic energy principle are not exactly similar in all respects, as Shafranov and Yurchenko[13] have suggested in a somewhat different context.

In these calculations the plasma radius was usually divided into 8 finite elements, this number being more than sufficient to ensure convergence of the solution. To test convergence a series of runs was made using from 2 to 20 radial line elements. It was found that the solution converged to a limiting value when 7 or more elements were used confirming the observations of Ohta, Shimomura and Takeda.[14]

13.4 Pinch with free boundary

When the plasma $(0 \leqslant r \leqslant a)$ is surrounded by a vacuum which is limited at $r = b$ by a perfectly conducting wall it is convenient to extend the definition of $\boldsymbol{\xi}$ into the vacuum region and to write the vector potential

$$\mathbf{A} = \boldsymbol{\xi} \times \mathbf{B}_v \tag{13.24}$$

so that the perturbed vacuum field

$$\mathbf{Q}_v = \nabla \times (\boldsymbol{\xi} \times \mathbf{B}_v) \tag{13.25}$$

has now the same form as that in the plasma. The boundary condition (13.4) at the plasma–vacuum interface is automatically satisfied since $\mathbf{B}(a) = \mathbf{B}_v(a)$ and the radial component of $\mathbf{B}_v$ is zero; we may satisfy the boundary condition at the conducting wall by requiring

$$\xi_r = 0 \qquad \text{at } r = b \tag{13.26}$$

The radius of the plasma–vacuum system is divided into N line elements in such a way that a node occurs at the plasma vacuum interface, for example, by points $r_0 = 0, r_1, r_2, \ldots, r_i = a, r_{i+1}, \ldots, r_N = b$ and nodal parameters $\xi_0, \xi_1, \ldots, \xi_N$ assigned. However, by considering the radial component Q_r

of the perturbed field we see that the varable

$$\xi_r = \frac{-rQ_r}{iB_\theta(m + nq)} \tag{13.27}$$

can become infinite when $m + nq = 0$. This possibility may occur since $nq = krB_z/B_\theta = (r^2/a^2)nq(a)$ varies across the vacuum. In situations of interest this does in fact happen, consequently ξ_r is not a suitable variable within the vacuum. A natural alternative not possessing this singularity is η defined by[12]

$$\eta = -(mB_\theta + krB_z)\xi_r = -B_\theta(m + nq)\xi_r \tag{13.28}$$

The boundary conditions to be satisfied by η are

$$\eta(a) = -B_a(m + nq(a))\xi_r(a) \tag{13.29}$$

and

$$\eta(b) = 0 \tag{13.30}$$

A solution exhibiting all the expected properties is now obtained.

Shafranov[12] has obtained an analytic solution to this problem for small values of k

$$\xi_r \propto r^{-m-1} \tag{13.31}$$

and

$$\gamma^2 = -\frac{B_a^2}{4\pi\rho a^2}\left[2(m + nq) + \frac{2(m + nq)^2}{1 - \left(\dfrac{a}{b}\right)^{-2m}}\right] \tag{13.32}$$

from which we deduce that the maximum growth rate is determined by

$$\gamma_{\max}^2 = -\frac{B_a^2}{8\pi\rho a^2}\left[1 - \left(\frac{a}{b}\right)^{-2m}\right] \tag{13.33}$$

and the range of instability is

$$-m - 1 + \left(\frac{a}{b}\right)^{-2m} < nq < -m \tag{13.34}$$

These helical instabilities have growth rates some two orders of magnitude greater than those of the flute instabilities found for the fixed boundary case, and so are correspondingly more important.

The results of the present numerical study are presented in Figures 13.3–13.5 where, as appropriate, they are compared with analytic values. The excellent agreement found is discussed in the following subsections.

13.4.1 Amplitudes

In Figure 13.3 the amplitudes of the nodal parameters ξ_r for the $m = -2$ mode are shown for a plasma–wall radius ratio $a/b = 0{\cdot}5$ and various axial wave numbers k. We see, for small k ($k = 0{\cdot}02, 0{\cdot}2$) that ξ_r has a linear radial distribution in agreement with Equation 13.31. However, as k is increased the amplitude graph changes. The displacement at the plasma–vacuum interface tends to zero and the curve approaches a limiting shape

$$\xi_r \propto \frac{1}{r} J_2\left(z_2 \frac{r}{a}\right)$$

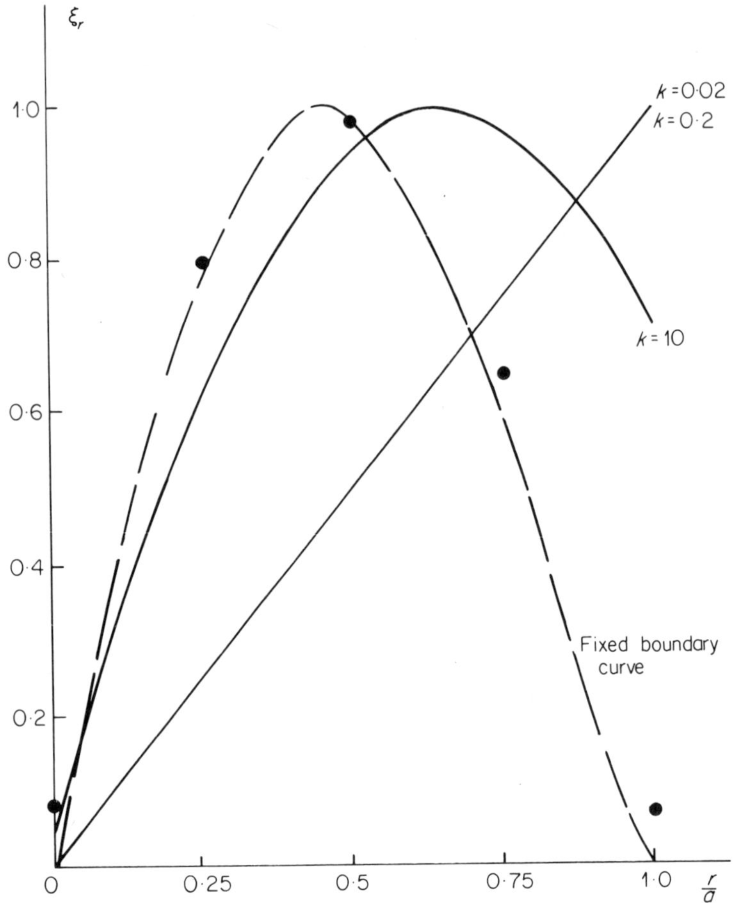

Figure 13.3 Radial variation of the perturbation ξ_r for $m = -2$ in a linear pinch with free boundary for various k and $a/b = 0{\cdot}5$. Points ● are for $k = 50$

where z_2 is the first zero of the Bessel function of order 2. This curve is the same as that obtained by Shafranov for a pinch with *fixed* boundary. In the free boundary case both flute and helical perturbations are possible as opposed to the fixed boundary pinch for which only flute deformations occur. By allowing k to increase in the free boundary case we increase $nq(a)$ so that we effectively stabilize the helical deformations[6] and retrieve the fixed boundary result in the limit of large k.

The amplitude distributions have also been determined for modes $m = -1$ and $m = -3$ for small k and an a/b ratio of 0·5. For $m = -1$ it was found that ξ_r is constant along a radius and for $m = -3$, ξ_r varies as r^2, in agreement with (13.31). No comparison with the results of Takeda and coworkers is possible since these authors have not presented any results for displacements.

13.4.2 Growth rates

For the $m = -2$ mode and axial wave number $k = 0{\cdot}2$ the square of the normalized growth rate $4\pi\gamma^2\rho a^2/B_a^2$ is plotted as a function of $nq(a)$, for various a/b ratios, in Figure 13.4. The instability ranges are found to agree exactly with (13.34). The maximum growth rates are also in good agreement. For small a/b ratios the agreement is excellent and becomes only slightly less so for ratios of $a/b > 0{\cdot}9$. A comparable dependence of growth rates on the position of the conducting wall has been presented in Reference 11, but for the case of a current density which shows a radial dependence as opposed to the constant current density chosen for Figure 13.4.

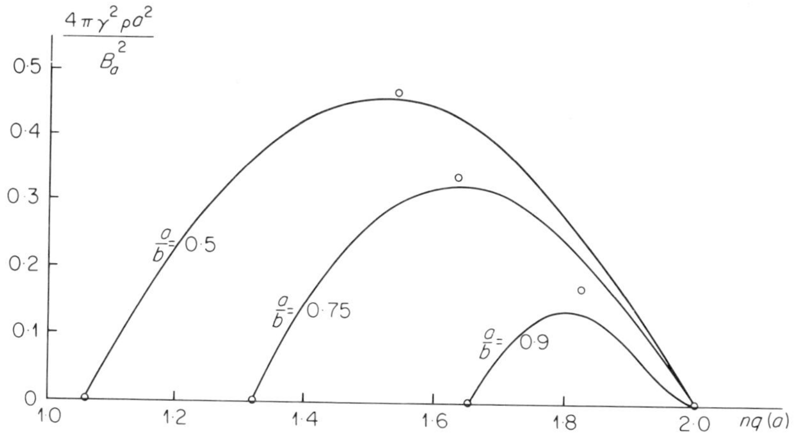

Figure 13.4 Square of the normalized growth rates for linear pinch with free boundary plotted against $nq(a)$ for $m = -2$ mode, $k = 0{\cdot}2$, for various values of a/b. The points ○ are Shafranov's analytic results

The variation of the square of the growth rate with axial wave number k has also been studied for the $m = -2$ mode. For small wave numbers k ($\leqslant 0{\cdot}2$) a limiting growth rate curve was obtained which agrees exactly with Equation 13.32. As k is increased the maximum growth increases in magnitude and the range of instability also increases with a limiting situation for which the instability range is approximately twice as large as for the small k case.

13.4.3 Current profile

In the preceding calculations the axial current flowing along the pinch has been constant. The situation in which the current density varies radially is of physical interest as Shafranov has shown that it is possible to stabilize the non-local helical perturbations provided the current density falls sufficiently rapidly with radius. We have examined this case numerically and have studied in particular the case of a current density profile given by

$$J = J_0\left[1 - \left(\frac{r}{a}\right)^{\nu}\right]$$

where J_0 is a constant. When current varies with radius it is found that 8 line elements across the plasma are no longer adequate to ensure convergence of the solution. From Figure 13.5 it can be seen that when $\nu = 2$, about 40 radial elements are necessary for convergence. This number of elements was therefore

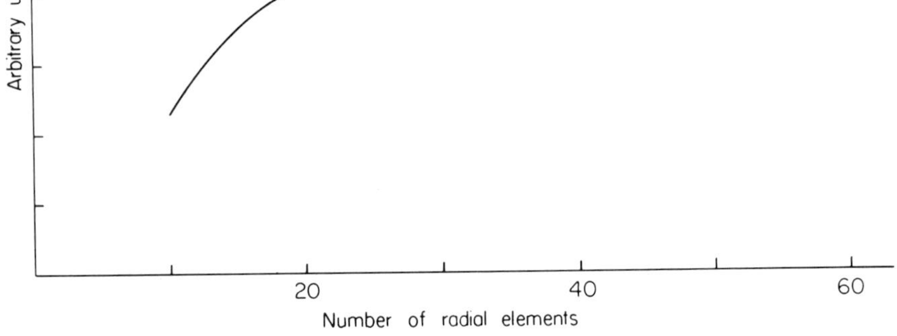

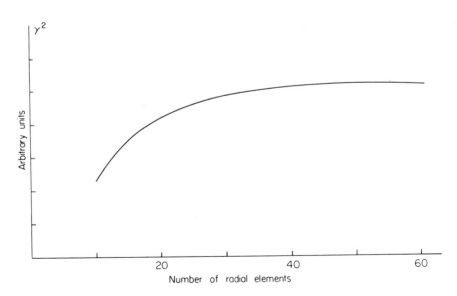

Figure 13.5 Variation in maximum value of γ^2 for $m = -2$, $k = 0{\cdot}2$ in pinch with current density $J = J_0[1 - (r/a)^2]$ when the number of radial elements is varied

used in the following calculations with $\nu \geqslant 2$. The effect of current density variation on growth rate and stability range is illustrated in Figure 13.6 for the case $m = -2$, $k = 0{\cdot}2$, $a/b = 0{\cdot}5$ and ν varying from 2 to 10. For comparison the constant current graph is also shown. The more rapid the fall off in current the smaller the range of instability and the lower the maximum growth rate.

Variations of the a/b ratio also affect stability. Figure 13.7 gives growth rate curves for the case $m = -2$, $k = 0{\cdot}2$, $\nu = 2$ and various a/b ratios. These results are in close agreement with Shafranov.[12]

It has been observed in the later stages of tokamak discharges[15] that the current density becomes peaked on the cylinder axis. We have therefore studied current densities of the form

$$J = J_0 \exp\left(-\frac{\alpha r}{a}\right)$$

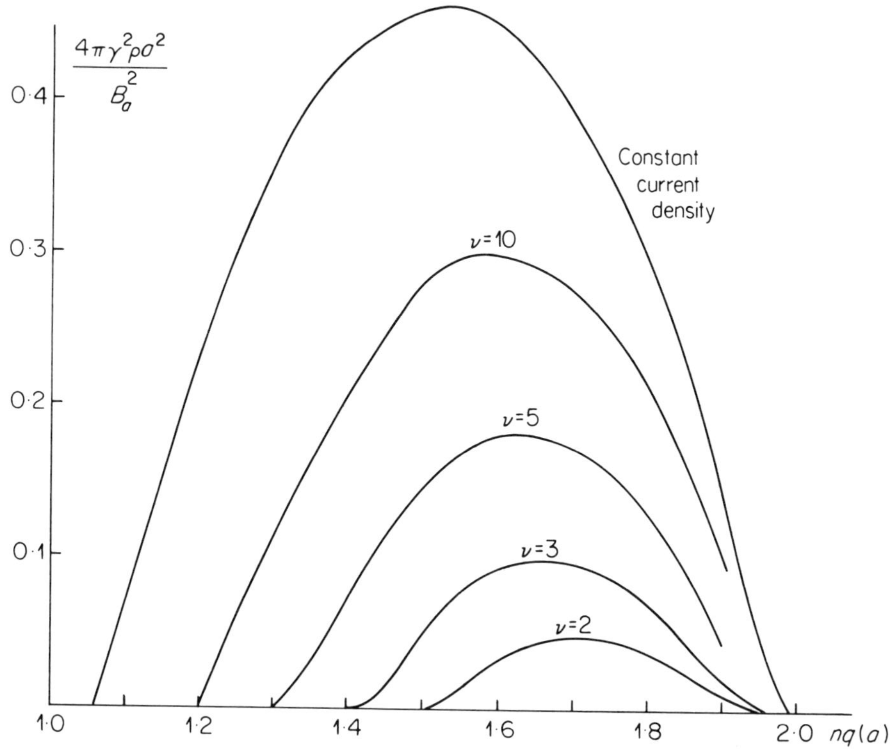

Figure 13.6 Variation in the square of the growth rate with nq for $m = -2$, $k = 0{\cdot}2$ mode in a linear pinch with $J = J_0[1 - (r/a)^\nu]$; ν varies from 2 to 10 and the curve for a constant current density is also plotted

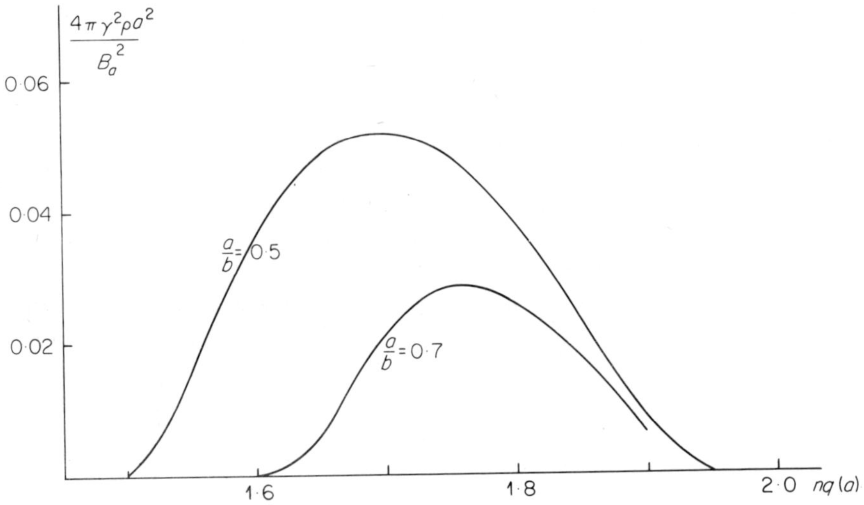

Figure 13.7 Variation in the square of the growth rate with $nq(a)$ for $m = -2$, $k = 0{\cdot}2$ mode in a linear pinch with $J = J_0[1 - (r/a)^2]$; $a/b = 0{\cdot}5$ and $0{\cdot}7$

with α having values 3 and 5. To illustrate our results we have plotted, in Figure 13.8, the growth rate of the $m = -1$ mode as a function of $nq(a)$ for

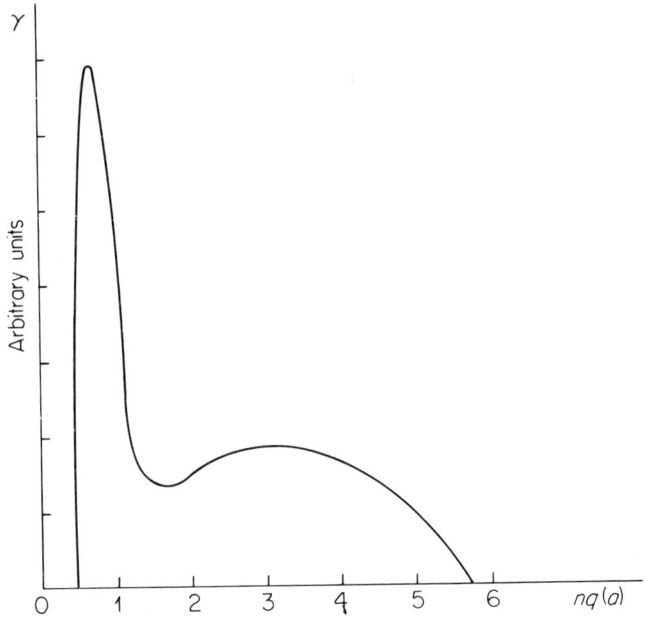

Figure 13.8 The growth rate of the $m = -1$ mode plotted as a function of $nq(a)$ for the current distribution $J = J_0 \exp(-3r/a)$

$\alpha = 3$ and in Figure 13.9 the square of the growth rate for modes $m = -2$ and $m = -3$ and $\alpha = 5$. The instability ranges observed and the positions of maximum growth rate agree with those reported by Grossmann and Ortolani.[15] We also confirm that the radial variation of the ξ_r perturbations is peaked on the cylinder axis, becoming more highly peaked for larger values of $nq(a)$.

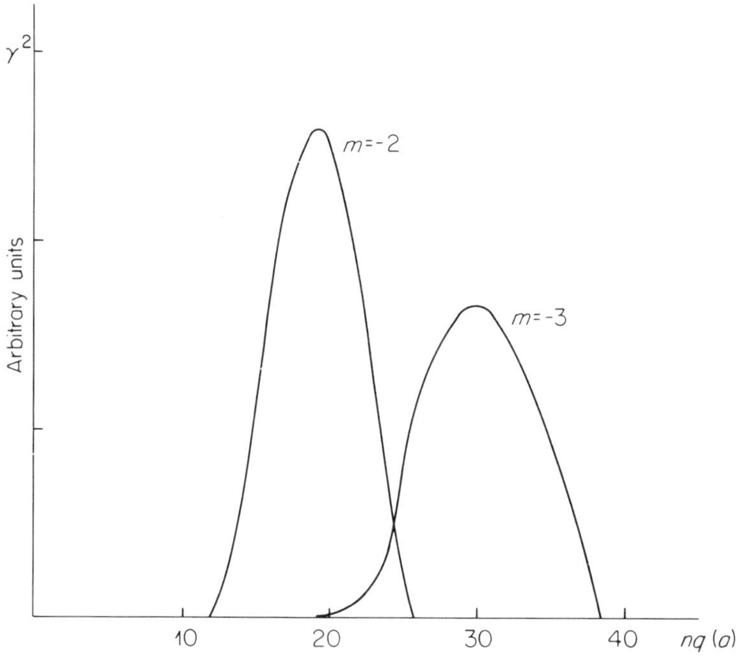

Figure 13.9 The square of the growth rate for the $m = -2$ and $m = -3$ modes plotted as functions of $nq(a)$ for the current distribution $J = J_0 \exp(-5r/a)$

13.5 Pinch with elliptic cross-section

Recently interest has been shown in pinches with non-circular cross section,[16,17,18] in particular in the so-called 'belt-pinch'. This is a plasma with elliptic cross-section carrying a longitudinal current with a magnetic field applied axially. We shall not be concerned with the details of such pinches other than to remark that the principal reason for interest in them is that increasing the ellipticity of the configuration affects the equilibria characteristics. As a first step towards applying a finite element analysis to belt-pinches in toroidal geometry which are important in practice, we have analysed the stability of a *straight* pinch with elliptic cross-section.

The pinch is again modelled as a column of plasma surrounded by an annular vacuum bounded by a perfectly conducting wall. The traces of the plasma–vacuum interface and the conducting wall on a cross-section, are confocal ellipses of focal distance a. Therefore it is convenient in this problem to use cylindrical elliptic coordinates (η, ψ, z), of focal distance a, with the z axis along the axis of the plasma column, so that the plasma–vacuum interface and conducting wall are the η = constant surfaces of Figure 13.10. The magnetic field in the plasma $\mathbf{B} = [B_\eta(\eta, \psi), B_\psi(\eta, \psi), B_z]$ is such that the axial field B_z is constant and the other two components are functions of η and ψ only. An axial current of density

$$J_z = \frac{1}{N^2}\{\partial_\eta(NB_\psi) - \partial_\psi(NB_\eta)\} \tag{13.35}$$

$$N = a(\cosh^2 \eta - \cos^2 \psi)^{\frac{1}{2}} \tag{13.36}$$

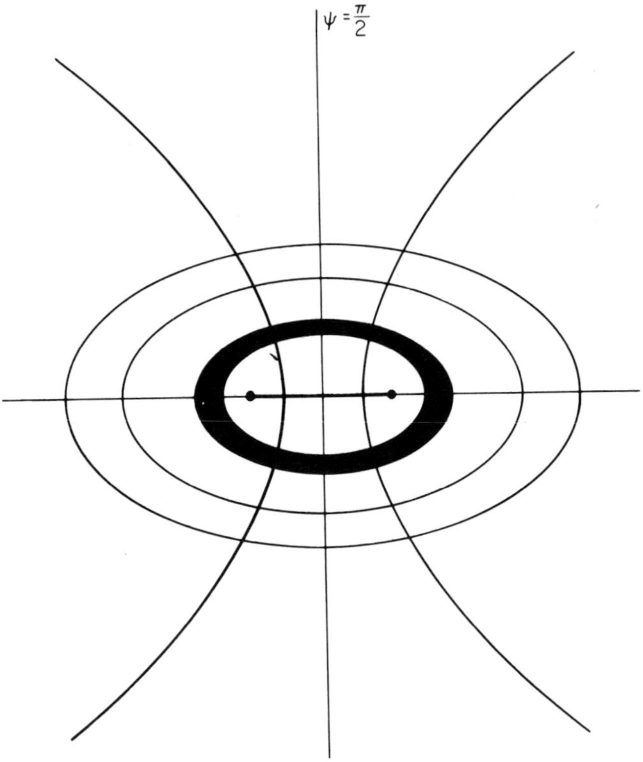

Figure 13.10 Cross-section of elliptic pinch showing elliptic coordinate system (η, ψ, z) where $0 \leqslant \eta < \infty$, $0 \leqslant \psi < 2\pi$, $-\infty < z < \infty$. The z axis is perpendicular to the plane of the figure. The focal distance is a, the eccentricity of the surface with major axis $2a \cosh \eta$, minor axis $2a \sinh \eta$ is $e = \operatorname{sech} \eta$. The ith finite element is shown shaded black

flows within the plasma. The perturbations $\boldsymbol{\xi}$ of the plasma will be expressed as Fourier modes

$$\boldsymbol{\xi} = \frac{1}{N}[\xi_\eta(\eta), \xi_\psi(\eta), \xi_z(\eta)] \exp(im\psi + ikz + i\gamma t) \tag{13.37}$$

where ξ_η, ξ_ψ, ξ_z are in general complex. The equation of continuity now takes the form

$$\partial_\eta \xi_\eta + im\xi_\psi + ikN\xi_z = 0 \tag{13.38}$$

and the perturbed fields are given by

$$\mathbf{Q} = \frac{1}{N}\begin{bmatrix} \Delta i\xi_\eta + \dfrac{B_\eta}{N}\partial_\eta\xi_\eta + \xi_\eta\,\partial_\psi\left(\dfrac{B_\psi}{N}\right) - \xi_\psi\,\partial_\psi\left(\dfrac{B_\eta}{N}\right) \\ \Delta i\xi_\psi + \dfrac{B_\eta}{N}\partial_\eta\xi_\psi + \xi_\psi\,\partial_\eta\left(\dfrac{B_\eta}{N}\right) - \xi_\eta\,\partial_\eta\left(\dfrac{B_\psi}{N}\right) \\ \Delta i\xi_z + \dfrac{B_\eta}{N}\partial_\eta\xi_z + \dfrac{\xi_z}{N}(\partial_\eta B_\eta + \partial_\psi B_\psi) \end{bmatrix} e^{i(m\psi + kz)} \tag{13.39}$$

where

$$\Delta = \frac{B_\psi m}{N} + B_z k \tag{13.40}$$

For the corresponding result in the vacuum we replace the components of B by those of $\mathbf{B}_v$.

The equilibrium transverse field $B_\perp$ components B_η, B_ψ and the current density J_z may be expressed in terms of a function ϕ by

$$B_\eta = \frac{-1}{N}\partial_\psi\phi, \qquad B_\psi = \frac{1}{N}\partial_\eta\phi \tag{13.41}$$

$$J_z = \frac{1}{N^2}(\partial_{\eta\eta}\phi + \partial_{\psi\psi}\phi) \tag{13.42}$$

where for a constant axial current density we have in the plasma[17]

$$\phi = \phi_I\left(\frac{\cosh^2\eta\cos^2\psi}{\cosh^2\eta_I} + \frac{\sinh^2\eta\sin^2\psi}{\sinh^2\eta_I}\right) \tag{13.43}$$

and in the vacuum

$$\phi_V = \phi_I + 2\phi_I\frac{\cosh 2\eta_I}{\sinh 2\eta_I}\left\{\eta - \eta_I - \frac{\sinh 2(\eta - \eta_I)}{2\cosh 2\eta_I}\cos 2\psi\right\} \tag{13.44}$$

On the plasma vacuum interface $\eta = \eta_I$ the function ϕ takes the (constant) value ϕ_I.

These expressions for the function ϕ are only valid when the conducting wall and plasma–vacuum interface are magnetic surfaces (constant pressure surfaces). These surfaces are only approximately confocal ellipses when the width of the vacuum is small compared with the dimensions of the plasma. This result also depends on the ellipticity of the configuration and the smaller the ellipticity the larger the ratio η_w/η_I may become for a valid computation.

Using Equations 13.41–13.43 we find that the current density is

$$J_z = \frac{2\phi_I}{a^2}\left(\frac{1}{\cosh^2 \eta_I} + \frac{1}{\sinh^2 \eta_I}\right)$$

and that β_η tends to zero at the plasma surface as required, so that the transverse field at the plasma–vacuum interface $B_{\perp I}$ is given by

$$B_{\perp I} = B_{\psi I} = \frac{2\phi_I N_I}{a^2 \cosh \eta_I \sinh \eta_I}$$

where $N_I = a(\cosh^2 \eta_I - \cos^2 \psi)^{\frac{1}{2}}$ is a function of ψ only. By analogy with the safety factor $nq(a)$ of the circular pinch we define a corresponding function for the elliptic case by

$$nq_I = \frac{kN_I B_z}{B_{\perp I}} = \frac{kB_z a^2 \cosh \eta_I \sinh \eta_I}{2\phi_I}$$

which is independent of both η and ψ. All growth rate calculations for the elliptic pinch will be expressed in terms of this variable.

The hydromagnetic energy of the pinch can be expressed in the form

$$W = \tfrac{1}{4}\int_0^{\eta_I}\int_0^{2\pi} IN^2\, d\psi\, d\eta + \tfrac{1}{4}\int_{\eta_I}^{\eta_w}\int_0^{2\pi} I_V N^2\, d\psi\, d\eta \qquad (13.45)$$

where $\eta = \eta_w$ at the conducting wall.

Expressions for I and I_V are found by using Equations 13.37 and 13.39 in the integrand of (13.1).

To set up a finite element subdivision of the pinch, the $\psi = 0$ radius of the plasma–vacuum system (FW in Figure 13.10) is divided into N line elements in such a way that a node occurs at the plasma–vacuum interface, for example by points $\eta_0 = 0, \eta_1, \eta_2, \ldots, \eta_i = \eta_I \eta_{i+1}, \ldots, \eta_N = \eta_w$. The ith finite element is thus an annulus bounded by the surfaces $\eta = \eta_{i-1}$ and $\eta = \eta_i$ Figure 13.10. However, since the variables $(\xi_\eta, \xi_\psi, \xi_z)$ are functions only of η the problem is essentially reduced to one dimension and the finite elements may be thought of as line elements with the complex qualities $(\xi_\eta, \xi_\psi, \xi_z)$ as the six parameters at each node. After each calculation we check that the resulting nodal parameters satisfy the equation of continuity (13.38). This must follow if the minimization of the Lagrangian is carried out successfully since it has been shown analytically that the minimizing displacements are incompressible.[19]

Alternative finite element formulations of the problem are possible. For example the equation of continuity can be used to eliminate the variables ξ_z and so ensure that the calculated displacements are incompressible. This introduces the second order derivatives $\partial_{\eta\eta}\xi_\eta$ so that it is no longer sufficient to obtain the distribution of ξ_η by linearly interpolating between nodal values, for second order derivatives will then always be zero. This problem can be avoided by, for example, using as nodal parameters $(\xi_\eta, \partial_\eta \xi_\eta, \xi_\psi)$. There are again six nodal parameters per element.

It has been shown that the expression for the hydromagnetic energy of an elliptic pinch, (13.45), reduces to the correct result for the pinch with circular cross-section in the limit of vanishing eccentricity. Computations now being carried out will allow us to compare our results with the work of Laval, Pellat and Soule[18] for a uniformly distributed current and with that of Marder[20] for a surface current.

13.6 Additional note

Subsequent to the writing of this paper in September 1973 our numerical results showed anomalies which led us to suspect the validity of our form for the displacement vector (cf. Equation 13.37). These suspicions were confirmed in work on the elliptic pinch by Dewar and others,[21] published in May 1974, in which it is shown that the perturbations are no longer represented by a single Fourier mode. These workers find that kink modes in elliptic tokamaks couple to Alfvén waves in such a way that the perturbation of the plasma–vacuum interface is proportional to

$$\tau_1 \exp[im\psi + ikz] + \tau_2 \exp[-im\psi + ikz]$$

The conclusions from their analysis show that the unstable regions for kink modes are widened by the ellipticity of the system. As a result of this we conclude that a proper finite element analysis of the stability of the elliptic pinch demands a representation of the plasma–vacuum boundary perturbation of the form

$$\boldsymbol{\xi} = \boldsymbol{\xi}(\eta, \psi) \exp[ikz + i\gamma t]$$

Acknowledgement

We wish to acknowledge helpful discussions with Dr. S. Kuhn, Institute for Theoretical Physics, University of Innsbruck.

References

1. V. D. Shafranov, 'The stability of a plasma column with a distributed current', *Plasma Physics and the Problem of Controlled Thermonuclear Reactions*, Vol. 4, Pergamon, Oxford, 1958.
2. R. J. Tayler, 'Hydromagnetic instabilities of an ideally conducting fluid', *Proc. Phys. Soc.*, **B70**, 31 (1957).
3. R. J. Tayler, 'The influence of an axial magnetic field on the stability of a constricted gas discharge', *Proc. Phys. Soc.*, **B70**, 1049 (1957).
4. I. B. Bernstein, E. A. Frieman, M. D. Kruskal and R. M. Kulsrud, 'An energy principle for hydromagnetic stability problems', *Proc. Roy. Soc.*, **A244**, 17 (1958).
5. W. A. Newcomb, 'Hydromagnetic stability of a diffuse linear pinch', *Ann. Phys.*, **10**, 232 (1960).
6. L. A. Artsimovich, 'Tokamak devices', *Nuclear Fusion*, **12**, 215 (1972).
7. M. J. Turner, R. W. Clough, H. C. Martin and L. J. Topp, 'Stiffness and deflection analysis of complex structures', *J. Aero Science*, **23**, 805 (1956).
8. O. C. Zienkiewicz, *The Finite Element Method in Engineering Science* (2nd ed.), McGraw-Hill, London, 1971.
9. J. T. Oden, *Finite Elements of Non-Linear Continua*, McGraw-Hill, New York, 1972.
10. J. S. Przemieniecki, *Theory of Matrix Structural Analysis*, McGraw-Hill, New York, 1968.
11. T. Takeda, Y. Simomura, M. Ohta and M. Yoshikawa, 'Numerical analysis of MHD instabilities by the finite element method', *Phys. Fluids*, **15**, 2193 (1972); T. J. M. Boyd, G. A. Gardner and L. R. T. Gardner, 'Numerical study of MHD stability using the finite element method', *Nuclear Fusion*, **13**, 764 (1973).
12. V. D. Shafranov, 'Hydromagnetic stability of a current carrying pinch in a strong longitudinal magnetic field', *Sov. Phys. Tech. Phys.*, **15**, 175 (1970).
13. V. D. Shafranov and E. I. Yurchenko, 'Condition for flute instability of a toroidal geometry plasma', *Sov. Phys., JETP*, **26**, 682 (1968).
14. M. Ohta, Y. Shimomura and T. Takeda, 'Analysis of hydromagnetic plasma stability by the finite element method', *Nuclear Fusion*, **12**, 271 (1972).
15. W. Grossmann and S. Ortolani, 'MHD model analysis of tokamak like plasma with various current density profiles', *European Conference on Controlled Fusion and Plasma Physics, Moscow 1973*, Vol. 1, p. 83.
16. L. A. Artsimovich and B. D. Shafranov, 'Tokamak with non-round section of the plasma loop', *JETP Letters*, **15**, 51 (1972).
17. R. Gajewski, 'MHD equilibrium of an elliptical plasma cylinder', *R. Phys. Fluids*, **15**, 70 (1972).
18. G. Laval, R. Pellat and J. L. Soule, 'Kink instabilities in a cylindrical plasma with elliptical cross section', *Fifth European Conference on Controlled Fusion and Plasma Physics, Grenoble: 1972: Proceedings*, Vol. 1, p. 25.
19. D. J. Rose and M. Clark, *Plasma and Controlled Fusion*, M.I.T. Press 1975, p. 343.
20. B. M. Marder, 'Kink instabilities in the belt pinch', *Phys. Fluids*, **17**, 447 (1974).
21. R. L. Dewar, R. C. Grimm, J. L. Johnson, E. A. Frieman, J. M. Greene and P. H. Rutherford, 'Long-wavelength kink instabilities in low-pressure, uniform axial current, cylindrical plasmas with elliptic cross sections', *Phys. Fluids*, **17**, 930 (1974).

Author Index

Subject Index